Sensors: Science and Engineering

Sensors: Science and Engineering

Edited by Marvin Heather

CLANRYE
INTERNATIONAL
www.clanryeinternational.com

Clanrye International,
750 Third Avenue, 9th Floor,
New York, NY 10017, USA

ISBN: 978-1-63240-618-7

Cataloging-in-Publication Data

Sensors : science and engineering / edited by Marvin Heather.
    p. cm.
Includes bibliographical references and index.
ISBN 978-1-63240-618-7
1. Detectors. 2. Detectors--Scientific applications. 3. Sensor networks.
4. Wireless sensor networks. I. Heather, Marvin.
TK7872.D48 S46 2017
681.2--dc23

For information on all Clanrye International publications
visit our website at www.clanryeinternational.com

*C*LANRYE
INTERNATIONAL

Printed in the United States of America.

Contents

Preface

Detecting events or changes in machines, equipment or environment is becoming highly critical. Such equipment needs reliable, robust and accurate sensors in order to avert risks that could endanger human lives, environment etc. This book discusses the theories and concepts of sensors. Most of the topics introduced herein cover new technologies and applications of sensors. The various studies that are constantly contributing towards advancing technologies and evolution of this field are examined in detail. The extensive content of this book provides the readers with a thorough understanding of working of sensors and its application.

The information contained in this book is the result of intensive hard work done by researchers in this field. All due efforts have been made to make this book serve as a complete guiding source for students and researchers. The topics in this book have been comprehensively explained to help readers understand the growing trends in the field.

I would like to thank the entire group of writers who made sincere efforts in this book and my family who supported me in my efforts of working on this book. I take this opportunity to thank all those who have been a guiding force throughout my life.

Editor

Fault-Tolerant Algorithms for Connectivity Restoration in Wireless Sensor Networks

Yali Zeng *, Li Xu * and Zhide Chen

Academic Editor: Leonhard M. Reindl

Fujian Provincial Key Laboratory of Network Security and Cryptology, School of Mathematics and Computer Science, Fujian Normal University, Fuzhou 350007, China; zhidechen@fjnu.edu.cn
*   Correspondence: yalizeng90@gmail.com (Y.Z.); xuli@fjnu.edu.cn (L.X.)

Abstract: As wireless sensor network (WSN) is often deployed in a hostile environment, nodes in the networks are prone to large-scale failures, resulting in the network not working normally. In this case, an effective restoration scheme is needed to restore the faulty network timely. Most of existing restoration schemes consider more about the number of deployed nodes or fault tolerance alone, but fail to take into account the fact that network coverage and topology quality are also important to a network. To address this issue, we present two algorithms named Full 2-Connectivity Restoration Algorithm (F2CRA) and Partial 3-Connectivity Restoration Algorithm (P3CRA), which restore a faulty WSN in different aspects. F2CRA constructs the fan-shaped topology structure to reduce the number of deployed nodes, while P3CRA constructs the dual-ring topology structure to improve the fault tolerance of the network. F2CRA is suitable when the restoration cost is given the priority, and P3CRA is suitable when the network quality is considered first. Compared with other algorithms, these two algorithms ensure that the network has stronger fault-tolerant function, larger coverage area and better balanced load after the restoration.

Keywords: wireless sensor networks; connectivity restoration; fault tolerance

---

1. Introduction

Wireless sensor networks (WSNs) are known for their wide use in industry, military, and environmental monitoring applications [1]. They are usually deployed in harsh environments, where nodes are subjected to failures and the networks are easy to be partitioned into disjoint segments. Therefore, fault tolerance becomes a critical issue for WSNs and numerous restoration algorithms are proposed [2–6] to address this issue. In order to achieve fault tolerance when restoring a faulty WSN, one approach is to deploy additional relay nodes to provide k ($k > 1$) vertex-disjoint paths (hereinafter referred to as k-connectivity) between every pair of network nodes (segments and relay nodes). In this way, the restored network can survive the failure of fewer than k nodes, which is more practical for WSNs. In this paper, we adopt this approach to repair the faulty network which is divided into many segments.

However, deploying additional relay nodes for network restoration brings us two conflicting requirements: On the one hand, it needs to spend some money to purchase the equipment. In order to save the cost, it is required to place as few nodes as possible to repair the faulty network. On the other hand, as a wireless sensor network is easy to fail, the network after the restoration is required to be with fault-tolerant function so that it can resist the attack and damage in the future. The network, which is constructed by using as few nodes as possible, may not be fault-tolerant, but the network with fault-tolerant function needs to deploy more relay nodes and costs more money. Hence, these two requirements are contradictory. In addition, as for a network, network coverage and topology quality are also important to a network. Therefore, when designing the restoration scheme, we should consider not only the cost and network fault tolerance, but also the other aspects. Only in this way can the network after the restoration be more practical.

1.1. Our Contributions

In this paper, we comprehensively consider the restoration cost, fault tolerance, network coverage and topology quality. We seek to use fewer nodes to establish a network with fault-tolerant function under the premise of multiple segments that are unable to communicate with each other. Meanwhile, except for the restoration cost and fault tolerance, we also consider the network coverage, the quality of topology and others in this paper, so as to ensure that the network can not only has better fault tolerance, but also has stronger robustness and higher coverage after the restoration. Certainly, these performances are not considered fully in the existing literature. The algorithms we propose in this paper are summarized as follows:

(1)   Full 2-Connectivity Restoration Algorithm (F2CRA) provides two vertex-disjoint paths between every pair of network nodes. This algorithm is suitable when the cost is considered first.

(2)   Partial 3-Connectivity Restoration Algorithm (P3CRA) provides three vertex-disjoint paths between every pair of segments and at least two vertex-disjoint paths between every pair of relay nodes. This algorithm is suitable when the fault tolerance, network coverage and topology quality are considered first.

1.2. Paper Organization

The remainder of this paper is organized as follows. Section 2 reviews some related works. Section 3 proposes the system model and preliminaries. Our algorithms are introduced in Section 4. Sections 5 and 6 conduct the theory and simulation analysis for our algorithms, respectively. Finally, we conclude this paper in Section 7.

2. Related Work

WSNs are prone to failures due to the hostile environments where they are deployed. How to recover a faulty WSN is an important issue that has attracted numerous researches. We summarize some existing restoration algorithms in Table 1. In the connected relay node placement problem, the aim is to ensure the network is connected ($k = 1$) [7–13], while in the survivable relay node placement problem, the aim is to ensure k-connectivity ($k > 1$) [2–6,14–18]. k-connectivity can be either full or partial [19]. Full k-connectivity implies that k node-disjoint paths exist between every pair of nodes, while partial fault-tolerance requires k-connectivity between original nodes (segments) only.

Table 1. Relay placement algorithms.

Algorithms	k	Deployment Locations	Fault-Tolerance	Network Types
Lloyd [9]	$k = 1$	Unconstrained	No	Homogeneous
Li [10]	$k = 1$	Unconstrained	No	Heterogeneous
Bhattacharya [13]	$k = 1$	Constrained	No	Homogeneous
Yang [11]	$k = 1, 2$	Constrained	Full	Hierarchical
Hao [2]	$k > 1$	Unconstrained	Partial	Hierarchical
Zhang [3]	$k = 2$	Unconstrained	Full	Hierarchical
Han [4]	$k > 1$	Unconstrained	Full, Partial	Heterogeneous
Senel [5]	$k = 2$	Unconstrained	Full	Homogeneous
Our algorithms	$k = 2, 3$	Unconstrained	Full, Partial	Homogeneous

In connectivity problems, most algorithms restore a faulty network by finding the minimum spanning tree or Steiner tree. Lin and Xue [7] show that the STP-MSP problem is NP-hard. They also show that the approximation obtained from the minimum spanning tree has a worst-case performance ratio at most 5, while Chen et al. [8] point out that this approximation has a performance ratio exactly 4. Chen *et al.* also present a new polynomial-time approximation with a performance ratio at most 3. Yang *et al.* [11] study two-tiered constrained relay node placement problems and propose polynomial time approximation algorithms with O(1)-approximation ratios. Lloyd *et al.* [9] study two versions of relay node placement problems, but the same objective of these two versions is to deploy the minimum number of relay nodes. Li *et al.* [10] also has the same objective as [9], but they study the placement problem in a heterogeneous WSN. Although easy to implement, these algorithms are usually not efficient when a failure occurs in the network.

In survivable problems, most algorithms aim to construct a fault-tolerant network topology in a WSN. Hao and Tang *et al.* [2,14] study a fault-tolerant relay node placement problem in a two-tiered network, while Zhang *et al.* [3] study the problem in both single and two tiered networks. Smith *et al.* [2] is further extended to cover *k*-connectivity in heterogeneous wireless sensor networks in [4] where sensor nodes possess different transmission radii. The same as [4], Misra *et al.* [15] study the placement problem in heterogeneous wireless sensor networks, but [15] studies a constrained version in which relay nodes can only be placed at a set of candidate locations. As many algorithms do in connectivity problem, many restoration algorithms in survivable problem also try to place fewest number of relay nodes in a WSN like [6,16,17].

In a word, most of the aforementioned algorithms try to place minimum relay nodes in a WSN. However, none of them take network quality into account which is also crucial in terms of application-level performance. Therefore, Senel and Lee *et al.* [5,18] opt to reestablish connectivity using the least number of relays while ensuring a certain quality in the formed topology. However, their algorithms produce many overlapped areas and cannot be practical in multiple node failures caused by aftermath. To address these issues, we jointly consider establishing fault-tolerant connectivity and providing large coverage area which has not been studied.

3. System Model and Preliminaries

3.1. System Model

WSN is often deployed in the hostile environment, and sometimes it may suffer from the large-scale damage, resulting in the entire network being divided into multiple segments which cannot communicate with each other. In this paper, the problem we consider is how to repair the faulty WSN composed of multiple segments. As mentioned above, our scheme is to deploy relay nodes between each segment, but this scheme brings us two contradictory requirements: One is to minimize the number of nodes, and the other is to construct a fault-tolerant network. If the segments are regarded as a node, the set of these nodes is defined as S and the set of the deployed relay nodes is defined as P, then our problem can be transformed into the following.

Given a set of nodes (segments) S on a plane with a random distribution, the nodes in the set S cannot communicate with each other. After the set of relay nodes P being added on the plane for the restoration, all the nodes $S \cup P$ can communicate with each other. It requires that (1) the number of relay nodes is minimized; and (2) the network is fault-tolerance after the restoration.

3.2. Preliminaries

Definition 1. *Minimum Convex Hull: Given a set of nodes S on a plane with a random distribution, the minimal convex polygon which contains all the points is called minimum convex hull (Figure 1a). In this paper, we need to construct two convex hulls. To describe them conveniently, we call this minimum convex hull as outer convex hull here. The outer convex hull is composed of the isolated segments in the network.*

Definition 2. *Inner Convex Hull: In this paper, when the minimum convex hull is found, in order to ensure the network has fault tolerance after the restoration, the other convex hull is built inside the minimum convex hull. This convex hull is called inner convex hull (Figure 1b).*

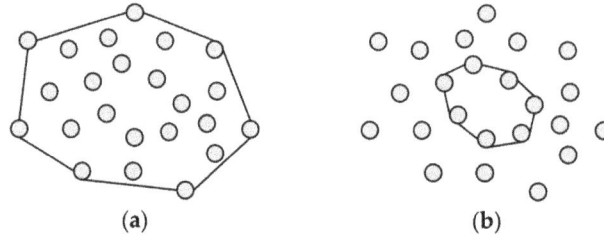

Figure 1. (a) Minimum convex hull; and (b) Inner convex hull.

Definition 3. *Corner Point of Convex Hull: The points to make up convex hull are called corner points. In outer convex hull, corner points are the isolated segments, while in inner convex hull, corner points are relay nodes.*

The notations used in this paper are as follows (Table 2).

Table 2. Notations.

Notation	Description
OCH	Outer Convex Hull
ICH	Inner convex hull
CP	Corner point
O	Center of OCH
R	Radius of relay node
P	Set of relay nodes
S	Set of n segments, $S = \{s_1, s_2, \ldots, s_n\}$. The corresponding coordinate set of these n segments is $\{(x_1, y_1), (x_2, y_2), \ldots, (x_n, y_n)\}$.

4. Algorithms

This section illustrates two proposed algorithms in detail.

4.1. Full 2-Connectivity Restoration Algorithm

Many schemes place too many relay nodes for improving network fault tolerance performance, but other performances like network coverage and topology quality have not been optimized. To address this issue, we propose a new restoration algorithm named Full 2-Connectivity Restoration Algorithm (F2CRA). This algorithm aims to deploy the minimum number of relay nodes to form a full 2-vertex connected network. Meanwhile, the restored network has a larger coverage area and a more balanced load than other schemes. The flow chart of F2CRA is shown in Figure 2.

The steps of F2CRA are as follows (Algorithms 1):

Given the scattered segments set S on the plane, in Step 1 the minimum convex hull composed of these segments is found by the method of Graham scan algorithm. The time complexity of Graham scan algorithm is $O(n \log n)$. By calculating the length from CPs to the center of OCH, in Step 2, we obtain the number of nodes and the accurate deployment position between each CP and O. Step 2 enables the nodes on ICH to form 3-connectivity, and Steps 3 and 4 enable the remaining segments on the plane to form 2-connectivity. Steps 2–4 make the network topology have the fan-shaped structure after the restoration. Compared with other algorithms, the network topology with such structure has better fault tolerance, larger coverage and more balanced load. The time complexity from Step 2 to Step 4 is $O(n)$; therefore, the time complexity of F2CRA algorithm is $O(n \log n)$.

Algorithms 1 F2CRA

INPUT: R, $S = \{s_1, s_2, \ldots, s_n\}$ and $\{(x_1, y_1), (x_2, y_2), \ldots, (x_n, y_n)\}$. P is null.

OUTPUT: A set of relay nodes P.

Step 1. Find OCH.

  (1) Adopt the Graham scan algorithm to find OCH in S.

    Suppose that CPs set of OCH is $\{s_1, s_2, \ldots, s_m\} \subset S$ and their corresponding coordinate is

    $\{(x_1, y_1), (x_2, y_2), \ldots, (x_m, y_m)\}$. The remaining segments set inside OCH is $\{s_{m+1}, s_{m+2}, \ldots, s_n\}$.

  (2) Calculate the coordinate (x_0, y_0) of O.

$$x_0 = \frac{1}{m} \sum_{i=1}^{m} x_i$$

$$y_0 = \frac{1}{m} \sum_{i=1}^{m} y_i$$

  (3) Calculate each side length of OCH.

    for $i = 1$ to m do

      $j = i + 1$

      if $i = m$ then

        $j = 1$

      end if

      $side_i = \sqrt{\left(x_i - x_j\right)^2 + \left(y_i - y_j\right)^2}$

    end for

  Then the set of the side lengths of OCH is $\{side_1, side_2, \ldots, side_m\}$.

Step 2. Find ICH.

  (1) Line each CP with O, respectively, and calculate the length of each line:

    for $i = 1$ to m do

  $l_i = \sqrt{(x_i - x_0)^2 + (y_i - y_0)^2}$

    Then the set of these lines is $\{l_1, l_2, \ldots, l_m\}$

  (2) Calculate the angle between two adjacent lines:

    for $i = 1$ to m do

      $j = i + 1$

        if $i = m$ then

          $j = 1$

        end if

      $\theta_i = \arccos\left(\dfrac{l_i^2 + l_{i+1}^2 - side_i^2}{2 \times l_i \times l_{i+1}}\right)$

    Calculate the number of relay nodes to be deployed in each line:

$$x_i' = x_0 \pm \frac{R \times |x_0 - x_i|}{2 \times l_i \times \sin\frac{\theta_i}{2}},$$

$$y_i' = y_0 \pm \frac{R \times |y_0 - y_i|}{2 \times l_i \times \sin\frac{\theta_i}{2}}$$

$$n_i' = \left\lfloor \frac{\sqrt{(x_i - x_i')^2 + (y_i - y_i')^2}}{R} \right\rfloor$$

    Start with a CP of OCH s_i, and deploy relay nodes towards O with one relay node every

    distance R, then get the corresponding deployment position of n_i' nodes. Add these nodes

    into the set of P and set the last node p_i, the coordinate (x_i'', y_i'').

    end for

  Then the CPs set of ICH is $\{p_1, p_2, \ldots, p_m\}$.

  (3) Deploy the nodes along with the edge of ICH.

    for $i = 1$ to m do

        if $i = m$ then

          $j = 1$

      end if

    Deploy relay nodes between p_i and p_{i+1}, and add these nodes into the set of P.

  end for

Step 3. Establish 2-connectivity for the segments on OCH.

      if $m\%2 = 0$ then

    Select $\frac{m}{2}$ non-adjacent sides of shortest total length from the side set $\{side_1, side_2, \ldots, side_m\}$,

    deploy relay nodes along with these sides, and add these nodes into the set of P.

  else

  There will be a vertex $u \in S$ not forming 2-connectivity. At this time, find a node v on ICH that is

  else

  closest to u but not collinear with u, deploy the nodes uniformly between v and u, and add these nodes into the set of P.

  end if

Step 4. Establish 2-connectivity for the isolated segments on the plane

    for $k = m + 1$ to n do

    Find the nearest two nodes u' and v' for s_k, $u', v' \in P$. Deploy relay nodes uniformly in s_k and u', s_k and v',

    and add these nodes into set P.

  end for

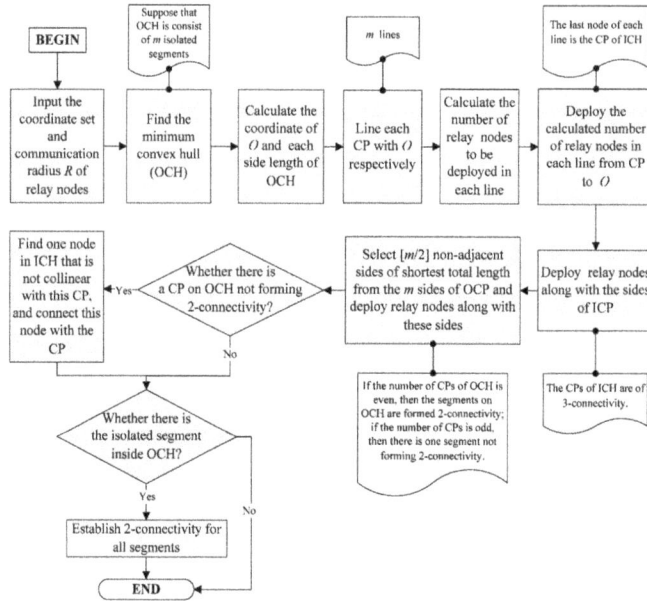

Figure 2. Flow chart of Full 2-Connectivity Restoration Algorithm.

4.2. Partial 3-Connectivity Restoration Algorithm

F2CRA uses fewer nodes to establish a network topology with fault tolerance. Therefore, F2CRA is suitable for the case when the number of available relay nodes is small. When the number of available nodes is sufficient, we can extend F2CRA, so that the network topology can have the stronger fault tolerance after the restoration. Here, we propose an improved algorithm Partial 3-Connectivity Restoration Algorithm (P3CRA). P3CRA is similar to F2CRA, but the network restored by P3CRA will have partial 3-connectivity structure. Partial 3-connectivity means that after the restoration, all the segments have 3-connectivity at least, and the deployed relay nodes have 2-connectivity at least. The network restored by P3CRA has larger coverage and better fault tolerance than that by F2CRA. However, P3CRA needs to deploy more nodes; therefore, P3CRA is suitable when the network quality is taken into consideration first, and F2CRA is suitable when the cost is in consideration. P3CRA flow chart is shown in Figure 3.

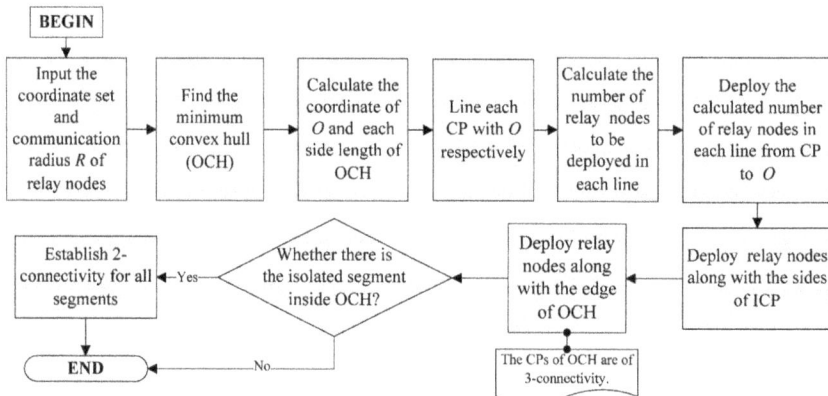

Figure 3. Flow chart of Partial 3-Connectivity Restoration Algorithm.

The steps of P3CRA are as follows (Algorithms 2):

Algorithms 2 P3CRA

INPUT: $R, S = \{s_1, s_2, \ldots, s_n\}$ and $\{(x_1, y_1), (x_2, y_2), \ldots, (x_n, y_n)\}$. P is null.
OUTPUT: A set of relay nodes P.
Step 1 and Step 2 are the same with F2CRA's.
Step 3. Establish 3-connectivity for the segments on OCH.
    for $i = 1$ to m **do**
Deploy relay nodes between s_i and s_{i+1}. (if $i = m$, then $i + 1 = 1$)
      end for
Step 4. Establish 3-connectivity for the isolated segments on the plane.
    for $k = m + 1$ to n **do**
        Find the nearest three nodes u', v' and w' for s_k. (u', v', w' are not on the same line)
        Deploy nodes uniformly in s_k and u', s_k and v', s_k and w'.
    end for

The first two steps are consistent in F2CRA algorithm and P3CRA algorithm, but in Step 3, P3CRA algorithm directly deploys nodes along the edge of OCH. At that time, all CPs (segments) on OCH form 3-connectivity. In Step 4, all segments on the plane eventually form 3-connectivity. Like F2CRA algorithm, the time complexity of P3CRA algorithm is also $O(n\log n)$.

To summarize, the network topology repaired by F2CRA algorithm has 2-connectivity. As it needs fewer nodes, this algorithm is suitable when the cost is considered first. Compared with F2CRA algorithm, P3CRA algorithm needs to deploy more nodes. Due to the stronger fault tolerance, larger coverage and more balanced load, P3CRA algorithm is applicable when the performance of network is considered first.

5. Algorithm Analysis

It is known that the coordinates of CPs and the center coordinate of OCH are $\{(x_1, y_1), (x_2, y_2), \ldots, (x_n, y_n)\}$ and (x_0, y_0), respectively, and the value of communication radius of relay nodes is R. Assume that nodes are deployed every distance R. When the CP coordinate of ICH is $(x_0 \pm \dfrac{R \times |x_0 - x_i|}{2 \times l_i \times \sin\frac{\theta_i}{2}}, y_0 \pm \dfrac{R \times |y_0 - y_i|}{2 \times l_i \times \sin\frac{\theta_i}{2}})$, the restoration algorithm will use the minimum number of relay nodes.

Proof. As shown in Figure 4, we assume the coordinates of point A, B, and O are (x_1, y_1), (x_2, y_2), and (x_0, y_0), respectively. Here point A and point B represent the different CPs of OCH, and point O represents the center of OCH. Our algorithm deploys relay nodes from the CPs (A and B) to the central point (O). As the values of AB, AO, BO and R are fixed, to minimize the nodes, it requires the total length of AD and BE to be the shortest, that is, the total length of OD and OE is the longest. Consequently, the problem is transformed into: Seeking the coordinate values of point D and E when the total length of OD and OE is the longest.

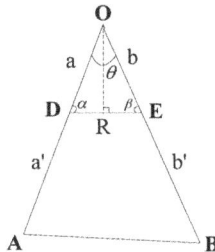

Figure 4. Diagram of triangle.

Set $\angle AOB = \theta$, the lengths of DE, OD and are R, a and b, respectively. $\angle ODE = \alpha$, $\angle OED = \beta$, a, b, α, β are unknown. When a is equal to b, $a + b$ reaches the maximum value, which means the total length of OD and OE is the longest. The detailed argument is relegated to the Appendix.

When $a = b$, by trigonometric function, we have:

$$a = b = \frac{\frac{R}{2}}{\sin\frac{\theta}{2}} = \frac{R}{2\sin\frac{\theta}{2}} \tag{1}$$

Assume the coordinate of point D is (x_1', y_1'), then:

$$\sqrt{(x_1' - x_0)^2 + (y_1' - y_0)^2} = \frac{R}{2\sin\frac{\theta}{2}} \tag{2}$$

As D is on line AO, then we have:

$$\frac{y_0 - y_1}{x_0 - x_1} = \frac{y_0 - y_1'}{x_0 - x_1'} \tag{3}$$

By Equations (2) and (3), we have:

$$x_1' = x_0 \pm \frac{R \times |x_0 - x_1|}{2 \times l_1 \times \sin\frac{\theta}{2}} \tag{4}$$

$$y_1' = y_0 \pm \frac{R \times |y_0 - y_1|}{2 \times l_1 \times \sin\frac{\theta}{2}} \tag{5}$$

If $x_1 < x_0$, then:

$$x_1' = x_0 - \frac{R \times |x_0 - x_1|}{2 \times l_1 \times \sin\frac{\theta}{2}} \tag{6}$$

Otherwise:

$$x_1' = x_0 + \frac{R \times |x_0 - x_1|}{2 \times l_1 \times \sin\frac{\theta}{2}} \tag{7}$$

y_1' is similar to x_1', and the method to get the coordinate of E is similar to that of D. That is, when the coordinates of D and E are, respectively, $(x_0 \pm \frac{R \times |x_0 - x_1|}{2 \times l_1 \times \sin\frac{\theta}{2}}, y_0 \pm \frac{R \times |y_0 - y_1|}{2 \times l_1 \times \sin\frac{\theta}{2}})$ and $(x_0 \pm \frac{R \times |x_0 - x_2|}{2 \times l_2 \times \sin\frac{\theta}{2}}, y_0 \pm \frac{R \times |y_0 - y_2|}{2 \times l_2 \times \sin\frac{\theta}{2}})$, the total length of OD and OE is the longest, the total length of AD and BE is the shortest, and the number of nodes is the least.

To summarize, the number of the nodes can be the least when the CP coordinate of ICH is

$$(x_0 \pm \frac{R \times |x_0 - x_i|}{2 \times l_i \times \sin\frac{\theta_i}{2}}, y_0 \pm \frac{R \times |y_0 - y_i|}{2 \times l_i \times \sin\frac{\theta_i}{2}})$$

6. Algorithm Comparison and Simulation Analysis

6.1. Algorithm Comparison

Figure 5a is the distribution of segments before the restoration, where there is no mutual communication between the isolated segments. Figure 5b is the diagram of network topology structure which is restored by 2C-SpiderWeb algorithm. From Figure 5b, we can see that this network topology has a large overlapping coverage. Because of this, the network has high average degree after the restoration. However, in this case, high average degree do not represent the network has better fault tolerance. If the localized fault occurs near the CP of OCH, such as the fire, the network restored by 2C-SpiderWeb may be divided into several segments again with high probability. Therefore, although the network has 2-connectivity after the restoration by 2C-SpiderWeb, it does not have good fault tolerance.

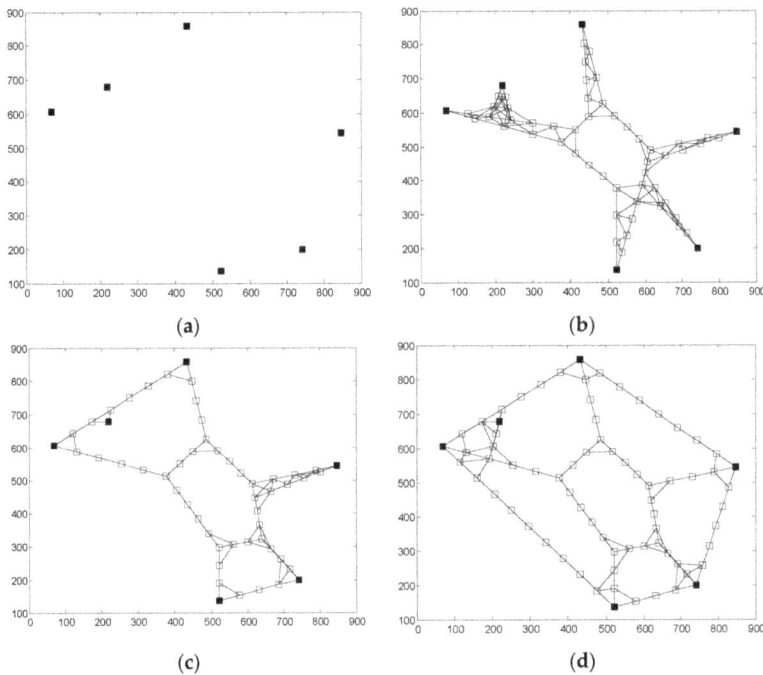

Figure 5. (a) The distribution of segments before the restoration; (b) 2C-SpiderWeb; (c) F2CRA; and (d) P3CRA.

Figure 5c,d are, respectively, the graphs of network topology structure of F2CRA and P3CRA after the restoration. From Figure 5c, we can see that all the network nodes after the restoration have at least 2-connevtivity. With such topology structure, the network can continue to run stably when one node fails. From Figure 4d, we can see that the network topology structure formed by P3CRA has larger coverage and better fault tolerance than others'. When the localized fault occurs, both F2CRA and P3CRA can maintain the stability of the network, and the network is not easy to be divided into many segments again.

6.2. Simulation Analysis

In this part, we will make a comparison among Hamilton Path algorithm, 2C-SpiderWeb algorithm, F2CRA and P3CRA from the four aspects: the number of nodes (Figure 6), total coverage of nodes (Figure 7), average coverage of each node (Figure 8) and average degree (Figure 9), so as to verify the feasibility and superiority of the proposed algorithm. In this simulation, the segments are distributed on the 2D plane of 1000×1000 m^2 randomly. Besides, the node communication range in Figures 6a, 7a, 8a and 9a is fixed with the value of 50 m, and the number of segments in Figures 6b, 7b, 8b and 9b is fixed at 8.

6.2.1. The Number of Relay Nodes

From Figure 6a, we can see that when the communication radius of nodes is fixed, the number of nodes used in these four algorithms will increase with the growth of the number of segments. The more segments and the longer the total path length of the segments are, the more nodes that need to be deployed; therefore, more nodes will be used totally. It can be seen from the figure that no matter how many segments the network being divided, the nodes used in F2CRA is fewer than those in 2C-SpiderWeb algorithm, but more than those in Hamilton Path algorithm. This is determined by different topology structure of various algorithms. As P3CRA has partial 3-connectivity, the number of nodes used in this algorithm is higher than that in the other three algorithms.

From Figure 6b, we can see that the number of segments being fixed, the relay nodes used in these four algorithms are reduced with the increase of the communication radius. This is because the number of nodes is determined by the communication range of nodes when the position of segments and the distance between the segments are fixed. When the node radius is enlarged, the number of nodes between the segments is less and then the total number of nodes will be less. From Figure 6b, we can see that no matter how the radius of nodes changes, the number of nodes in P3CRA is larger than that of the other three algorithms. In addition, with the increase of the node radius, the number of nodes used in F2CRA will be more close to that in 2C-SpiderWeb algorithm. This is because the network topology formed by 2C-SpiderWeb algorithm is more similar to the one formed by F2CRA when the node radius is enlarged. As a result, the number of nodes used in these two algorithms is close.

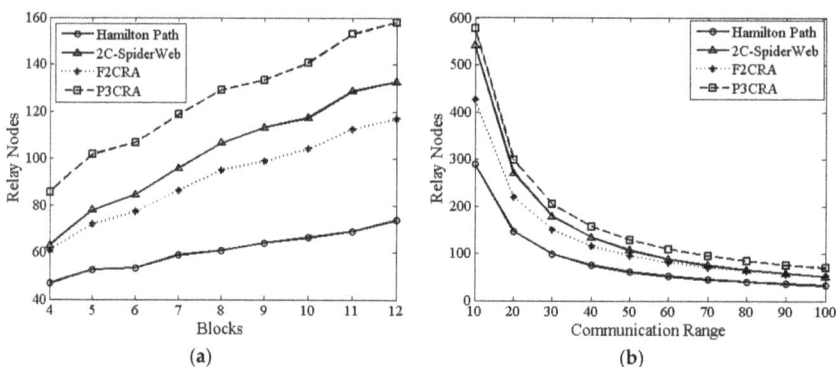

Figure 6. (a) Relay Nodes *vs.* Segments; and (b) Relay Nodes *vs.* Communication Radius.

6.2.2. Total Coverage

From Figure 7, we can see that with the increase of the number of segments and the communication radius of nodes, the total coverage of these four algorithms increases. The coverage area of F2CRA is larger than that of 2C-SpiderWeb algorithm and Hamilton Path algorithm, while the coverage area of P3CRA is much larger than other three algorithms'. Although 2C-SpiderWeb algorithm has more nodes than F2CRA, the coverage area of F2CRA is always larger than that of 2C-SpiderWeb algorithm, no matter how the number of segments or the communication radius of nodes changes. From Figure 5, we can know that compared with 2C-SpiderWeb algorithm, F2CRA has a smaller overlapping area. Therefore, no matter how the number of segments or the communication radius of nodes changes, F2CRA has larger coverage area than 2C-SpiderWeb algorithm. Similarly, P3CRA uses more nodes and has smaller coverage area than F2CRA. Hence, no matter how the number of segments or the communication radius of nodes changes, P3CRA has larger coverage area than other algorithms.

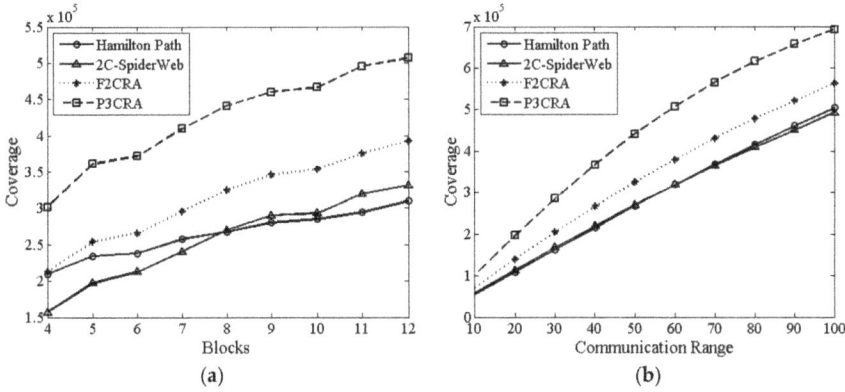

Figure 7. (a) Coverage Area *vs.* Segments; and (b) Coverage Area *vs.* Communication Radius.

6.2.3. Average Coverage

From Figure 8a, we can see that the average coverage of F2CRA is basically the same with that of P3CRA, less than that of each node by Hamilton Path algorithm, but more than that of 2C-SpiderWeb algorithm. From Figure 5, we can visually know that 2C-SpiderWeb algorithm has a large overlapping area. Because of the

large overlapping area, the actual coverage area of network topology formed by 2C-SpiderWeb algorithm becomes small. As a result, the average coverage area of each node becomes small. Compared with other three algorithms, the network topology formed by Hamilton Path algorithm has the smallest network coverage area. Consequently, the average coverage area of each node of Hamilton Path algorithm is the largest.

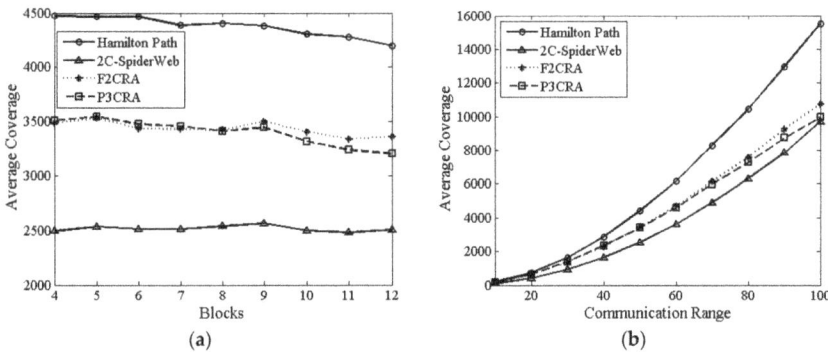

Figure 8. (a) The Average Coverage of Each Node *vs.* Segments; and (b) The Average Coverage of Each Node *vs.* Communication Radius.

From Figure 8b, we can see that the average coverage area of each node will increase with the growth of the communication radius. When the position of segments is fixed, the nodes deployed between the segments depend on the communication radius of nodes. The larger the communication radius is, the smaller the number of nodes will be, but the larger the communication radius is, the larger the coverage area will be, so is the average coverage of all the nodes.

6.2.4. Average Degree

From Figure 9a, we can see that the average degrees of both F2CRA and P3CR are lower than that of 2C-SpiderWeb algorithm because 2C-SpiderWeb algorithm has larger coverage overlap, and the nodes in the overlapping part have larger degree. Therefore, the average degree of this algorithm is greater than that of the other three algorithms. Moreover, the topology of Hamilton Path algorithm can be regarded as a ring, where the degree of the node is nearly 2. As some nodes overlap between them, the final average degree is slightly larger than 2. The average degrees of the two algorithms we proposed are between 2C-SpiderWeb algorithm's and Hamilton Path algorithm's, the reason of which can be seen clearly from the algorithm topology diagram. Compared with 2C-SpiderWeb algorithm, the two algorithms we proposed have a smaller overlapping coverage; while compared with Hamilton Path algorithm, they have a larger overlapping coverage. Hence, the final average degrees of the two algorithms we proposed are between 2C-SpiderWeb algorithm's and Hamilton Path algorithm's. Moreover, from Figure 9a, we can see that the average degree of P3CRA is slightly lower than that of F2CRA. The reason is that compared with F2CRA, the network topology formed by P3CRA has the smaller overlapping area; therefore, the average degree of P3CRA is slightly lower than that of F2CRA.

The case in Figure 9b is similar with that in Figure 8a, but from Figure 9b, we can see that the average degree increases with the increase of the communication radius. When the deployment location of the node is determined, the larger the communication radius of the node is, the larger the overlapping area between the nodes will be. As a result, the average degree will be larger.

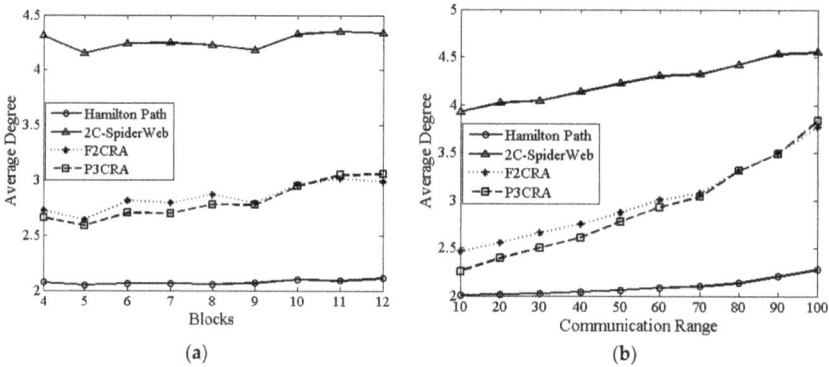

Figure 9. (a) Average Degree *vs.* Segments; and (b) Average Degree *vs.* Communication Radius.

7. Conclusions

Due to the deployment environment, WSN is prone to large-scale failure; therefore, the effective algorithm is needed for timely recovery so that the network can run normally and stably. In this paper, we propose two fault restoration algorithms, respectively, solving the WSN fault restoration problem from different points. F2CRA is suitable when cost is considered first; and P3CRA is suitable when the performances of the network are considered first. Compared with other algorithms, these two algorithms ensure that the network has the stronger fault-tolerant function, larger coverage area and more balanced load after the restoration. In future work, we plan to consider other factors of deployment environment in our algorithms, such as obstacles and rough terrain, so that the proposed algorithms can be more in line with the actual situation.

Acknowledgments: The authors wish to thank National Natural Science Foundation of China (Grant No. U1405255). Fujian Normal University Innovative Research Team (No. IRTL1207).

Author Contributions: Yali Zeng raised the research problem, directed the algorithm design, matlab implementation, simulation execution and drafted the manuscript. Li Xu and Zhide Chen organized the manuscript structure, conducted algorithm design, data analysis, result presentation and polished the writing.

Conflicts of Interest: The authors declare no conflict of interest.

Appendix

Proof of "When a is equal to b, $a + b$ reaches the maximum value, which means the total length of OD and OE is the longest".

As shown in Figure A1, we assume the coordinates of point A, B, and O are (x_1, y_1), (x_2, y_2), and (x_0, y_0), respectively. Set $\angle AOB = \theta$, the lengths of DE, OD and OE are R, a and b, respectively. $\angle ODE = \alpha$, $\angle OED = \beta$, a, α, β are unknown.

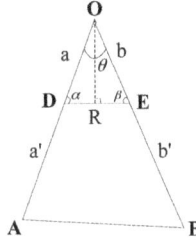

Figure A1. Diagram of triangle.

From the known, we know AB, AO and BO are, respectively:

$$side_1 = \sqrt{(x_1 - x_2)^2 + (y_1 - y_2)^2} \tag{A1}$$

$$l_1 = \sqrt{(x_1 - x_0)^2 + (y_1 - y_0)^2} \tag{A2}$$

$$l_2 = \sqrt{(x_2 - x_0)^2 + (y_2 - y_0)^2} \tag{A3}$$

From Equations (A1)–(A3) and the cosine law, we can obtain the value of θ:

$$\theta = \arccos\left(\frac{l_1^2 + l_2^2 - side_1^2}{2 \times l_1 \times l_2}\right) \tag{A4}$$

By trigonometric function, we have:

$$\frac{R}{\sin\theta} = \frac{a}{\sin\alpha} = \frac{b}{\sin\beta} = \frac{a+b}{\sin\alpha + \sin\beta} \tag{A5}$$

So

$$a + b = \frac{\sin\alpha + \sin\beta}{\sin\theta} R
\begin{aligned}
&= \frac{2\sin\dfrac{\alpha + \beta}{2}\cos\dfrac{\alpha - \beta}{2}}{\sin\theta} R \\
&= \frac{2\sin\dfrac{\pi - \theta}{2}\cos\dfrac{\alpha - \beta}{2}}{\sin\theta} R \\
&= \frac{2\cos\dfrac{\theta}{2}\cos\dfrac{\alpha - \beta}{2}}{\sin\theta} R \\
&= \frac{\cos\dfrac{\alpha - \beta}{2}}{\sin\dfrac{\theta}{2}} R
\end{aligned} \tag{A6}$$

As θ and R are known, when $\alpha = \beta$, $\cos\dfrac{\alpha - \beta}{2}$ has the maximum value, then $a + b$ reaches the maximum value. That is, when $a = b$, $a + b$ reaches the maximum value, which means the total length of OD and OE is the longest.

References

1.   Malaver, A.; Motta, N.; Corke, P.; Gonzalez, F. Development and integration of a solar powered unmanned aerial vehicle and a wireless sensor network to monitor greenhouse gases. *Sensors* **2015**, *15*, 4072–4096. [CrossRef] [PubMed]

2.   Hao, B.; Tang, J.; Xue, G. Fault-Tolerant Relay Node Placement in Wireless Sensor Networks: Formulation and Approximation. In Proceedings of the International Conference on High Performance Switching and Routing, Phoenix, AZ, USA, 19–21 April 2004; pp. 246–250.

3.   Zhang, W.; Xue, G.; Misra, S. Fault-Tolerant relay Node Placement in Wireless Sensor Networks: Problems and Algorithms. In Proceedings of the IEEE International Conference on Computer Communications, Anchorage, AK, USA, 6–12 May 2007; pp. 1649–1657.

4.   Han, X.; Cao, X.; Lloyd, E.L.; Shen, C. Fault-tolerant relay node placement in heterogeneous wireless sensor networks. *IEEE Trans. Mob. Comput.* **2010**, *9*, 643–656.

5.   Senel, F.; Younis, M.F.; Akkaya, K. Bio-inspired relay node placement heuristics for repairing damaged wireless sensor networks. *IEEE Trans. Veh. Technol.* **2011**, *60*, 1835–1848. [CrossRef]

6.   Sitanayah, L.; Brown, K.N.; Sreenan, C.J. A fault-tolerant relay placement algorithm for ensuring k vertex-disjoint shortest paths in wireless sensor networks. *Ad Hoc Netw.* **2014**, *23*, 145–162. [CrossRef]

7.   Lin, G.H.; Xue, G. Steiner tree problem with minimum number of Steiner points and bounded edge-length. *Inf. Process. Lett.* **1999**, *69*, 53–57. [CrossRef]

8.   Chen, D.; Du, D.Z.; Hu, X.D.; Lin, G.; Wang, L.S.; Xue, G.L. Approximations for Steiner trees with minimum number of Steiner points. *J. Glob. Optim.* **2000**, *18*, 17–33. [CrossRef]

9.   Lloyd, E.L.; Xue, G. Relay node placement in wireless sensor networks. *IEEE Trans. Comput.* **2007**, *56*, 134–138. [CrossRef]

10. Li, S.; Chen, G.; Ding, W. Relay Node Placement in Heterogeneous Wireless Sensor Networks with Basestations. In Proceedings of the International Conference on Communications and Mobile Computing, Kunming, China, 6–8 January 2009; pp. 573–577.

11. Yang, D.; Misra, S.; Fang, X.; Xue, G.; Zhang, J. Two-tiered constrained relay node placement in wireless sensor networks: Computational complexity and efficient approximations. *IEEE Trans. Mob. Comput.* **2012**, *11*, 1399–1411. [CrossRef]

12. Senel, F.; Younis, M. Relay node placement in structurally damaged wireless sensor networks via triangular Steiner tree approximation. *Comput. Commun.* **2011**, *34*, 1932–1941. [CrossRef]

13. Bhattacharya, A.; Kumar, A. A shortest path tree based algorithm for relay placement in a wireless sensor network and its performance analysis. *Comput. Netw.* **2014**, *71*, 48–62. [CrossRef]

14. Tang, J.; Hao, B.; Sen, A. Relay node placement in large scale wireless sensor networks. *Comput. Commun.* **2006**, *29*, 490–501. [CrossRef]

15. Misra, S.; Hong, S.D.; Xue, G.; Tang, J. Constrained Relay Node Placement in Wireless Sensor Networks to Meet Connectivity and Survivability Requirements. In Proceedings of the IEEE International Conference on Computer Communications, Phoenix, AZ, USA, 13–18 April 2008; pp. 281–285.

16. Kashyap, A.; Khuller, S.; Shayman, M.A. Relay Placement for Higher Order Connectivity in Wireless Sensor Networks. In Proceedings of the IEEE International Conference on Computer Communications, Barcelona, Spain, 23–29 April 2006; p. 12.

17. Bredin, J.; Demaine, E.D.; Hajiaghayi, M.T.; Rus, D. Deploying Sensor Networks with Guaranteed Capacity and Fault Tolerance. In Proceedings of the 6th ACM International Symposium on Mobile Ad Hoc Networking and Computing, Cologne, Germany, 28 August–2 September 2005; pp. 309–319.

18. Lee, S.; Younis, M.; Lee, M. Connectivity restoration in a partitioned wireless sensor network with assured fault tolerance. *Ad Hoc Netw.* **2015**, *24*, 1–19. [CrossRef]

19. Pu, J.; Xiong, Z.; Lu, X. Fault-tolerant deployment with k-connectivity and partial *k*-connectivity in sensor networks. *Wirel. Commun. Mob. Comput.* **2009**, *9*, 909–919. [CrossRef]

2

An Electrically Tunable Zoom System Using Liquid Lenses

Heng Li [1,†], Xuemin Cheng [2,†] and Qun Hao [1,*]

Academic Editor: Manuela Vieira

[1]   School of Optoelectronics, Beijing Institute of Technology, Beijing 100081, China; liheng@bit.edu.cn
[2]   Graduate School at Shenzhen, Department of Precision Instrument, Tsinghua University, Shenzhen 518055, China; cheng-xm@mail.tsinghua.edu.cn
[*]   Correspondence: qhao@bit.edu.cn
[†]   These authors contributed equally to this work.

Abstract: A four-group stabilized zoom system using two liquid lenses and two fixed lens groups is proposed. We describe the design principle, realization, and the testing of a 5.06:1 zoom system. The realized effective focal length (EFL) range is 6.93 mm to 35.06 mm, and the field of view (FOV) range is $8°$ to $40°$. The system can zoom fast when liquid lens 1's (L_1's) optical power take the value from 0.0087 mm^{-1} to 0.0192 mm^{-1} and liquid lens 2's (L_2's) optical power take the value from 0.0185 mm^{-1} to -0.01 mm^{-1}. Response time of the realized zoom system was less than 2.5 ms, and the settling time was less than 15 ms. The analysis of elements' parameters and the measurement of lens performance not only verify the design principle further, but also show the zooming process by the use of two liquid lenses. The system is useful for motion carriers e.g., robot, ground vehicle, and unmanned aerial vehicles considering that it is fast, reliable, and miniature.

Keywords: optical imaging; liquid lens; stabilized zoom system; Gaussian bracket

1. Introduction

The optical zoom system used for imaging or projection is widely applied to many fields, such as surveillance, medicine, aviation, and aerospace [1–5]. Optical zoom systems used in imaging must satisfy two basic conditions: adjustable focal length and a fixed image plane. Traditional zoom systems usually comprise several lens groups and the separation between adjacent groups is allowed to vary. Focal length of the zoom system is varied continuously by the displacements of lens groups, the individual lens group must move along the precise track, and the action of multiple lens groups must ensure synchronization [6]. As a result, a complex mechanical camera system is required in the process of zooming. As the conventional zoom system is applied to high-speed moving vehicles, the reliability of the conventional zoom systems will be reduced due to vibration, and the response time is hard to meet the requirement of high-speed as well.

A novel optical zoom system based on focal length variable optical components has been presented. Focal length of such optical components can be adjusted by varying its surface shape or refractive index [7,8]. Different from traditional optical zoom systems, this new type of optical zoom system can operate by adjusting the focal length variable optical components without involving any motorized movements. This novel optical zoom system is more attractive because of the low complexity, high reliability, and the ability to be easily miniaturized, which make it suitable for being utilized in robots or other moving carriers. Many researchers designed zoom systems based on such focal length variable optical components, which reduced the dependence on motorized components, therefore effectively decreasing the complexity of the system [9–17]. However, there still remain difficulties in the realization of stabilized zoom systems using optical power as a zoom variable. For example, it would be ideal if the focal length variable elements would compensate for the image shift while taking over the variations of the focal length of the zoom system. This is one reason why most of the existing designs still have a small amount of motorized components. Requirements in large zoom ratio, high zoom precision, high imaging quality, and excellent electrical performance are also giving lens design greater challenges.

In this paper, we propose to realize a four-group stabilized zoom system based on two liquid lenses. Firstly, we analyze the solution of image displacement compensation, and discuss the design method based on the theory of Gaussian brackets. Stable image plane provides help for improving the imaging quality. The method has been presented in our previous work [18]. It solved the problem of increasing zoom ratio with a mathematical method. Optimization of parameters was realized by mathematically solving extreme value and maximum gradient of the function. Application of the method has been extended in this paper. In view of the specific application, we designed and optimized a realizable zoom system, and the zoom ratio is designed to be 5:1. Secondly, we realized the zoom device and set up an experiment platform to test the performance of the zoom system. We test zoom ratio, zoom precision, and imaging quality. The results show that the realized system has good electrical performance and high imaging quality. Zoom ratio and precision of the system have reached the design target.

2. Theory Analysis and Design

2.1. Imaging System Design

We presented a design method that is applicable to four-group stabilized zoom system in the previous work [18]. The optical layout of the four-group stabilized zoom system is shown in Figure 1, where Φ_1 and Φ_3 are the optical power of the fixed groups, Φ_2 and Φ_4 are optical power of the stabilized focal-length-variable groups. The terms e_1, e_2, e_3 are the separations between the consecutive principle planes in each group, e_4 is the separation between the last group and the image plane, m_i is magnification of each group.

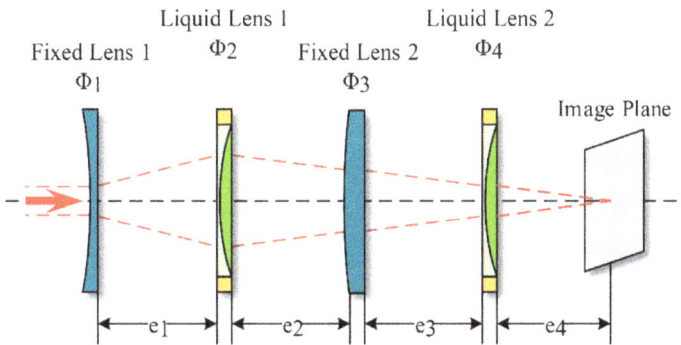

Figure 1. Structure of the zoom system.

In previous work, we analyzed this four-group stabilized zoom systems using the optical power of the optical element as the independent variables in the zoom equation. Simplified differential function is used to describe the relationship between the focal length variation of varifocal elements and the focal length variation of zoom system. The differentiable condition optimization can be used to get the analytic solution. In the zooming process, the function of the optical power of the focal length variable groups is monotonic, and the function of magnification has a breakpoint. The optical layout can be miniaturized as magnification has a larger derivative near the breakpoint. If the sign of m_3 is changeable during zooming, a larger derivative of m_3 can be found near the breakpoint. Then we optimize the Gaussian parameters of the fixed lens groups, make $\dfrac{e_4}{(1 - e_1\phi_1) \cdot {}^2B_4}$ and $\dfrac{1 - e_1\phi_1}{e_4 \cdot {}^2B_4}$ ($^2B_4 = [-e_2, \phi_3, -e_3] = e_2 \cdot e_3 \cdot \phi_3 - e_2 - e_3$ is Gaussian brackets parameter) converge to a minimum to reach the requirement of zoom ratio.

In this paper, the optical design of a 5:1 stabilized zoom system is investigated. The zoom range is designed to be 7 mm to 35 mm. Design parameters are determined based on the requirement of a kind of ground vehicle being deployed on a bumpy road. The values of the Gaussian parameters are listed in Table 1, $\left|\dfrac{e_4}{(1 - e_1\phi_1) \cdot {}^2B_4}\right| = 0.12$, and $\left|\dfrac{1 - e_1\phi_1}{e_4 \cdot {}^2B_4}\right| = 0.001$.

The relationship between m_2, m_3 and m_4 is shown in Figure 2, a larger derivative of m_3 can be found near the breakpoint since the sign of m_3 is changeable.

Table 1. Gaussian parameters of the zoom system.

	$1/\Phi$ (mm)	Φ_2 (mm^{-1})	Φ_4 (mm^{-1})	m_2	m_3	m_4
Wide	7.27	0.008	0.019	1.28	−0.28	0.56
Middle	26.14	0.019	0	2.31	0.11	1.03
Tele	36.34	0.02	−0.01	2.53	0.19	1.24

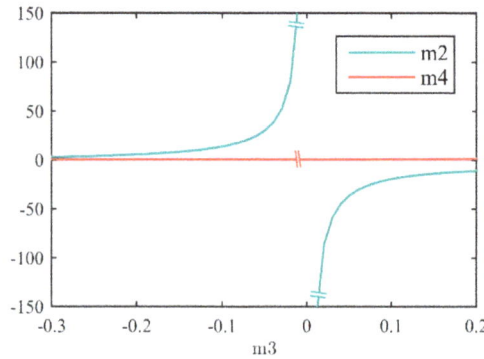

Figure 2. Relationship between m_2, m_3, and m_4. A larger derivative of m_3 can be found near the breakpoint since the sign of m_3 is changeable.

We use two liquid lenses (L_1 and L_2) as the focal length variable groups (Φ_2, Φ_4), the variation of Φ_2 and Φ_4 can be obtained as follows:

$$\Phi_2 = \frac{e_4\Phi - [\phi_1, -(e_1 + e_2), \phi_3, -e_3]}{(1 - e_1\phi_1) \cdot {}^2B_4} \tag{1}$$

$$\Phi_4 = \frac{1 - e_1\phi_1}{e_4\Phi \cdot {}^2B_4} + \frac{{}^2B_4 - e_4[-e_2, \phi_3]}{e_4 \cdot {}^2B_4} \tag{2}$$

$$\Delta\Phi_2 = \frac{e_4}{(1 - e_1\phi_1) \cdot {}^2B_4}\Delta\Phi \tag{3}$$

$$\Delta\Phi_4 = \frac{1 - e_1\phi_1}{e_4 \cdot {}^2B_4}\Delta\left(\frac{1}{\Phi}\right) \tag{4}$$

where $\Delta\Phi = \Phi_s - \Phi_l$ is the variation of the system optical power, $\Delta\left(\dfrac{1}{\Phi}\right) = \dfrac{1}{\Phi_s} - \dfrac{1}{\Phi_l}$, Φ_s and Φ_l represent the system optical power at the wide-angle end and the telephoto end respectively, $\Delta\Phi_2$ and $\Delta\Phi_4$ are the variation needed to achieve a given zoom range $\dfrac{1}{\Phi_s}$ to $\dfrac{1}{\Phi_l}$.

The general parameters of the zoom system are shown in Table 2. The initial lens structure of the optical system using these Gaussian parameters is simulated and optimized with the commercial optical design software named CODE V (Optical Research Associates, Tucson, AZ, USA), the result is shown in Figure 3, as the wide-angle end, the middle phase, and the telephoto end are shown in (a), (b), and (c), respectively.

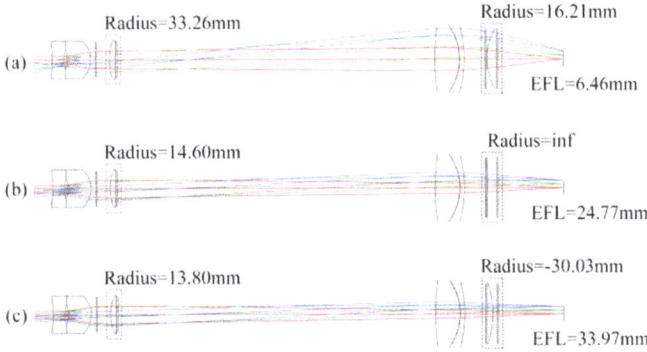

Figure 3. Initial lens structure of the optical system.

Table 2. General parameters of the zoom system.

Surface	#	Curvature Radius	Thickness	Glass
Object		Infinity	Infinity	
	1	Infinity	−9.7160	
	2	−27.4979	3.7885	HLAF4_CDGM
Fixed Lens1	3	7.5970	1.4003	
	4	−9.0978	7.8405	ZK9_CHINA
	5	−8.1539	1.5000	
	6	Infinity	0.1000	
	7	Infinity	0.5000	BK7_SCHOTT
	8	Infinity	zoom	
Liquid Lens1	9	zoom	zoom	"OL1024"
	10	Infinity	0.5000	BK7_SCHOTT
	11	Infinity	1.1000	
	12	Infinity	25.9971	
	13	Infinity	81.4883	
	14	77.7631	8.2500	ZK9_CHINA
Fixed Lens2	15	−18.8293	0.1000	
	16	−18.5589	1.5000	ZF50_CDGM
	17	−41.6000	7.0723	
	18	Infinity	0.0500	
	19	Infinity	0.5000	BK7_SCHOTT
	20	Infinity	zoom	
Liquid Lens2	21	zoom	zoom	"OL1024"
	22	Infinity	0.5000	BK7_SCHOTT
	23	Infinity	3.4500	
	24	Infinity	0.8943	
	25	Infinity	17.9482	
Image		Infinity	0.0000	

2.2. Operating Range and Precision

The zoom system operates by changing the optical power of two liquid lenses (L_1's optical power Φ_2, and L_2's optical power Φ_4). Two focus tunable lenses (Optotune, Swizerland) are used in the experiments, a 10 mm aperture commercial lens EL-10-30 for L_1, and a 20 mm aperture custom-design lens for L_2. Both L_1 and L_2 are liquid-filled membrane lenses, the refractive index of the lenses is 1.30, and Abbe number is 100. Φ_2 and Φ_4 are controlled by adjusting L_1's control current (I_1) and L_2's control current (I_2), respectively. Range of Φ_2 is 0.008 mm^{-1} to 0.022 mm^{-1}, control precision is 4.67×10^{-5} mm^{-1}. Range of Φ_4 is -0.0142 mm^{-1} to 0.02 mm^{-1}, the control precision is 3×10^{-4} mm^{-1}, zero optical power is attainable. The relationships between optical power and control currents are shown in Figure 4. As shown in figure, precision of L_1 is much higher than L_2, thus we use L_2 as active element, and use L_1 as servo element to compensate the image displacement.

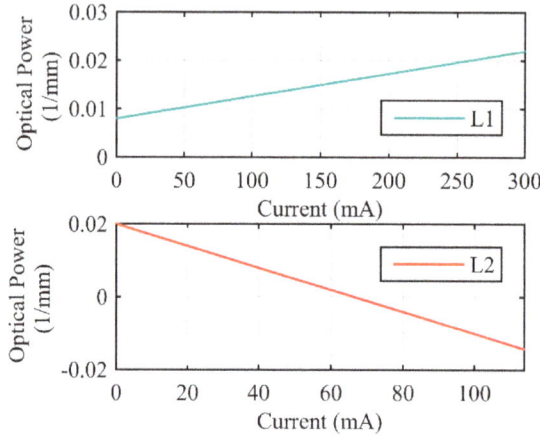

Figure 4. Optical power of the liquid lenses as functions of control currents.

Our purpose is to design a zoom system with effective focal length (EFL) of 7 mm to 35 mm. In Section 2.1, we calculated the Gaussian parameters of the system, and optimized the system with CODE V. From Equation (2), we can calculate EFL of the zoom system theoretically. The calculated range of EFL is 7.27 mm to 36.34 mm, the zoom ratio is 5:1. Then we simulate the designed system, select 20 sampling points, set I_2 from 5 mA to 100 mA, respectively. Relationships between EFL and I_2 are shown in Figure 5. The simulated EFL range is 6.46 mm to 33.97 mm, and the zoom ratio is 5.26:1.

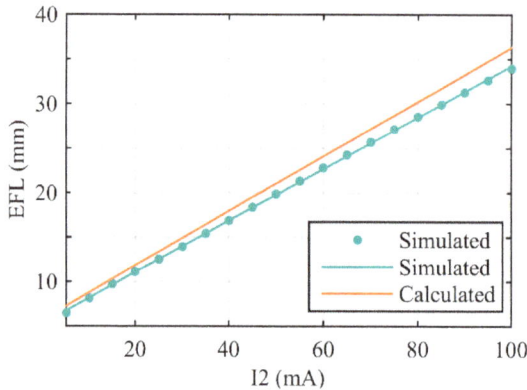

Figure 5. Calculated and simulated EFL as functions of I_2. The "simulated dots" mark the EFL of sampling points that be obtained by CODE V. The "simulated line" is the fitting curve of the dots.

In this paper, zoom precision is defined as the minimum variation of EFL that can be adjusted *i.e.*, the variation of EFL when I_2 increase or decrease 1 mA. From Equation (4), we can theoretically calculate the zoom precision to be 0.3 mm. Then we simulate the zoom precision with CODE V. We firstly select 20 sampling points when I_2 takes different values listed in Table 3, then make I_2 increase 1 mA, corresponding values of Φ_4 can be obtained from Figure 4. The simulated results are shown in Table 3.

Table 3. Simulated zoom precision of the zoom system.

I_2 (mA)	Zoom Precision (mm)	I_2 (mA)	Zoom Precision (mm)
5	0.2911	55	0.2945
10	0.2731	60	0.2924
15	0.2612	65	0.2898
20	0.2762	70	0.2867
25	0.2862	75	0.2832
30	0.292	80	0.2792
35	0.295	85	0.2749
40	0.2964	90	0.2703
45	0.2966	95	0.2652
50	0.2959	100	0.2601
Average	0.2830	RMS	0.0122

Therefore, we designed and simulated a zoom system with 5:1 zoom ratio. The calculated EFL range is 7.27 mm to 36.34 mm while the simulated range is 6.46 mm to 33.97 mm. The calculated zoom precision is 0.3 mm while the average simulated value is 0.2830, and root-mean-square (RMS) of the simulated zoom precision is 0.0122 mm.

3. Experimental and Analysis

3.1. Experimental Setup

In order to test the performance of the zoom system, we set up an experimental platform shown in Figure 6. Test targets were backlit with white light and two forms of targets were used in measurement. The size of the target was 240 mm × 170 mm. The zoom system was placed between the test target and the CCD camera, and the object distance was set to be 265 mm. The CCD was 1/4 inch, pixel size of which was 7.4 μm. In fact, the zoom system and the CCD camera were fixed connected through a standard C lens mount. CCD camera captured the images and then sent them to the computer. We obtained the information on imaging quality and magnification through image processing, and then sent new control parameters to the zoom system. A dc power supply was used to complete the D/A conversion.

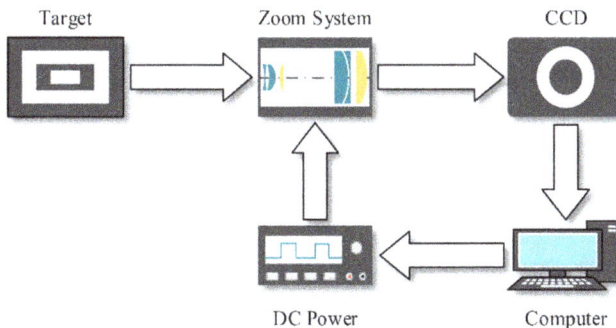

Figure 6. Structure of the experimental platform.

There were two major parameters we needed to measure: EFL and modulation transfer function (MTF). EFL was used to evaluate the zoom ratio and zoom precision, and MTF was used to evaluate the imaging quality. During measurement, we selected 20 sampling points as we had done in simulation when adjusted I_2 from 5 mA to 100 mA. The forms of targets and the methods of measuring EFL and MTF would be discussed in the following sections.

3.2. Initial Structure

In their initial state, I_1 and I_2 both took 0 mA. MTF was used to evaluate the quality of imaging. The method of calculating MTF of the system would be provided in Section 3.6, we currently used the calculation result. Initial parameters of the zoom system are shown in Table 4. MTF value when spatial frequency took 20 lp/mm was used to evaluate the image quality.

Table 4. Initial parameters of the zoom system.

I_2 (mA)	EFL (mm)	FOV (Degree)	MTF Value at 20l p/mm
0	6.56	41.7	0.01
5	6.93	39.7	0.28

Image quality was poor because L_1 had not compensated for L_2 and the optical aberration of L_2 was high when Φ_4 took values near the limit position.

In order to achieve high image quality in the whole zoom process, we set $I_2 = 5$ mA and adjusted L_1 to compensate for L_2. MTF diagram is shown in Figure 7. Parameters of the zoom system when I_2 takes 5 mA are shown in Table 4.

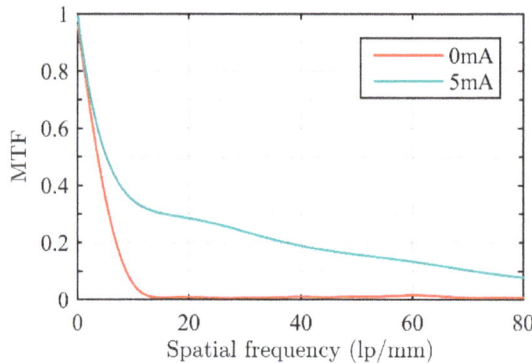

Figure 7. MTF curve of the zoom system when I_2 took 0 mA and 5 mA, respectively.

In the following discussion, we take the state when $I_2 = 5$ mA as the initial state.

3.3. Electrical Performance Testing

In the process of measuring, L_2 was adjusted actively to change EFL of the zoom system when L_1 supplied compensation to get clear images. The relationship between I_1 and I_2 can be calculated from Equations (1) and (2).

We set I_2 to a value, coarsely adjusted I_1 till we got clear image on computer, and then tightly adjusted I_1. Range of tight adjustment was usually less than ± 4 mA in actual measurement. We could obtain the real-time MTF of the zoom system with the software written by ourselves on computer. In fact, the MTF curve had no obvious change when we adjusted I_1 in ± 2 mA range. For I_1, this was an acceptable error range. Then we chose the value of MTF when spatial frequency was 20l p/mm as the image quality factor (Q). We considered the zoom system to achieve highest imaging quality when Q got maximum, and I_1 would be recorded. The results of simulation and measurement are shown in Figure 8. Although there were differences between simulated curve and measured curve, we could easily obtain I_1 as function of I_2 by curve fitting. This enabled the zoom system to realize precise electric control. Both the simulated and the measured curves could be fitted well by three-order polynomials, and the correlation coefficients were larger than 0.999. This meant that we only needed a small amount of addition and multiplication operations to obtain I_1 by I_2.

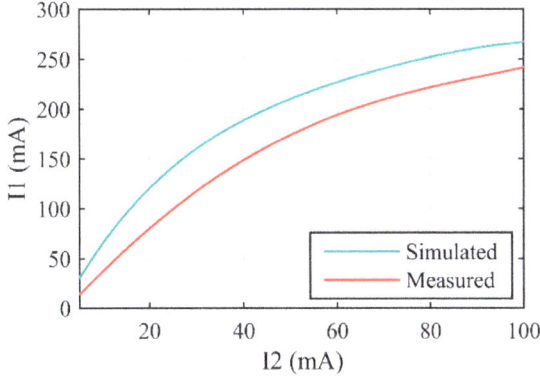

Figure 8. L_1's control current (I_1) as function of L_2's control current (I_2).

Optical power of the liquid lens was the only variant in the zoom system. Although in actual application on ground vehicle, it needed time to calculate I_1 and I_2 for specific EFL, the calculation time was usually at a microsecond level for most of the processors since there were only multiply and add operations in the fitting function. Therefore, the zoom speed depended only on the performance of the liquid lens. The response time of the system was less than 2.5 ms, and the settling time was less than 15 ms due to fluid viscosity.

3.4. Zoom Ratio Testing

Relationships between EFL, field of view (FOV), object height (OH), paraxial image height (PIH), and object distance (OD) could be obtained refer to Equations (5) and (6).

$$PIH = EFL \cdot \tan(\frac{1}{2}FOV) \tag{5}$$

$$OH = OD \cdot \tan(\frac{1}{2}FOV) \tag{6}$$

We obtained the magnification (m) of the zoom system by imaging a fixed length (l_o) target. The magnification could be calculated refer to Equation (7) by counting the pixels (n) when the pixel size (μ) of the CCD camera was known.

$$m = n \cdot \mu/l_o = PIH/OH \tag{7}$$

Thus from Equations (5)–(7), we can obtain EFL refer to Equation (8).

$$EFL = m \cdot OD \tag{8}$$

The measured and simulated results of EFLs are shown in Table 5, and the EFL comparison diagram of calculated value, simulated value and measured value is shown in Figure 9. An approximate linear relationship exists between EFL and I_2. Thus, we can easily determine I_2 to achieve specific EFL in actual application. As presented in Section 3.3, I_1 could be calculated out by I_2 through the fitting function. Thus far, issues of determining control parameters to achieve specific EFL are completely solved.

Table 5. Comparisons between simulated EFL and measured EFL.

I_2 (mA)	Simulated EFL (mm)	Measured EFL (mm)	Difference (mm)
5	6.46	6.93	0.47
10	8.14	7.81	−0.33
15	9.78	8.58	−1.20
20	11.12	9.77	−1.35
25	12.52	10.85	−1.67
30	13.97	12.15	−1.82
35	15.43	13.38	−2.05
40	16.91	14.80	−2.11
45	18.39	16.28	−2.11
50	19.88	17.95	−1.93
55	21.35	19.67	−1.68
60	22.82	21.40	−1.42
65	24.28	23.19	−1.09
70	25.72	24.91	−0.81
75	27.15	26.7	−0.45
80	28.56	28.55	0.01
85	29.94	30.77	0.83
90	31.31	32.31	1.00
95	32.65	33.46	0.81
100	33.97	35.06	1.09

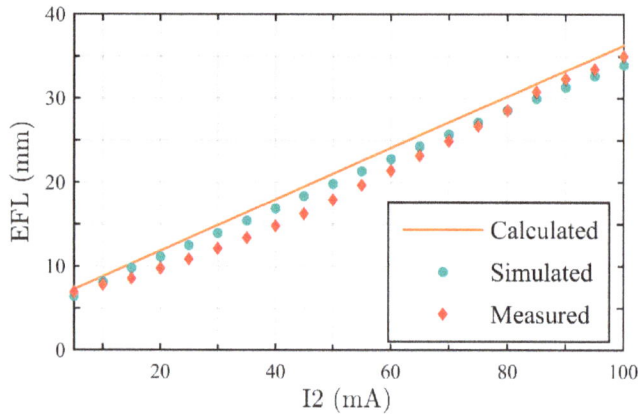

Figure 9. EFL comparison among calculated value, simulated value, and measured value.

Therefore, the measured EFL range of the zoom system is 6.93 mm to 35.06 mm, the zoom ratio is 5.06:1. When $I_2 = 5$ mA, EFL takes minimum, at this moment $\Phi_2 = 0.0087$ mm^{-1} and $\Phi_4 = 0.0185$ mm^{-1}, the FOV is 40°. When $I_2 = 100$ mA, EFL takes maximum, $\Phi_2 = 0.0192$ mm^{-1} and $\Phi_4 = -0.0100$ mm^{-1} at this moment, the FOV is 8°.

Images captured in different EFL are shown in Figure 10a–c show the wide end, the middle phase, and the tele end respectively. In wide end, EFL is 6.93 mm, the zoom system has the widest FOV, and there are 18 pixels between two red lines shown in Figure 10a. In tele end, EFL is 35.06 mm, the zoom system is most detailed, and the number of pixels between two red lines shown in Figure 10c increases to 91. We can also get the conclusion of a 5.06:1 zoom ratio. A middle phase is presented too, in order to make readers observe the zoom process from wide end to tele end visually.

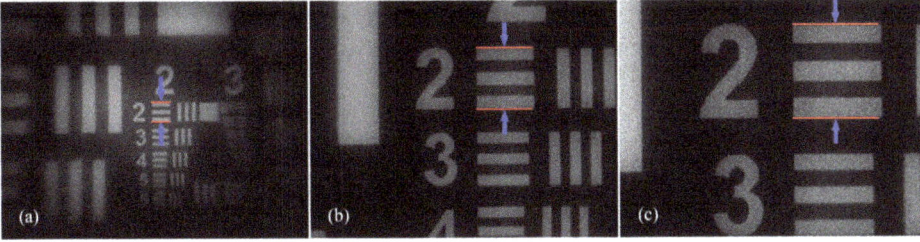

Figure 10. Images captured in different EFL. (**a**) wide end, when $I_2 = 5$ mA, 18 pixels between two red lines; (**b**) middle phase, when $I_2 = 60$ mA, 59 pixels between two red lines; (**c**) tele end, when $I_2 = 100$ mA, 91 pixels between two red lines.

3.5. Zoom Precision Testing

In order to measure the zoom precision of the system, we set I_2 to a value firstly, measured EFL at this moment. Then we adjusted I_2 to increase 1 mA, and measured the variation of EFL. In the above, we had simulated the test process. The simulated results had been shown in Table 3. The measured results are shown in Table 6, and diagram of comparison among calculated, simulated, and measured results is shown in Figure 11.

Table 6. Measured zoom precision.

I_2 (mA)	Zoom Precision (mm)	I_2 (mA)	Zoom Precision (mm)
5	0.2975	55	0.3125
10	0.2725	60	0.2900
15	0.3000	65	0.2750
20	0.2750	70	0.2650
25	0.2700	75	0.3000
30	0.2675	80	0.2625
35	0.3150	85	0.2825
40	0.2750	90	0.3050
45	0.3150	95	0.3050
50	0.3075	100	0.2775
Average	0.2885	RMS	0.0181

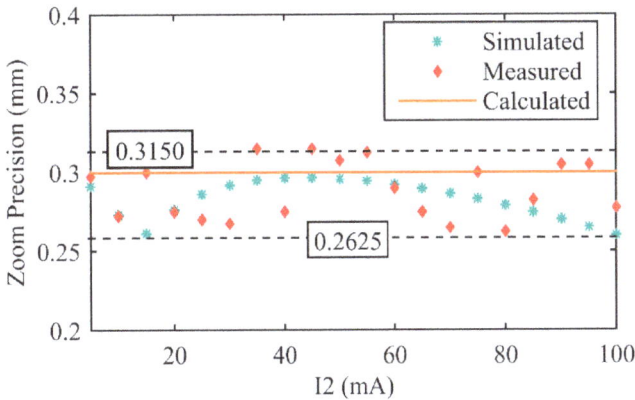

Figure 11. Zoom precision comparison among calculated, simulated, and measured values.

Zoom precision of the realized zoom system is approximately uniform in the whole zoom process. The average of measured zoom precision is 0.2885 mm, near to the calculated value 0.3 mm, and all of measured values are in the range of ± 0.0265 mm to average zoom precision, RMS of the measured zoom precision is 0.0181 mm. Parameters of the realized lens are consistent with the calculated and simulated values.

Therefore, there are 96 discrete EFL states in the entire zoom range, spaces between adjacent states are approximately uniform to be 0.2885 mm. Because the EFL states are so compact, we can consider the realized system as a continuous zoom system.

3.6. Imaging Quality Testing

We used MTF to evaluate the imaging quality of the zoom system. The test target used in imaging quality testing is shown in Figure 12, similar to a black and white chessboard. The tilt of edge was set to $5°$ according to ISO12233.

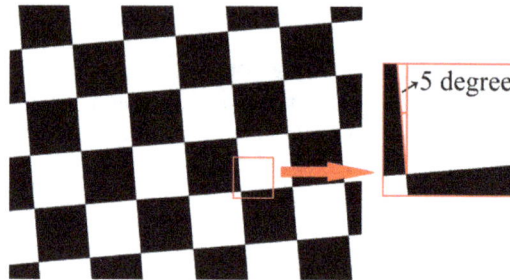

Figure 12. Structure of the target used to test the imaging quality.

In the test process, we selected a region-of-interest (ROI) that contains a slanted-edge in the beginning, then obtained the Edge Spread Function (ESF) by the use of knife-edge method and got the Line Spread Function (LSF) through the differential of ESF. The MTF could be computed by a Fast Fourier Transform (FFT) of LSF. MTFs were measured when I_2 took different values. The results are shown in Figure 13a.

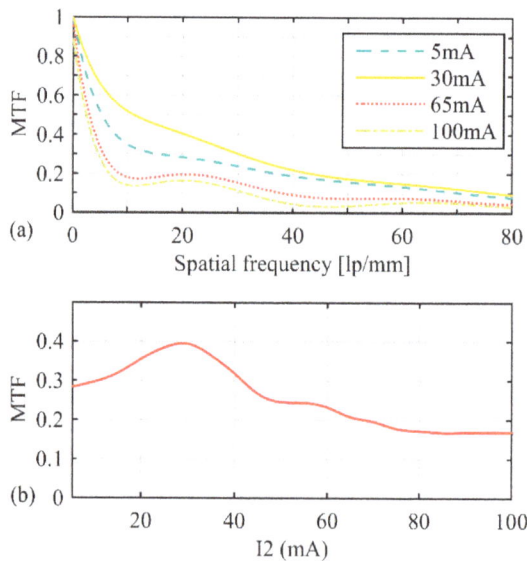

Figure 13. MTF of the zoom system. (**a**) MTFs were measured when I_2 took different values; (**b**) relationship between MTF and I_2 at 20 lp/mm.

At 20 lp/mm, the relationship between MTF and I_2 is shown in Figure 13b. The zoom system got the highest imaging quality when $I_2 = 30$ mA, MTF reached 0.40 at 20 lp/mm. When $I_2 > 90$ mA, MTF reduced to 0.17 at 20 lp/mm.

4. Conclusions

In this paper, we designed and realized an electrically tunable four-group stabilized zoom system by using two liquid lenses. The zoom equations were established by using Gaussian bracket method. L_1 and L_2 were used as the second and the fourth lens group that were focal length variable. The calculated EFL range was 7.27 mm to 36.34 mm while the zoom precision was calculated to be 0.3 mm. We simulated the zoom system, got the simulated EFL range was 6.46 mm to 33.97 mm and the zoom precision was 0.283 mm. Then we made a series of experiments to verify the performance of the realized system. Response time of the realized zoom system was less than 2.5 ms, and the settling time was less than 15 ms. The measured EFL range was 6.93 mm to 35.06 mm, the zoom ratio was 5.06:1. When $\Phi_2 = 0.0087$ mm^{-1} and $\Phi_4 = 0.0185$ mm^{-1}, EFL got the minimum value, the FOV was 40°. When $\Phi_2 = 0.0192$ mm^{-1} and $\Phi_4 = -0.01$ mm^{-1}, EFL got the maximum value, the FOV was 8°. The measured EFL adjust precision was 0.2885 mm, the RMS of precision was 0.0181 mm. MTF was used to evaluate the imaging quality of the zoom system. The range of MTF was 0.17 to 0.40 at 20 lp/mm. The lens realized in this paper had only four groups of elements, and none of them was motorized, thus it had good anti-shake performance. It was applicable to many fields e.g., ground vehicles used in a field-environment, and was also suitable for applications in high speed vehicles and fast moving target tracking as a result of the good electrical performance.

However, the gravity influence of liquid lens was not considered in this paper. Although in the experiments, no apparent gravity effect had been observed, it would exist in theory. In fact, gravity effect in the stationary state had been contained in the parameter "wavefront error" in the datasheet of the liquid lens, the wavefront error was less than 0.5λ. In practice, we need to consider the system working in doubling even tripling gravitational acceleration environments. We will do more research on this issue in future works.

Acknowledgments: The work was supported by the National Natural Science Foundation of China (NSFC) (61275003, 51327005, 91420203).

Author Contributions: Qun Hao and Xuemin Cheng conceived and designed the experiments; Heng Li performed the experiments and analyzed the data; Heng Li and Xuemin Cheng wrote the paper.

Conflicts of Interest: The authors declare no conflict of interest.

References

1. Herzberger, M. *Modern Geometrical Optics*; Interscience Publishers, Inc.: New York, NY, USA, 1958.
2. Lin, Y.H.; Chen, M.S. A pico projection system with electrically tunable optical zoom ratio adopting two liquid crystal lenses. *J. Disp. Technol.* **2012**, *8*, 401–404. [CrossRef]
3. Foresti, G.L.; Micheloni, C.; Piciarelli, C.; Snidaro, L. Visual sensor technology for advanced surveillance systems: Historical view, technological aspects and research activities in Italy. *Sensors* **2009**, *9*, 2252–2270. [CrossRef] [PubMed]
4. Zou, T.; Tang, X.; Song, B.; Wang, J.; Chen, J. Robust feedback zoom tracking for digital video surveillance. *Sensors* **2012**, *12*, 8073–8099. [CrossRef] [PubMed]
5. Zhao, P.P.; Ataman, C.; Zappe, H. An endoscopic microscope with liquid-tunable aspheric lenses for continuous zoom capability. In Proceedings of the SPIE Micro-Optics 2014, Brussels, Belgium, 14–16 April 2014.
6. Wick, D.V.; Martinez, T.; Payne, D.M.; Sweatt, W.C.; Restaino, S.R. Active optical zoom system. In Proceedings of the SPIE Conference on Spaceborne Sensors II, Orlando, FL, USA, 28–29 March 2005.
7. Wick, D.V.; Martinez, T. Adaptive optical zoom. *Opt. Eng.* **2004**, *43*, 8–9.
8. Ren, H.W.; Wu, S.T. Variable-focus liquid lens. *Opt. Express* **2007**, *15*, 5931–5936. [CrossRef] [PubMed]

9. Lin, Y.S.; Chen, M.S.; Lin, H.C. An electrically tunable optical zoom system using two composite liquid crystal lenses with a large zoom ratio. *Opt. Express* **2001**, *19*, 4714–4721. [CrossRef] [PubMed]

10. Zhang, W.; Tian, W. A novel micro zoom system design with liquid lens. In Proceedings of the 2008 International Conference on Optical Instruments and Technology—Optical Systems and Optoelectronic Instruments, Beijing, China, 16–19 November 2008.

11. Peng, R.; Chen, J.; Zhu, C.; Zhuang, S. Design of a zoom lens without motorized optical elements. *Opt. Express* **2007**, *15*, 6664–6669. [CrossRef] [PubMed]

12. Fang, Y.C.; Tsai, C.M.; Chung, C.L. A study of optical design and optimization of zoom optics with liquid lenses through modified genetic algorithm. *Opt. Express* **2011**, *19*, 16291–16302. [CrossRef] [PubMed]

13. Miks, A.; Novak, J. Analysis of two-element zoom systems based on variable power lenses. *Opt. Express* **2010**, *18*, 6797–6810. [CrossRef] [PubMed]

14. Park, S.C.; Lee, W.S. Design and analysis of a one-moving-group zoom system using a liquid lens. *J. Korean Phys. Soc.* **2013**, *62*, 435–442. [CrossRef]

15. Peng, R.; Chen, J.; Zhuang, S. Electrowetting-actuated zoom lens with spherical-interface liquid lenses. *J. Opt. Soc. Am. A* **2008**, *25*, 2644–2650. [CrossRef]

16. Sun, J.H.; Hsueh, B.R.; Fang, Y.C.; MacDonald, J.; Hu, C.C. Optical design and multiobjective optimization of miniature zoom optics with liquid lens element. *Appl. Opt.* **2009**, *48*, 1741–1757. [CrossRef] [PubMed]

17. Zhang, W.; Li, D.; Guo, X. Optical Design and Optimization of a Micro Zoom System with Liquid Lenses. *J. Opt. Soc. Korea* **2013**, *17*, 447–453. [CrossRef]

18. Hao, Q.; Cheng, X.; Du, K. Four-group stabilized zoom lens design of two focal-length-variable elements. *Opt. Express* **2013**, *21*, 7758–7767. [CrossRef] [PubMed]

3

An Analog-Digital Mixed Measurement Method of Inductive Proximity Sensor

Yi-Xin Guo *,†, Zhi-Biao Shao † and Ting Li †

Academic Editor: Vittorio M. N. Passaro

The School of Electronic and Information Engineering, Xi'an Jiaotong University, No.28, Xianning West Road, Xi'an 710049, China; zbshao@mail.xjtu.edu.cn (Z.-B.S.); agike@sina.com (T.L.)
* Correspondence: macray@126.com
† These authors contributed equally to this work.

Abstract: Inductive proximity sensors (IPSs) are widely used in position detection given their unique advantages. To address the problem of temperature drift, this paper presents an analog-digital mixed measurement method based on the two-dimensional look-up table. The inductance and resistance components can be separated by processing the measurement data, thus reducing temperature drift and generating quantitative outputs. This study establishes and implements a two-dimensional look-up table that reduces the online computational complexity through structural modeling and by conducting an IPS operating principle analysis. This table is effectively compressed by considering the distribution characteristics of the sample data, thus simplifying the processing circuit. Moreover, power consumption is reduced. A real-time, built-in self-test (BIST) function is also designed and achieved by analyzing abnormal sample data. Experiment results show that the proposed method obtains the advantages of both analog and digital measurements, which are stable, reliable, and taken in real time, without the use of floating-point arithmetic and process-control-based components. The quantitative output of displacement measurement accelerates and stabilizes the system control and detection process. The method is particularly suitable for meeting the high-performance requirements of the aviation and aerospace fields.

Keywords: inductive proximity sensor; analog-digital mixed measurement method; two-dimensional look-up table; built-in self-test; position detection

1. Introduction

Most proximity sensors use non-contact methods to detect the distance or the proximity event, and their advantages include a resistance to fouling and abrasion, water tightness, a long service life, and low mechanical system maintenance cost [1]; as well as a high mean time between failure (MTBF) value [2]; and strong magnetic immunity [3,4]. Comparing with capacitive proximity sensors, inductive proximity sensors (IPSs) have better sensitivity with a target of alloy steels. IPS is applicable in the industry field of position detection, especially in the aviation and aerospace fields [5].

The primary transducer of IPS is a coil, and its inductance component correlates to the distance between IPS and the target [1]. However, the resistance component of coil changes significantly as the temperature changes due to the effects of the coil structure and materials [1,6,7]. The resistance component also constrains the detection precision of IPS severely [8,9]. Besides, the traditional IPS detects the nearness of the target, but it cannot create quantitative output. Mizuno *et al.* [10] proposed the use of magneto-plated wire to fabricate coil because this wire can decrease AC resistance and reduce temperature drift [11,12]. However, this method cannot solve the temperature drift of the DC resistance element fundamentally. The performance of IPS depends on the coil structure and the processing circuit. In certain conditions, the greater the inductance component is, the better the IPS sensitivity will be. However, the temperature drift will be severe as the resistance component increases simultaneously, and modeling calculation will be complicated by the simultaneous increase in distributed capacitance [13]. The sensor signal processing circuit is important to the IPS performance [14]. Many methods have been developed to reduce temperature drift by improving the processing circuit, including the analog [6,15] and digital measurement methods [16].

Analog measurement procedures include applying pulse excitation to a sensor coil, comparing the thresholds of the r-L discharge waveform via a comparator to determine the inductance value, and evaluating if a target is approaching. This method is simple and popular. Due to the influence of temperature drift, however, a quantitative output is difficult to realize. Ferican [6] and Nabavi [15] proposed thermistor and differential coils to reduce temperature drift, respectively. These methods improve the IPS temperature characteristic but cannot produce a quantitative output because of the precision and consistency restriction in analog compensation or offsetting.

Digital measurement procedures include applying a sine wave excitation to a sensor coil, sampling voltage and current waveforms, using a Fourier transform algorithm to identify the phase difference between the voltage and the current, and calculating the inductance and resistance components of coil [7]. The inductance component is directly related to proximity distance; therefore, temperature drift can be reduced and quantitative output produced [16]. Digital signal processing is complicated; this process uses random-access memory (RAM) and process-control-based components such as micro-controller unit (MCU) or digital signal processor (DSP). The procedure may induce large volume and high power consumption; thus, its application in the aviation field is constrained [17].

Leons *et al.* [18] proposed a measurement method that applied pulse excitation to the sensor coil and obtained discharge waveform samples, and measured an integral average value of a discharge waveform for reducing the noise impact and calculated the distance using a one-dimensional look-up table. That system can measure distances in the range of 0~5 mm at 89% accuracy under room temperature. However, the reference also indicated that "We also plan to study the accuracy of our design with modifications in temperature".

This paper combines the characteristics of the analog and digital measurements; pulse excitation and sampling are commonly used for these techniques, respectively. Given the characteristics of the two methods, this study proposes a new method and processing circuit that obtains the discharge waveform samples of an IPS coil twice as well as separates the inductance and resistance components through discharge model calculation. The measurement errors caused by temperature drift are reduced, and a quantitative output is realized. This work also proposes a two-dimensional look-up table-based method, and online data calculation is simplified. Without the need for complex floating point computation, process control units are omitted, such as the MCU-class. The look-up table is effectively compressed by maximizing the distribution characteristics of the sample value. Furthermore, memory can be significantly decreased. A real-time built-in self-test (BIST) is designed with reference to the partition and analysis of abnormal sample data. This method has the advantages of the traditional analog and digital measurement methods; it also improves temperature stability, reliability, and real-time measurement. The servo system applies proportional-integral-differential (PID) control through the IPS quantitative output; this accelerates and stabilizes the process in addition to expanding the IPS application function.

2. System

2.1. Front-End Interface

IPS mainly consists of a sensing head and a processing circuit, as shown in Figure 1. The processing circuit can be either embedded into the sensing head or spread outside the sensing head via a cable connection. The distance between the sensing head and the target is calculated and outputted by driving and detecting the coil.

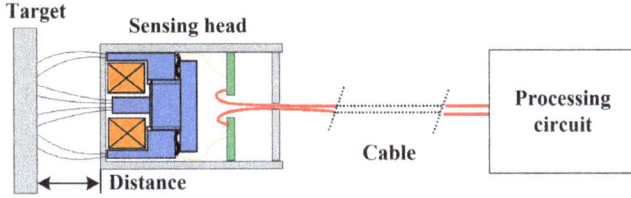

Figure 1. The schematic diagram of inductive proximity sensors.

In an equivalent circuit, as shown in Figure 2, L_i is the inductance component of the coil in the sensing head, r_i is the resistance component of the coil, L_w is the equivalent distributed inductance of cable, r_w is the equivalent distributed resistance of the cable, and C_P is the equivalent parasitical capacitance. The general series resistance r_{DC} (equal to $r_i + r_w$) is 13.2 Ω based on the measurement of the sensing head in this paper (on the condition that 1.0 V at 1 kHz, The equipment of E4980A (Agilent, Santa Clara, CA, United States) is used, and the cable is 200.0 mm long). When the distance between the sensing head and the target is greater than 10.0 mm, the general series inductance L_S (equal to $L_i + L_w$) is 4.9 mH; when the distance is less than 0.1 mm, L_S is approximately 9.4 mH. When an impedance analyzer (HP4195A, Agilent, Santa Clara, CA, United States) is employed to scan the sensing head, C_P is 262.7 pF. The effect of distributed capacitance on the system parameter is considerably less than 1‰; the distribution parameter can be thus ignored.

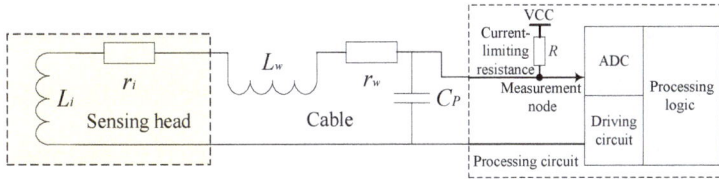

Figure 2. The equivalent circuit of the IPS.

Experimental data shows that r_{DC} changes significantly with temperature but is weakly correlated with the distance between the sensing head and the target. Moreover, L_S varies significantly with distance but is slightly correlated with temperature [1,7,8]; thus, L_S reflects the distance between the sensing head and the target. The measurement is affected by cable length and temperature; as for the linear time-invariant parameter system, the influence of cable length on measurement is fixed and can be eliminated through calibration. The influence of temperature on measurement is dynamic; therefore, separating the resistance component r_{DC} and the inductance component L_S of the IPS coil is key to eliminating the effect of temperature variation on measurement precision.

The sensing head is connected to the processing circuit via a cable. When driving circuit generates an excitation of rising edge, a signal passes through current-limiting resistance and drives the sensing head to discharge. As shown in Figure 3, the discharge waveforms vary significantly in the time domain as the distance changes. Calculating and processing logic are employed to obtain a sample and to analyze the discharge waveform through an analog-to-digital converter (ADC). Subsequently, the distance information can be deduced.

This study focuses on a common integrative IPS; its processing circuit is embedded in the sensing head and is directly connected to the coil. The distribution parameter of the cable that connects the sensing head and the processing circuit can be neglected. The driving and sampling circuits are designed as shown in Figure 4.

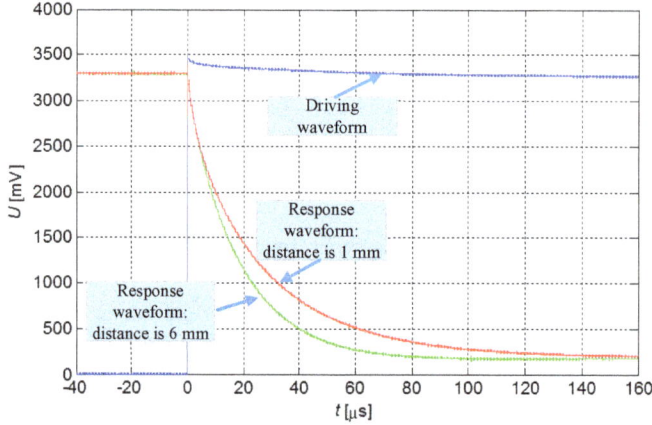

Figure 3. Driving and response waveforms of the sensing head.

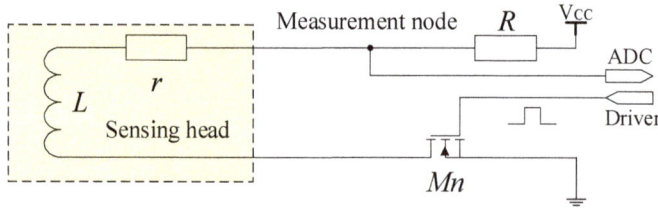

Figure 4. Driving and sampling of the integrative IPS.

Once the switch transistor M_n is activated, the inductance L discharges slowly through r and R as shown in Figure 4. The voltage of the measurement node is determined as follows:

$$U = \frac{V_{CC}}{R+r}\left[r + Re^{-\frac{R+r}{L}t}\right] \tag{1}$$

This paper presents a unique measurement method, the ADC in the processing circuit samples twice at t_1 and t_2. The samples U_1 at t_1 and U_2 at t_2 can be represented as follows:

$$\begin{cases} U_1 = \dfrac{V_{CC}}{R+r}\left[r + Re^{-\frac{R+r}{L}t_1}\right] \\[3mm] U_2 = \dfrac{V_{CC}}{R+r}\left[r + Re^{-\frac{R+r}{L}t_2}\right] \end{cases} \tag{2}$$

Through equipment measurement mentioned above, the scopes of coil parameters are determined. When the target moves, the variation range of L is from 4.0 mH to 10.0 mH; when the temperature changes, the variation range of r is from 8 Ω to 20 Ω. The aforementioned scopes are regarded as the definition domain of (r, L).

When the sample value of the 12-bit ADC is applied to express U_1 and U_2, V_{CC} is 4095 times the least significant bit (LSB) of ADC.

The discharge curve of the IPS is determined when the parameters take the mean value of the scope of resistance r and inductance L. t_1 is the moment in which the measurement node discharges to 60% V_{CC}, whereas t_2 is the moment in which the measurement node discharges to 30% V_{CC}. t_1 and t_2 are regarded as constants; at the two specific moments, ADC is controlled to conduct sampling twice. Based on the current-limiting condition, current-limiting resistance R takes a value of 300 Ω.

V_{CC}, t_1, t_2, and R are determined; L and r can be calculated by substituting U_1 and U_2 into Equation (2). The distance between the sensing head and the target can be indicated by L.

2.2. Construction of a Look-up Table

Sensor proximity status is obtained by building a look-up table and deriving the inductance value L from the sample (U_1, U_2). This process avoids complicated calculation; nevertheless, the following requirement should be fulfilled. Inductance value L can be determined through reasonable sampling (U_1, U_2), and the searching result is unique and correct. Thus, the following lemma should be met.

Lemma 1. Equation (2) has one and only one solution.

Proof. $U(r, L)$ is a monotone increasing function about r and L. As $U(r, L)$ is a monotone and continuous function, an inverse function exists, and Equation (2) generates a solution.

Given the value (U_1, t_1), r and L are subject to restriction conditions. Space p is defined as a set of (r_p, L_p), which enables function $U(t)$ to pass through the point (U_1, t_1) as shown in Figure 5; space q is defined as a set of (r_q, L_q), which enables function $U(t)$ to pass through the point (U_2, t_2) as shown in Figure 6.

Hence, Equation (2) generates one and only one solution.   □

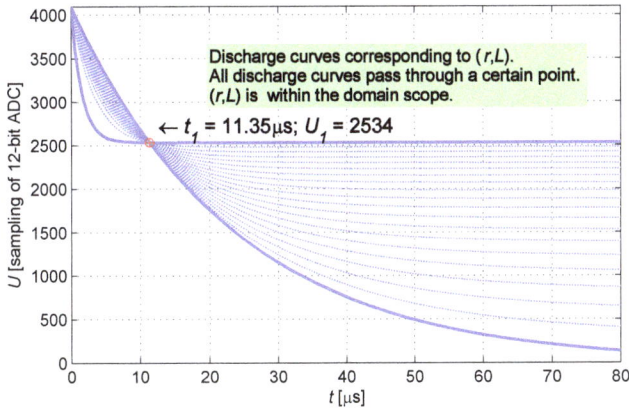

Figure 5. Cluster of curves passing through (U_1, t_1).

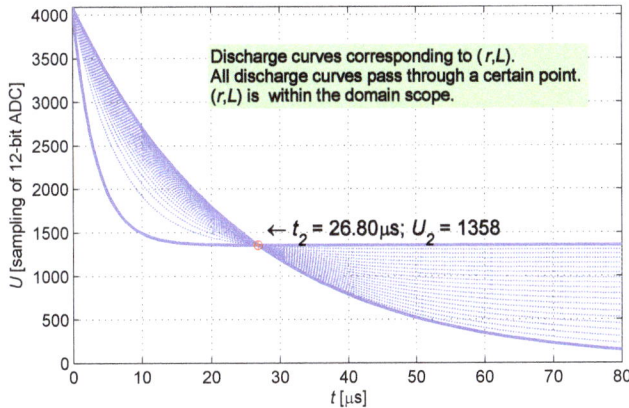

Figure 6. Cluster of curves passing through (U_2, t_2).

The spaces p and q are drawn in r-L coordinates, as shown in Figure 7. The intersection of the two curves is the solution of the equation because it passes through (U_1, t_1) and (U_2, t_2) synchronously. $L(r)$ is monotonic; therefore, a maximum of only one intersection exists. Equation (2) thus produces only a single solution at most.

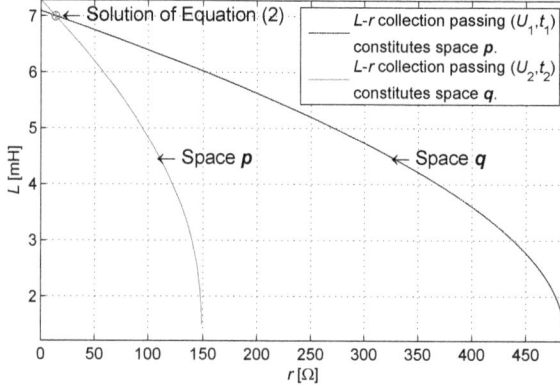

Figure 7. The intersection of spaces p and q as the solution of Equation (2).

Equation (2) is a transcendental equation, and obtaining an analytical solution is difficult. According to Lemma 1, a numerical solution that satisfies an engineering accuracy requirement can be obtained through iteration.

Subjecting to (s.t.) the given conditions, the values of L and r that minimize (min) the below formula are the numerical solution of Equation (2):

$$\begin{aligned} &\min && [U_1 - U(L, r, t_1)]^2 + [U_2 - U(L, r, t_2)]^2 \\ &\text{s.t.} && r > 0 \quad , \quad L > 0 \\ & && 0 < t_1 < t_2 \quad , \quad t_1 \text{ and } t_2 \text{ are determined} \end{aligned} \tag{3}$$

The digital solution of (r, L) that corresponds to (U_1, U_2) is obtained by traversing the sample value (U_1, U_2), and a look-up table is established. In practice, the value of L is obtained with U_1 and U_2 to determine the target proximity status. The optimization algorithm can ensure that the precision of the solved L meets the requirement. L reflects the distance between the sensing head and the target.

2.3. Compression of the Look-up Table

For instance, the size of a complete two-dimensional look-up table is $2^{12} \times 2^{12}$ units when the 12-bit ADC is used; this size indicates poor practicability. Therefore, an effective method to compress the look-up table should be developed to ensure the practicability of this method. The look-up table can be compressed effectively by analyzing the distribution characteristics of the sample value (U_1, U_2) in a real circuit.

In practice, (r_i, L_i) is calculated by the given (U_{1i}, U_{2i}) value. The solution scope can be narrowed by the restriction of the physical model.

Supposing $U_1 < U_2$ (when $t_1 < t_2$), the monotony of $U(t)$ changes. Thus, the equation solution is negative in value and is physically unreasonable.

Equation (1) is converted into an inverse function of U for L:

$$L = -\frac{(R+r)t}{\lg\left[\dfrac{U}{4095R}(R+r) - \dfrac{r}{R}\right]} \tag{4}$$

L belongs to a positive real domain, so the following constraint should be fulfilled:

$$0 < \left[\frac{U}{4095R}(R+r) - \frac{r}{R} \right] < 1 \tag{5}$$

That means the sample value should meet the following constraint:

$$\frac{4095r}{R+r} < U < 4095 \tag{6}$$

If (U_{1i}, U_{2i}) is improperly given, the solution (r_i, L_i) may consist of a negative or even a complex number. In fact, the sampled (U_1, U_2) is certainly reasonable. A restriction condition exists between U_1 and U_2 such that (r, L) belongs to a positive real domain. The intersection of two curves depicted in Figure 7 is observed at the first quadrant of the real domain coordinate. Beyond the restriction condition, (U_1, U_2) will not be obtained through practical sampling and need not be recorded in the look-up table.

Given samples U_1 at t_1 and U_2 at t_2, the following equation is obtained.

$$\begin{cases} L = -\dfrac{(R+r)t_1}{\lg\left[\dfrac{U_1}{4095R}(R+r) - \dfrac{r}{R} \right]} \\[2em] L = -\dfrac{(R+r)t_2}{\lg\left[\dfrac{U_2}{4095R}(R+r) - \dfrac{r}{R} \right]} \end{cases} \tag{7}$$

Furthermore, the constraint relationship of U_1 and U_2 is obtained.

$$\left[\frac{U_1}{4095R}(R+r) - \frac{r}{R} \right]^{t_2/t_1} = \left[\frac{U_2}{4095R}(R+r) - \frac{r}{R} \right] \tag{8}$$

If a set of (U_1, U_2) satisfies both Equations (6) and (8), then Equation (2) generates a solution with a positive real number.

A restriction condition of $U_2(r)$ is obtained when U_1 is given. $U_2(r)$ is proven as a monotone function, and the range of its values is determined by the definition domain of r. Under the range of $U_2(r)$ in combination with the given U_1, Equation (2) generates positive real number solutions, as shown in Figure 8.

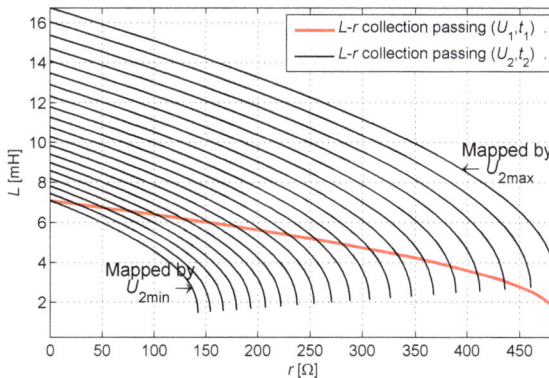

Figure 8. Cluster of curves mapped by U_2, which have positive real intersections with a given curve mapped by U_1.

The essence of a look-up table lies in the mapping of the points from the $U_1 - U_2$ coordinate to the $r - L$ coordinate. All points in the $r - L$ coordinate can be mapped into the $U_1 - U_2$ coordinate. Some points in the $U_1 - U_2$ coordinate can be mapped back into the $r - L$ coordinate. A few of the other points in the $U_1 - U_2$ coordinate will fall under the other quadrants (negative number) or a four-dimensional complex space (complex number) when mapping to the $r - L$ coordinate.

Through equipment measurement mentioned in Subsection 2.1, the scopes of coil parameters are determined. The engineering parameter definitions of the coil are $r \in [8.0, 20.0]$ Ω and $L \in [4.0, 10.0]$ mH. The definition domain of U_1 can be obtained based on the restriction condition of Equation (1) and the monotony of $U(r, L)$, as expressed in Equation (9):

$$
\begin{cases}
U_{1\,\text{min}} = \dfrac{4095}{R + r_{\text{min}}} \left[r_{\text{min}} + Re^{-\frac{R + r_{\text{min}}}{L_{\text{min}}} t_1} \right] \\[3mm]
U_{1\,\text{max}} = \dfrac{4095}{R + r_{\text{max}}} \left[r_{\text{max}} + Re^{-\frac{R + r_{\text{max}}}{L_{\text{max}}} t_1} \right]
\end{cases}
\tag{9}
$$

Given the restriction condition of Equation (8), a range of U_2 values corresponding to r within a physical changing range can be determined with each given U_1, as expressed in Equation (10):

$$
\begin{cases}
U_{2\,\text{min}} = 4095R \dfrac{\left[\dfrac{U_1}{4095R}(R + r_{\text{min}}) - \dfrac{r_{\text{min}}}{R} \right]^{t_2/t_1} + \dfrac{r_{\text{min}}}{R}}{(R + r_{\text{min}})} \\[5mm]
U_{2\,\text{max}} = 4095R \dfrac{\left[\dfrac{U_1}{4095R}(R + r_{\text{max}}) - \dfrac{r_{\text{max}}}{R} \right]^{t_2/t_1} + \dfrac{r_{\text{max}}}{R}}{(R + r_{\text{max}})}
\end{cases}
\tag{10}
$$

The definition domain of U_2 is determined in sequence by traversing the definition domain of U_1, and the definition domain of (U_1, U_2) is determined as shown in Figure 9.

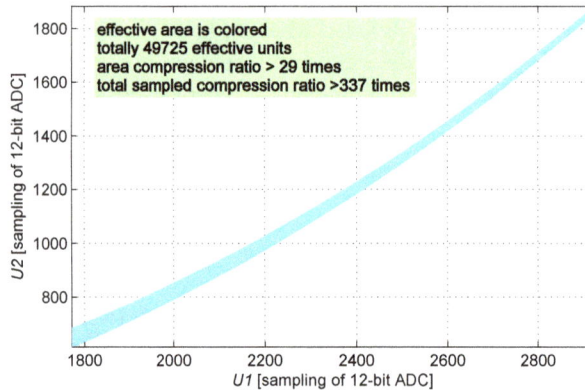

Figure 9. Effective table area according to the domain of the engineering definition.

The size of the look-up table is compressed from 4096×4096 into 49,725 units; therefore, the look-up table is compressed 337 times.

2.4. Real-Time BIST

In general, the traditional scheme implements the BIST process after the system powers on and enters the working state. Therefore, BIST detection cannot be performed simultaneously during the system working process.

The position information is obtained by analyzing and calculating the sample value of the discharge waveform. When coil disorder or driving circuit failure occur and the sample value is beyond the definition domain, failure statuses and modes can be assessed and outputted in real time.

As mentioned previously, look-up table information is established in the parameter scope of the IPS coil. Sample datum that are not recorded in this table can be obtained during IPS system failure. Different failure modes can be obtained without the need for an additional circuit by analyzing the unrecorded sample data. As shown in Figure 10, if the sample value falls under domain **N**, then the system searches for position information in the look-up table. If the sample value falls under domain **S**, then the system evaluates the coil as short-circuited; if the sample value falls under domain **O**, then the system regards the coil as an open circuit or as subject to driving circuit failure.

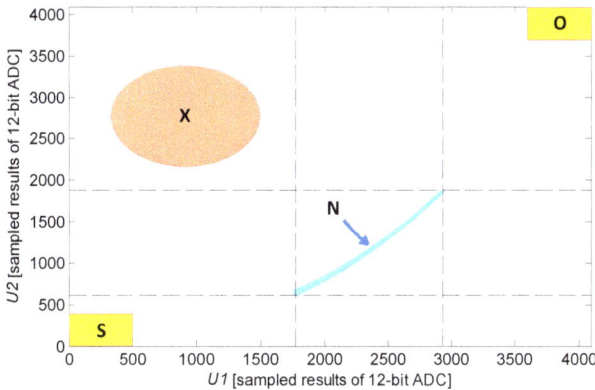

Figure 10. Built-in self-test vector diagram.

The increased accumulation of engineering applications in the future will be conducive to improving the integrality of failure modes and to establishing the relationship of these failure modes with the **X** region. In this case, the area covered by BIST expands, the lifetime of IPS can be forecasted, and an effective online intelligent monitoring function can be established for the sensor IPS.

3. Implementation

As shown in Figure 11, the integrative micro-assembly module of the processing circuit includes a gate driver, look-up table logic, and the interpolation and filtering algorithms. On the excitation of the gate driver, the sensing head generates an analog input for ADC. Given the sample value, the look-up table logic calculates a proximity distance. The interpolation algorithm extracts an additional linear resolution of the proximity distance. The filtering algorithm is used to analyze abnormal samples and to generate outputs in the following format: the quantitative output is exhibited by a universal asynchronous receiver/transmitter (UART); the switching output is exhibited by a one-bit digital signal, and the BIST output is exhibited by a two-bit digital signal. The module requires external conditions, including a current-limiting resistor, probe assembly (coil), external analog-filtering circuit, flash, oscillator, reset, and power circuit.

The engineering prototype of the IPS based on integrative micro-assembly is depicted in Figure 12. The IPS coil is a $\phi 5$ mm $\times$ 12 mm cylinder that is wound by $\phi 0.14$ mm of enameled wire. The core is integrated into a processing circuit that is composed of two circuit boards: the power board is inputted with an external power supply and outputs the voltages for the digital board; the digital board calculates and outputs information on distance and BIST. Those two circuit boards are integrated into a threaded barrel that measures φ14 mm × 59 mm.

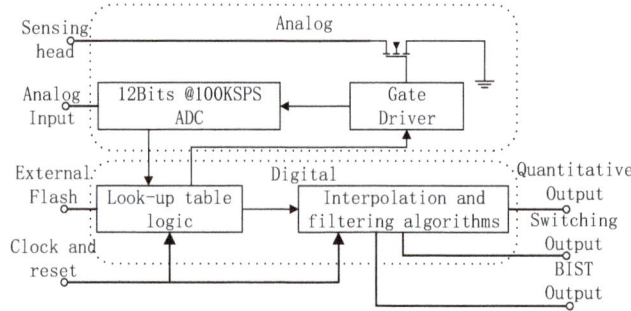

Figure 11. Block diagram of the integrative micro-assembly module.

Figure 12. The engineering prototype of the inductive proximity sensor.

This study is implementing a procedure for the realization of an application-specific integrated circuit (ASIC) used to substitute for the micro-assembly module. Besides, system integration contributes to the improvement of stability.

4. Results And Discussion

This section establishes the relationship between the sample value (U_1, U_2) and the inductance component L of the IPS coil based on the discharge model. Moreover, a quantitative error analysis is conducted on the look-up table. A prototype experimental platform is designed and built as described in this section. The relationship between the inductance component L of the IPS coil and proximity distance is established with this platform, and the look-up table of proximity distance that is related to sample value (U_1, U_2) is constructed. Finally, the look-up table is programmed into the IPS engineering prototypes, and a testing procedure is implemented.

According to the definition of the (U_1, U_2) domain illustrated in Figure 9, the corresponding L is solved by applying Equation (3), and the look-up table is established. An image of the function $L(U_1, U_2)$ within the definition domain is obtained by taking L as the Z-axis, as shown in Figure 13.

The quantitative error evaluation testifies that the maximum quantization error of L is 12.08 μH. This error is detected when $U_1 = 2925$, $U_2 = 1882$, and $L = 9984.51$ mH. The IPS system reaches a theoretical resolution of 1.21‰ when the system uses a 12-bit ADC.

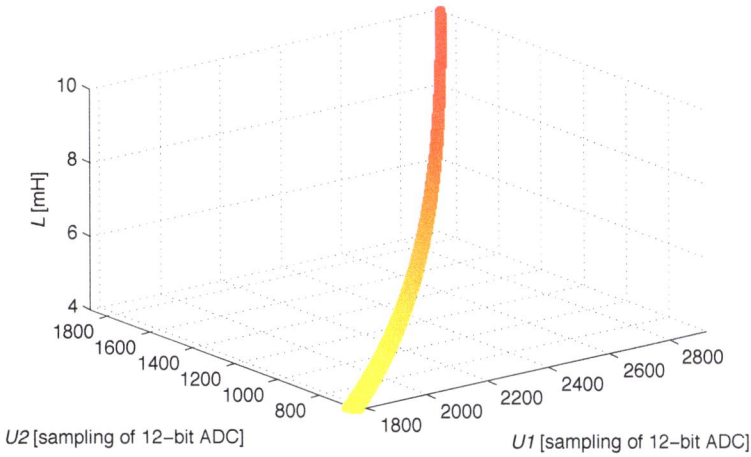

Figure 13. The solution of function $L(U_1, U_2)$.

This study emphasizes the $L_S - r_{DC}$ parameters. It also establishes the relationship of $L_S - r_{DC}$ with proximity distance through an experimental method given that production and assembly are inconsistent and that the difference of coil is less than 2%, which is far beyond the theoretical resolution [19,20].

An experimental platform is designed and built to control the distance between the sensing head and the target automatically, as shown in Figure 14. According to an optics motorized precision translation stage, the re-orientation accuracy of the experimental platform is ±20 μm, which is ensured by the related metrology accreditation.

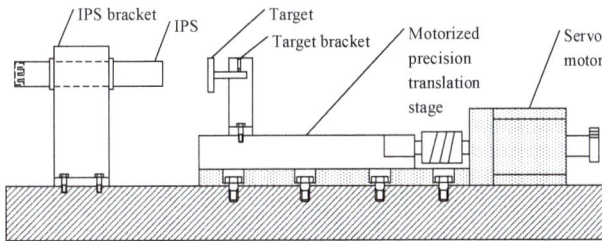

Figure 14. Experimental platform.

The materials of the target are related to the magnetic environment of the IPS working process. The IPS can detect a target of steels. Moreover, slight sensitivity differences are observed among various alloy steels. The alloy steel of $0Cr17Ni5Cu4Nb$ is commonly selected as the material of the target in the industry; thus, this material comprises the target on the experimental platform in this study. The target is produced in the shape of a coin that is 16.0 mm in diameter and is 2.5 mm thick. The IPS assembled with only a coil component is installed on the platform; this IPS is used to establish the relationship of $L_S - r_{DC}$ with proximity distance. Under this circumstance, the relationship of the observed distance (between the sensing head and the target) and the L_S of the coil is determined, as shown in Figure 15.

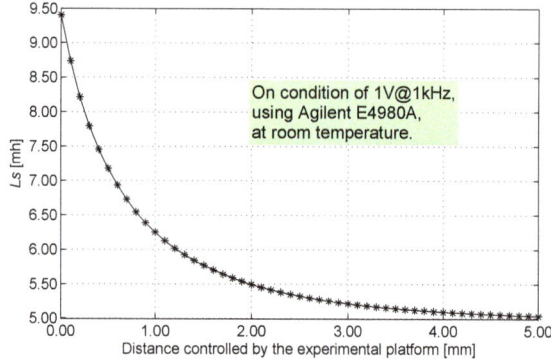

Figure 15. Distance *vs.* L_S.

The constraint relationship of distance and L_S is nonlinear [7]; therefore, the measurement precision in relation to distance is uneven. The effective resolution of IPS worsens if distance increases [18]. Sensors of this type are constrained by this characteristic, and this feature is applicable only in short-distance (less than 8 mm) detection.

According to the calibrated data, the distance between the sensing head and the target can be searched from the sample values (U_1, U_2) when the inductance value of the look-up table is replaced with distance.

To conduct the temperature experiment, the servo motor on the mechanical part of the experimental platform is connected to the control box via cables. In this experiment, the mechanical part is placed into a high-temperature box. We can set the distance between the IPS and the target through a human-machine interface of the control box outside of the high-temperature box.

Figures 16–19 show the measurement errors and results of the engineering prototype at 20 °C, 50 °C, 80 °C and 110 °C respectively. The experimental data indicates two rules: first, the maximum and mean errors increase as temperature increases. They are mainly caused by the temperature drift of the experimental platform and the inductance component. Second, the error increases with distance because the sensitivity of the sensing head drops rapidly if distance increases. The maximum error is 0.18 mm, which is below 4%.

Four engineering prototypes are included in the aforementioned experiment. No significant difference is observed among these prototypes based on the analysis of the experimental data. The max distance error of the four engineering prototypes is detected at 0–5 mm, as shown in Figure 20.

Figure 21a,b exhibits the ADC samples U_1 and U_2 at different temperatures. The smoothness and monotony of the curves suggest that the system sampling is stable and that background noise is controlled below 1 LSB of ADC.

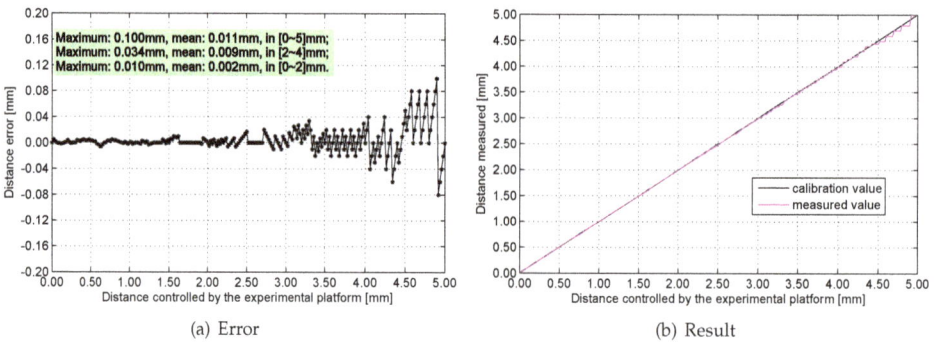

(a) Error

(b) Result

Figure 16. Measurement errors and results at 20 °C.

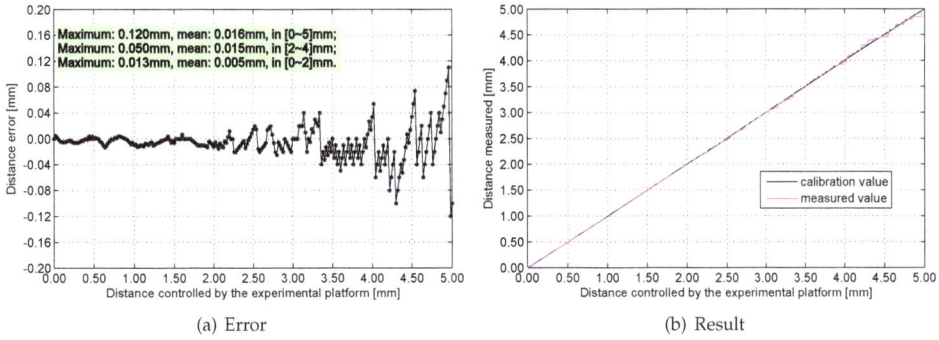

(a) Error

(b) Result

Figure 17. Measurement errors and results at 50 °C.

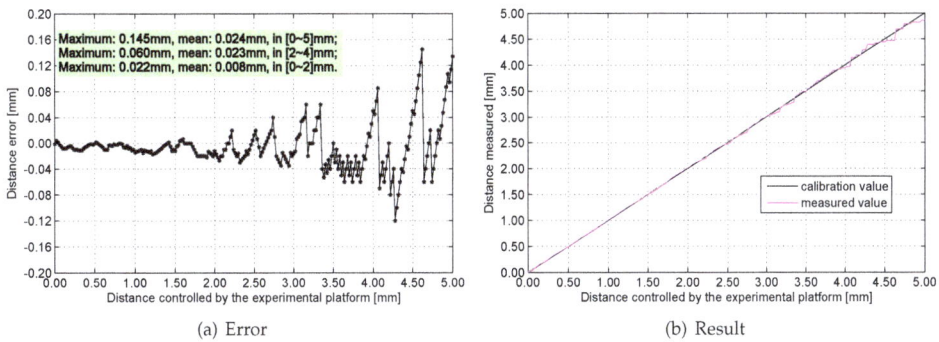

(a) Error

(b) Result

Figure 18. Measurement errors and results at 80 °C.

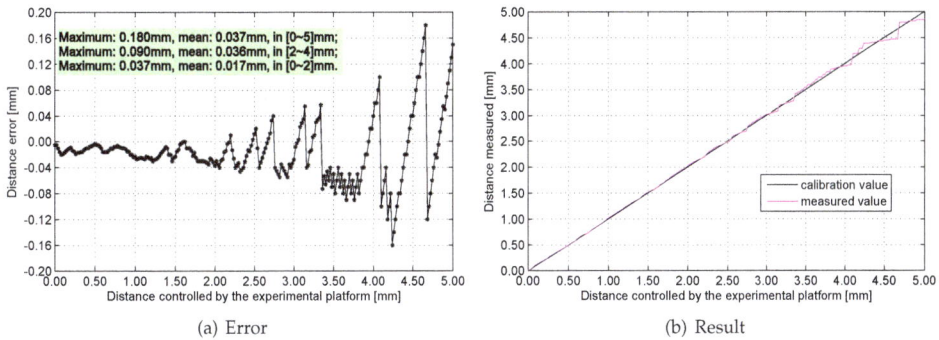

(a) Error

(b) Result

Figure 19. Measurement errors and results at 110 °C.

If the temperature effects on the sensing head are neglected, the four waveforms in Figure 21 would be coincident, and the twice-sampling method can be simplified into a single sampling method. As shown in Figure 21b, if the distance reaches 5 mm at 110 °C, the curve generated at 20 °C could explain the curve established at 110 °C (if the temperature effects are neglected). The computing distance value is 2.914 mm, and the error is 2.086 mm.

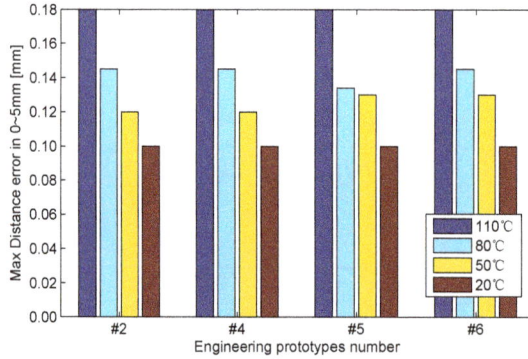

Figure 20. Max distance error of the four engineering prototypes at 0–5 mm.

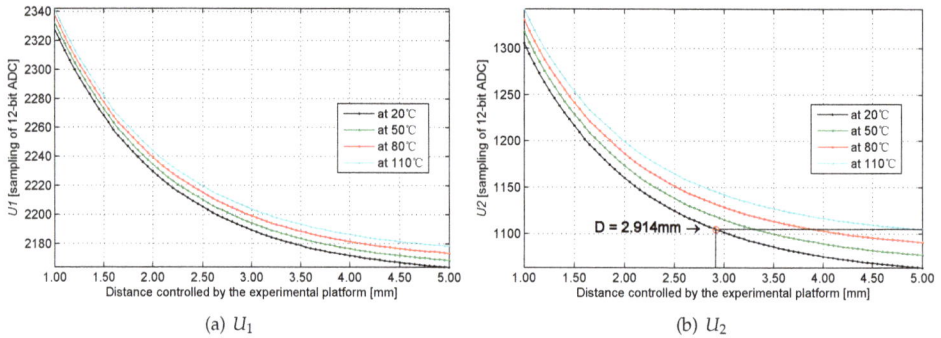

(a) U_1 (b) U_2

Figure 21. ADC samples U_1 and U_2 at different temperatures.

Unlike the study conducted by Leons *et al.* [18], this study used different processing circuits that can effectively control background noise in addition to obtaining samples of the discharge waveform of IPS coils twice; the inductance and resistance components are separated through the calculation process based on a two-dimensional look-up table, and the temperature drift is reduced. Experimental data shows that, if the temperature drift of the coil is disregarded, the temperature variance in the applicable scope strongly affects system precision.

Therefore, obtaining the discharge waveform samples of the IPS coil twice as well as separating the inductance and resistance components are the key points in reducing temperature drift and improving the precision of the IPS system.

The system measurement is implemented during the activation of the controlled switch; the external jamming out of the measurement window is thus unrelated to the measurement result. The system, therefore, has a unique anti-jamming capability.

This study established a testing framework that is similar to the voltage spike test of category A in section 17 of DO-160F. The engineering prototype has passed the related test of DO-160F, and the result confirms that input jamming is inadequate to cause a temporary failure. Instead of the common mode jamming applied in the test of DO-160F, differential mode jamming is input into the IPS discharge circuit. As shown in Figure 22, a pulse generator creates 10 V of a 200 Hz square wave and inputs jamming into the system via capacitance C_t and transformer L_t. C_t is 1 μF. The inductance in series with discharge circuit is 5 μH.

In the test, an oscilloscope is used to observe discharge waveforms with jamming, as shown in Figure 23; the jamming input is enough to cause a temporary failure. Even though jamming beyond the measurement window alters the coil recharge status and then the status is recovered quickly through discharge, the jamming effect does not influence the working status of the subsequent measurement window. Therefore, a narrow measurement window improves the anti-jamming capability.

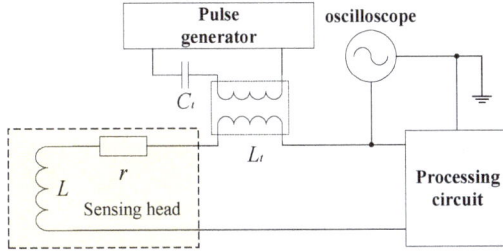

Figure 22. Jamming input method.

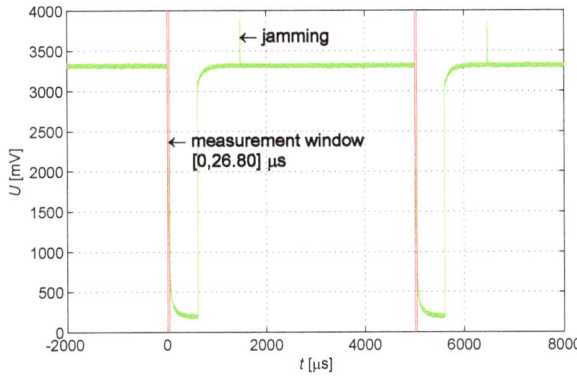

Figure 23. An anti-jamming process.

Some public released references were selected to make a comparison, as they have the similarities with this study in the following aspects: they have the same application background in the field of aviation; they are all integrative IPSs and have the similar parameters; they all avoid using process-control-based components. Table 1 presents the comparison and highlights the benefits of this study.

Table 1. Comparison of this work with related studies.

	This work	[17]	Honeywell ZS-00305 [21]
Temperature drift reduction	Self-adaptive, compensation is unnecessary	Customized thermal resistance, compensation is required	Unknown
Measurement method	Analog-digital mixed	Analog	Analog
Guaranteed actuation distance	4.50 ± 0.20 mm	3.28 ± 0.20 mm	4.00 ± 0.50 mm
Quantitative output	Yes	No	No
Built-in self-test	Yes, real-time	Yes, non-real-time	Yes, non-real-time
Electromagnetic compatibility	Narrow measurement window, strong anti-jamming capability	No measurement window, weak anti-jamming capability	No measurement window, weak anti-jamming capability
Input current (at 28 V)	6 mA	4 mA	10 mA
MCU or DSP	Non-adoptive	Non-adoptive	Non-adoptive

This work obtains two samples of the pulse discharge waveform of the IPS coil and separates the two variables L_S and r_{DC} through modeling calculation. Precision is improved, and the quantitative output of proximity distance is realized. The two-dimensional look-up table method reduces online computational complexity. Thus, process-control-based components, such as MCU, are saved.

Through an analysis of the characteristics of the look-up table and the exploitation of the physical parameter constraint of the coil, the look-up table is compressed 337 times, the processing circuit is simplified, and power consumption is reduced. This study establishes an abnormal sample partition for data beyond the look-up table, extracts IPS failure modes, and establishes a real-time BIST mechanism. Two failure models are assessed, namely, the short-circuit and open-circuit models. Based on more engineering application accumulation in future, BIST covering area will be comprehensive, and healthy online intelligent monitoring function will be realized [21].

In this work, the sample waveform includes the static component of the coil, which does not change with temperature or proximity status [22]. Under its influence, the dynamic range of ADC cannot be maximized [23]. The two sets of coils with symmetric parameters embedded into the sensor are proposed for future studies. The coil in front is close to the target, and can extract proximity variation; the rear coil is far from the target, and can be regarded as a parameter reference. The DC resistance of the rear coil changes with temperature as with the coil in front. System sensitivity and resolution can be improved by detecting the discharge waveform extracted from the differential circuit [24].

5. Conclusions

The IPS structure and test method were researched in this study. To reduce temperature drift and to improve the test performance, a new mixed method of analog and digital testing was presented that conducts twice sampling of the discharge waveform. Through discharge modeling calculation, the inductance and resistance components are separated, the measurement error caused by temperature drift is reduced, and the quantitative output is realized, moreover, the IPS application scope is expanded. The establishment and application of the two-dimensional look-up table reduce system online computational complexity. It also saves process-control-based components and improves measurement stability, reliability, and real-time performance. By analyzing the exploitation of the physical parameter constraints of the sensing head, the size of the look-up table is compressed 1/337 of the original processing circuit. Furthermore, power consumption is reduced. The realization of the real-time BIST extends the IPS detection capabilities.

Experiment data and comparison of related products indicate that the aforementioned method is especially suitable for fulfilling the high-performance requirements, such as displacement measurement in the fields of aerospace and aviation.

Acknowledgments: We would like to gratefully acknowledge Kai-Liang Xu for his mechanical technical support, and we are grateful for Xi'an Runxin Aviation Technology Co., Ltd, for providing the coil component of IPS and experimental environment.

Author Contributions: These authors contributed equally to this work. Yi-Xin Guo proposed the original idea, carried out the simulation, designed the processing circuit of IPS, and designed the control system of the experimental platform; Zhi-Biao Shao proposed the research direction and made a substantial contribution to the design of algorithms; Ting Li performed acquisition of experimental data and analyzed the experiment results. All authors contributed to the organization of the paper, writing, and proofreading. All authors read and approved the final manuscript.

Conflicts of Interest: The authors declare no conflicts of interest.

References

1.   Jagiella, M.; Fericean, S.; Dorneich, A. Progress and recent realizations of miniaturized inductive proximity sensors for automation. *IEEE J. Sens.* **2006**, *6*, 1734–1741.

2.	Lugli, A.B.; Rodrigues, R.B.; Santos, M.M.D. A New Method of Planar Inductive Sensor for Industrial Application. In Proceedings of the 39th Annual Conference Of the IEEE Industrial Electronics Society, Vienna, Austria, 10–13 November 2013; pp. 3858–3863.
3.	Masi, A.; Danisi, A.; Losito, R.; Perriard, Y. Characterization of Magnetic Immunity of an Ironless Inductive Position Sensor. *IEEE J. Sens.* **2013**, *13*, 941–948.
4.	Danisi, A.; Masi, A.; Losito, R.; Perriard, Y. Modeling and Compensation of Thermal Effects on an Ironless Inductive Position Sensor. *IEEE Trans. Ind. Appl.* **2014**, *50*, 375–382.
5.	Volpe, R.; Ivlev, R. A survey and experimental evaluation of proximity sensors for space robotics. In Proceedings of the IEEE International Conference on Robotics and Automation, San Diego, CA, USA, 8–13 May 1994; pp. 3466–3473.
6.	Fericean, S.; Droxler, R. New noncontacting inductive analog proximity and inductive linear displacement sensors for industrial automation. *IEEE J. Sens.* **2007**, *7*, 1538–1545.
7.	Kruger, H.; Ewald, H.; Frost, A. Multivariate data analysis for accuracy enhancement at the example of an inductive proximity sensor. In Proceedings of the IEEE International Conference on Sensors, Christchurch, New Zealand, 25–28 October 2009; pp. 759–763.
8.	Vyroubal, D. Eddy-Current Displacement Transducer With Extended Linear Range and Automatic Tuning. *IEEE Trans. Instrum Meas.* **2009**, *58*, 3221–3231.
9.	Bakhoum, E.G.; Cheng, M.H.M. High-Sensitivity Inductive Pressure Sensor. *IEEE Trans. Instrum Meas.* **2011**, *60*, 2960–2966.
10.	Mizuno, T.; Mizuguchi, T.; Isono, Y.; Fujii, T.; Kishi, Y.; Nakaya, K.; Kasai, M.; Shimizu, A. Extending the Operating Distance of Inductive Proximity Sensor Using Magnetoplated Wire. *IEEE Trans. Magn.* **2009**, *45*, 4463–4466.
11.	Shinagawa, H.; Suzuki, T.; Noda, M.; Shimura, Y.; Enoki, S.; Mizuno, T. Theoretical Analysis of AC Resistance in Coil Using Magnetoplated Wire. *IEEE Trans. Magn.* **2009**, *45*, 3251–3259.
12.	Mizuno, T.; Yachi, S.; Kamiya, A.; Yamamoto, D. Improvement in Efficiency of Wireless Power Transfer of Magnetic Resonant Coupling Using Magnetoplated Wire. *IEEE Trans. Magn.* **2011**, *47*, 4445–4448.
13.	Mizuno, T.; Enoki, S.; Hayashi, T.; Asahina, T.; Shinagawa, H. Extending the linearity range of eddy-current displacement sensor with magnetoplated wire. *IEEE Trans. Magn.* **2007**, *43*, 543–548.
14.	Jagiella, M.; Fericean, S.; Dorneich, A.; Eggimann, M. Theoretical progress and recent realizations of miniaturized inductive sensors for automation. In Proceedings of the IEEE International Conference on Sensors, Irvine, CA, USA, 30 October–3 November 2005; pp. 488–491.
15.	Nabavi, M.R.; Nihtianov, S. Stability considerations in a new interface circuit for inductive position sensors. In Proceedings of the IEEE International Conference on Electronics, Circuits, and Systems, Hammamet, Tunisia, 13–16 December 2009; pp. 932–935.
16.	Fericean, S.; Dorneich, A.; Droxler, R.; Krater, D. Development of a Microwave Proximity Sensor for Industrial Applications. *IEEE J. Sens.* **2009**, *9*, 870–876.
17.	Huang, W.; Wang, C.; Liu, L.; Huang, X.; Wang, G. A signal conditioner IC for inductive proximity sensors. In Proceedings of the IEEE 9th International Conference on ASIC (ASICON), Xiamen, China, 25–28 October 2011; pp. 141–144.
18.	Leons, P.; Yaghoubian, A.; Cowan, G.; Trajkovic, J.; Nazon, Y.; Abdi, S. On improving the range of inductive proximity sensors for avionic applications. In Proceedings of the 16th International Symposium on Quality Electronic Design (ISQED), Santa Clara, CA, USA , 2–4 March 2015; pp. 547–551.
19.	AnimAppiah, K.D.; Riad, S.M. Analysis and design of ferrite cores for eddy-current-killed oscillator inductive proximity sensors. *IEEE Trans. Magn.* **1997**, *33*, 2274–2281.
20.	Koibuchi, K.; Sawa, K.; Honma, T.; Hayashi, T.; Ueda, K.; Sasaki, H. Loss estimation and sensing property enhancement for eddy-current-type proximity sensor. *IEEE Trans. Magn.* **2006**, *42*, 1447–1450.
21.	Aerospace Proximity Sensors, by Honeywell Sensing & Control. Available online: http://sensing.honeywell.com/products/aero-prox4?N=0 (accessed on 20 August 2015).
22.	Kumar, P.; George, B.; Kumar, V.J. A Simple Signal Conditioning Scheme for Inductive Sensors. In Proceedings of the Seventh International Conference on Sensing Technology (ICST), Wellington, New Zealand, 3–5 December 2013; pp. 512–515.

23. Chen, S.C.; Le, D.K.; Nguyen, V.S. Inductive Displacement Sensors with a Notch Filter for an Active Magnetic Bearing System. *Sensors* **2014**, *14*, 12640–12657.
24. Liu, C.Y.; Dong, Y.G. Resonant Coupling of a Passive Inductance-Capacitance-Resistor Loop in Coil-Based Sensing Systems. *IEEE J. Sens.* **2012**, *12*, 3417–3423.

4

Study and Experiment on Non-Contact Voltage Sensor Suitable for Three-Phase Transmission Line

Qiang Zhou [1,*], Wei He [1], Dongping Xiao [1], Songnong Li [2] and Kongjun Zhou [2]

Academic Editor: Vittorio M. N. Passaro

[1]   State Key Laboratory of Power Transmission Equipment & System Security and New Technology, Chongqing University, Chongqing 400044, China; weihe2016@126.com (W.H.); dongpingxiao2016@sina.com (D.X.)
[2]   State Grid Chongqing Electric Power CO. Electric Power Research Institute, Chongqing 400015, China; songnongli2016@126.com (S.L.); kongjunzhou2016@sina.com (K.Z.)
*   Correspondence: 20131101031@cqu.edu.cn

Abstract: A voltage transformer, as voltage signal detection equipment, plays an important role in a power system. Presently, more and more electric power systems are adopting potential transformer and capacitance voltage transformers. Transformers are often large in volume and heavyweight, their insulation design is difficult, and an iron core or multi-grade capacitance voltage division structure is generally adopted. As a result, the detection accuracy of transformer is reduced, a huge phase difference exists between detection signal and voltage signal to be measured, and the detection signal cannot accurately and timely reflect the change of conductor voltage signal to be measured. By aiming at the current problems of electric transformation, based on electrostatic induction principle, this paper designed a non-contact voltage sensor and gained detection signal of the sensor through electrostatic coupling for the electric field generated by electric charges of the conductor to be measured. The insulation structure design of the sensor is simple and its volume is small; phase difference of sensor measurement is effectively reduced through optimization design of the electrode; and voltage division ratio and measurement accuracy are increased. The voltage sensor was tested on the experimental platform of simulating three-phase transmission line. According to the result, the designed non-contact voltage sensor can realize accurate and real-time measurement for the conductor voltage. It can be applied to online monitoring for the voltage of three-phase transmission line or three-phase distribution network line, which is in accordance with the development direction of the smart grid.

Keywords: voltage sensor; non-contact measurement; ansoft maxwell; electric field intensity; three phase transmission lines

1. Introduction

The voltage transformer is an important part of an electric power system and the detection performance of the transformer can directly influence the reliability of an electric power system voltage measurement and relay protection device motion as well as the accuracy of electric power measurement. Presently, voltage transformers extensively applied to electric power system mainly include potential transformer and capacitance voltage transformer [1–5].

Structure diagram of Capacitance Voltage Transformer (CVT) is shown in Figure 1. The high voltage terminal is directly connected to electrified body and the other terminal is connected to the ground; C1 and C2 are adopted for capacitance voltage division, and voltage signal after voltage division will be transformed into output signal of transformer through potential transformer [6–9].

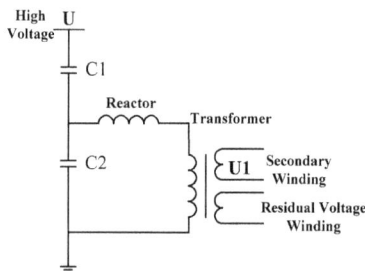

Figure 1. Structure diagram of CVT.

Potential transformer is shown in Figure 2, and the high voltage terminal is connected to electrified body.

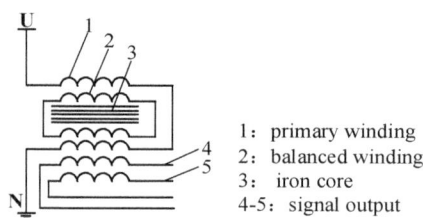

1: primary winding
2: balanced winding
3: iron core
4-5: signal output

Figure 2. Potential transformer.

According to analysis the Figures 1 and 2 the traditional voltage transformers shown in Figures 1 and 2 the high voltage terminal of currently used voltage transformer has direct electrical connection with conductor to be measured, and the other terminal of the transformer is connected to the ground. Therefore, insulation design of the transformer is difficult, and, meanwhile, it involves a large volume and heavyweight; due to the iron core design and multi-grade capacitance voltage division of the transformer, serious time delays occur during the detection signal and voltage signal to be measured in phase position. As a result, detection accuracy of the transformer is reduced and its application range is restricted. Moreover, three-phase voltage transformer is seldom adopted for the voltage grade above 35 kV in electric power systems. Therefore, traditional voltage transformers can no longer satisfy the development of the electric power system [10–12].

In consideration of the above-mentioned problems, this paper designed a non-contact voltage sensor on the basis of electric field coupling principle, as shown in Figure 3. In Figures 1–3 are non-contact voltage sensors installed on the three-phase transmission line. Positive and negative induction electrodes are designed on the sensor, and the electric field produced by electric charges on the transmission line will generate induced charges on the electrode of voltage sensor by way of electrostatic induction. The electric charges will distribute on the electrode to form electric potential, and the potential signal of positive and negative induction electrodes is output signal of the non-contact voltage sensor.

Compared with traditional voltage transformer, non-contact voltage sensor has no electrical connection with electrified body. Therefore, the insulation design is simple and the cost is low; the induction electrode of non-contact voltage sensor is optimized through simulation calculation. Miniature design of the sensor is realized, the voltage division ratio is increased, and phase error of sensor detection signal is reduced. An experimental platform that consisted of a 50 kV simulated three-phase transmission line was established in the laboratory to test the non-contact voltage sensor designed in this paper. According to the result, the designed voltage sensor can realize real-time detection for the voltage of three-phase transmission line and accurately measure the voltage and phase position.

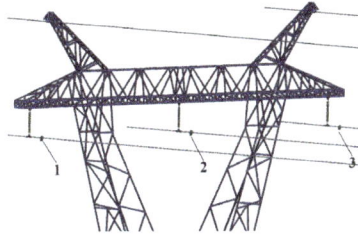

Figure 3. Application of voltage sensor in transmission line.

Non-contact voltage sensor used at the laboratory test stage adopts closed annular electrode, and it is mainly applied to detection performance and insulation performance test of the sensor. In the follow-up study, open-type annular electrode presented in Figure 4 will be designed. The sensor can be conveniently installed on the line, while the signal detection and processing circuit and wireless transmission circuit will be designed. In this way, the detection signal can be transferred to the control terminal in time. In Figure 4, 1 refers to non-contact voltage sensor; 2 is the insulation fixture—rubber with good insulation performance and toughness is used in the center fixture so that when the fixture is compressed, the centers of the line and sensor electrode can be maintained in the coaxial position, owing to the existence of non-conducting rubber; 3 indicates the signal processing circuit, wireless transmission module and battery; 4 is power transmission line; 5 is the insulator; and 6 denotes the connection fitting.

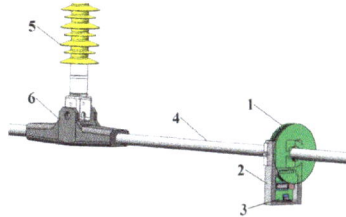

Figure 4. Non-contact open-type voltage sensor.

2. Detection Principle

The surface of any metal electrified conductor has electric charges that will generate electric field. In electric engineering, the power frequency electric field of 50 Hz can be treated as quasi static electric field. At this time, when metal electrode is introduced into this electric field, induced charges will be produced on the metal electrode, and the quantity of electric charges will change with the variation of electric field intensity. Distribution of electric charges on the metal surface will form electric potential of the metal conductor. When there is potential to ground in this environment, potential difference between these two is the induced voltage of metal electrode.

A non-contact voltage sensor was designed according to the above electrostatic induction principle. Different from the traditional sensor that directly measures the voltage of electrified conductor, this sensor couples electric field produced by electric charges of the conductor to be measured into voltage sensor, so as to form detection signal. Figure 5 is the schematic diagram for the principle of sensor measurement.

In Figure 5, S indicates the external surface boundary of electrified conductor in any shape; Ω means the computational domain of the whole electric field; ε refers to the electric medium constant in the area; r signifies the position vector of the calculation site; r′ denotes the position vector of original point; $E(r)$ is the electric field intensity of any point in the computational domain; $F(r')$ refers to the potential distribution on the external surface of electrified conductor; $\varphi_0(t)$ means the electric potential of electrified conductor; A indicates the external surface of metal electrode of the voltage sensor; B is the sensor support (connected to the ground); and R_m denotes the sampling resistance of voltage sensor.

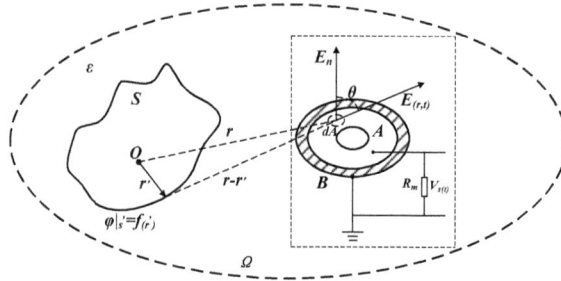

Figure 5. Principle of sensor measurement.

No free charges exist in the computational domain, i.e., $\rho(r') = 0$. The computational domain meets progressive boundary conditions, and any point in the computational domain meets Poisson equation of Equation (1):

$$\nabla^2 \varphi_{(r)} = -\frac{\rho_{(r)}}{\varepsilon} \tag{1}$$

Green function is introduced and the following result is gained by solving Equation (1) through the first boundary condition $\varphi \mid s' = f(r')$:

$$\varphi_0 = \frac{1}{\varepsilon F(r)} E(r) \tag{2}$$

In Equation (2), $F(r)$ is decided by the distance of field point from the conductor to be measured. According to this equation, in the computational domain Ω, the electric field intensity of any point is in direct proportion to electric potential of the conductor to be measured, and they have a linear relation.

Suppose that the electric potential change of electrified conductor with time is $\varphi_0(t)$ and the frequency is 50 Hz; at this time, the electric field produced by electrified body can be treated as quasi static field. Under the effect of this electric field, induced charge q will appear on the surface of metal electrode A due to electrostatic induction principle. Closed Gaussian surface is created on the electrode surface with the distance of r from the electrified conductor, and the differential element dA is taken [13–15]. The electric field intensity component at normal direction is E_n and the angular separation with $E(r,t)$ is θ. The following equation can be gained according to Gauss theorem:

$$\frac{dq}{dt} = \oint_A \varepsilon \frac{dE(r,t)}{dt} dA \tag{3}$$

According to Equation (3), the electric field intensity produced by electric charges on the energized conductor changes with the time t, and the electric field intensity will cause corresponding changes to the quantity of electric charges on induction electrode A. The induced charges will distribute on electrode A to form electric potential $\varphi_s(t)$. Sampling resistance R_m is used to connect induction electrode A and support B, and the voltage $V_s(t)$ at both ends of the sampling resistance is the detecting voltage:

$$V_s(t) = R_m \oint_A \varepsilon \frac{dE(r,t)}{dt} dA = \frac{\varepsilon_0 A r'}{r^2} R_m \frac{d}{dt} \varphi_0(t) \tag{4}$$

By organizing Equation (4), the following equation can be gained:

$$\varphi_0(t) = \frac{r^2}{\varepsilon A R_m r'} \int V_s(t)dt \tag{5}$$

When the electrified conductor is cylindrical transmission line conductor, r' is the section radius of the cylinder. At this time:

$$K = \frac{r^2}{\varepsilon A R_m r'} \tag{6}$$

In Equation (6), K is a constant, so:

$$\varphi_0(t) = K \int V_s(t)dt \tag{7}$$

According to Equation (7), the voltage value of electrified conductor can be calculated by conducting integral operation for the voltage signal of sampling resistance and multiplying the result by K; these two have a linear relation. According to the above discussion, a non-contact method can be found to realize non-contact measurement for the voltage of power frequency high voltage electrified body.

3. Design of Non-Contact Voltage Sensor

Figure 6 presents the lumped parameter equivalent circuit model of electric field coupling sensor based on electrostatic induction. $\varphi(t)$ refers to the equivalent voltage of the conductor to be measured; C_a means the equivalent mutual capacitance between the sensor electrode and conductor to be measured; C_{an} indicates the ground equivalent stray capacitance of sensor electrode; R_m signifies the detecting resistance; and R_1 and C_1 constitute integral circuit and the output signal of integral circuit is $V_0(t)$, respectively.

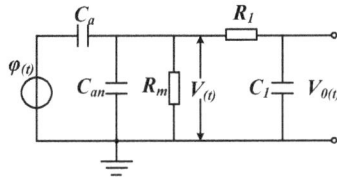

Figure 6. Equivalent circuit model.

Through Laplace transformation, transfer Equation (8) of sensor and transfer Equation (9) of passive integral circuit can be gained:

$$H_M(s) = \frac{C_a R_m s}{1 + (C_a + C_{an}) R_m s} \tag{8}$$

$$H_N(s) = \frac{1}{1 + C_1 R_1 s} \tag{9}$$

Suppose that ω_h and ω_l are upper limit and lower limit of sensor measurement bandwidth; the following equations can be gained:

$$\omega_H = \frac{1}{(C_a + C_{an}) R_m} \tag{10}$$

$$\omega_L = \frac{1}{C_1 R_1} \tag{11}$$

The characteristic amplitude-frequency response curve of the sensor, integral circuit and the entire measurement system can be obtained through Equations (8) and (9), as shown in Figure 7.

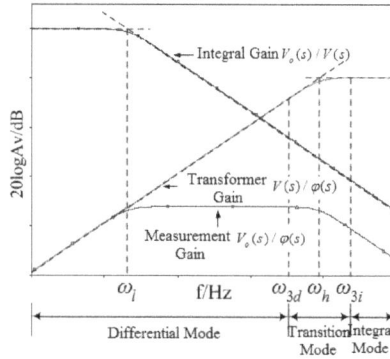

Figure 7. Amplitude-frequency response curve.

As shown in Equation (12), when $(C_a + C_{an})R_m \ll 1$, the sensor works under differential mode:

$$H_M(s) = \frac{V(s)}{\varphi(s)} = C_a R_m s \tag{12}$$

At this time, in order to meet the measurement requirements, integral circuit should be added, but integral circuit will cause troubles to signal-to-noise ratio design of sensor. Meanwhile, due to problems of component parameters in integral circuit, detecting waveform distortion and inaccurate measurement problems will be caused. Therefore, sensor should work under self-integration mode. In other words, Equation (13) is gained when $(C_a + C_{an})R_m \gg 1$. At this time, the required detection signal of sensor can be obtained without integral circuit.

$$H_M(s) = \frac{V(s)}{\varphi(s)} = \frac{C_a}{C_a + C_{an}} \tag{13}$$

In order to make the sensor work under self-integration mode at the power frequency of 50 Hz, the angular frequency should be $\omega_H < 2\pi \times 50$. Therefore, $(C_a + C_{an})R_m$ must meet the following condition:

$$(C_a + C_{an})R_m > 3.18 \times 10^{-3} \tag{14}$$

Non-contact measurement mode is adopted, so the distance between the sensor and conductor to be measured cannot be short. Thereby, the mutual capacitance C_a between the conductor to be measured and induction electrode cannot be large and the distance between the sensor and ground is long. Therefore, the ground stray capacitance C_{an} of the sensor is not large (generally speaking, C_a and C_{an} are at the level of pF). In order to meet the condition of Equation (14), the sampling resistance R_m should be at the level of GΩ, which will cause problems to equivalent internal resistance detection and impedance matching of the sensor. The small capacitance values of C_a and C_{an} also restrict the voltage division ratio of sensor. One terminal of the sampling resistance is connected to the ground, increasing the difficulty of the insulation design of the sensor. Most existing electric field coupling sensors based on electrostatic induction adopt the form of adding integral circuit, so as to gain the signal to be detected. Meanwhile, the sensor electrode often adopts plate or cylindrical design, which might restrict detection bandwidth, voltage division ratio, detection accuracy, and insulation design of the sensor.

By directing at the above problems, the equivalent circuit of non-contact voltage sensor designed in this paper is given, as shown in Figure 8. The voltage sensor of adopting differential input structure sets the difference of suspended potentials output by a pair of sensor electrodes working under self-integration mode as the output of sensor. In Figure 7, $\varphi_{(t)}$ means the equivalent voltage source of the conductor to be measured; C_a and C_b refer to the equivalent mutual capacitance between induction electrode and conductor to be measured; C_{an} and C_{bn} signify the ground equivalent stray capacitance of induction electrode; C_{ab} indicates the equivalent mutual capacitance between electrodes; Mark 1 denotes the suspended potential of voltage source; Marks 2 and 3 are suspended potentials output by two electrodes; and R_m refers to the detecting resistance [16].

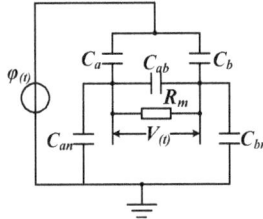

Figure 8. Equivalent circuit of voltage sensor.

Transfer function of the system is obtained through Laplace transformation:

$$H(s) = \frac{(C_a C_{an} - C_b C_{bn})R_m s}{(C_{ab} R_m s + 1)(C_a + C_b + C_{an} + C_{bn}) + (C_a + C_{an})(C_b + C_{bn})R_m s} \quad (15)$$

According to Figure 8, the non-contact voltage sensor designed in this paper adopts two induction electrodes and inputs suspended potentials of the two electrodes into the differential amplifier of signal processing circuit as detection signals. In this method, no terminal of the sensor is connected to the ground, greatly reducing the insulation design difficulty of the sensor. The signal processing method of differential type can effectively eliminate the interference of common-mode signal and increase the measurement accuracy of sensor.

By comparing Equation (8) with Equation (15), after design structure of this paper is introduced, transfer function of the sensor will not be changed. Only mutual capacitance C_{ab} between electrodes is introduced in the electrode point. Compared with the mutual capacitance C_a between the induction electrode and conductor to be measured that cannot be large capacitance, the distance between induction electrodes can be designed freely and the inter-electrode capacitance C_{ab} can be adjusted freely. Therefore, such circuit structure can meet the requirement of Equation (14), and the sensor is able to work under self-integration mode.

The amplitude-frequency characteristic function of sensor designed in this paper can be gained according to Equation (15):

$$|H(\omega)| = \frac{\dfrac{C_a C_{bn} + C_b C_{an}}{C_a + C_b + C_{an} + C_{bn}} R_m}{\sqrt{R_m^2 \left[\dfrac{(C_a + C_{an})(C_b + C_{bn})}{C_a + C_b + C_{an} + C_{bn}} + C_{ab}\right]^2 + \dfrac{1}{\omega^2}}} \quad (16)$$

Its phase-frequency characteristic function is:

$$\angle H(\omega) = \arctan \frac{1}{R_m \omega \left[\dfrac{(C_a + C_{an})(C_b + C_{bn})}{C_a + C_{an} + C_b + C_{bn}} + C_{ab}\right]} \quad (17)$$

According to Equation (16), the inter-electrode mutual capacitance C_{ab} exists in the denominator, so the transformation ratio of the sensor can be increased by magnifying the mutual capacitance C_{ab} between electrodes without changing other conditions. In Equation (17), C_{ab} also exists in the denominator only, so the phase difference between output signals of the electrified body to be measured and voltage sensor can be reduced by increasing the inter-electrode capacitance C_{ab}, thus measurement accuracy and response speed of the sensor can be further enhanced.

Through the above analysis, the sensor is required to work under self-integration mode at the power frequency of 50 Hz due to the following two reasons:

(1) If the sensor works under differential state, integral circuit should be designed. According to the analysis of circuit structure, the detection bandwidth upper limit of sensor is restricted by corner frequency of the sensor at this time, and detection bandwidth lower limit is restricted by corner frequency of the circuit. Therefore, bandwidth of the entire sensor detector is restricted.

(2) Capacitance and resistance in integral circuit are restricted by factors like temperature coefficient. As a result, the accuracy of simulating integral circuit can hardly reach the requirement. Meanwhile, due to stray parameters of components and parts in integral circuit, distortion might happen to the detection signal.

According to the analysis in this paper, in order to meet the above requirements, the mutual capacitance between sensors should be increased to the largest extent. This paper shows that the method of increasing the mutual capacitance between induction electrodes of the sensor is to treat suspended potential signal of sensor electrode as output signal of the sensor. The purpose of adopting differential amplifier is to restrain common-mode signal in suspended potential signal of sensor, and to amplify the differential signal at the same time.

In the future, solar panel will be adopted as power supply of signal processing circuit. The circuit board adopts the mode of DC +5 V power supply, and the differential amplifier uses the mode of +3 V and −3 V power supply. Signal processing circuit and solar panel are designed in the box, and installed in the tower of high-voltage transmission line, so as to realize voltage detection for high-voltage transmission line.

According to the above analysis, the non-contact voltage sensor designed in this paper is presented in Figure 9. Figure 9a shows the physical design of sensor. Annular copper electrodes are arranged on the front and back sides of PCB board (printed-circuit board) as induction electrodes. The sensor diameter is 70 mm and the pore diameter in the middle is 18 mm; there are 32 electrodes on the front side and the electrode spacing is 0.3 mm; there are 32 electrodes on the back side and the electrode spacing is 0.3 mm; and the detection signal is drawn forth by the screened wire. Figure 9b is the schematic diagram for sensor structure and the electric transmission line passes through the pore in the middle of the sensor. In order to enhance the mutual capacitance between electrodes and the sensor electrode ability of holding electric charges to the largest extent, multiple PCB boards are connected; all electrodes on the front side of each PCB board are connected and all electrodes on the back side of each PCB board are connected.

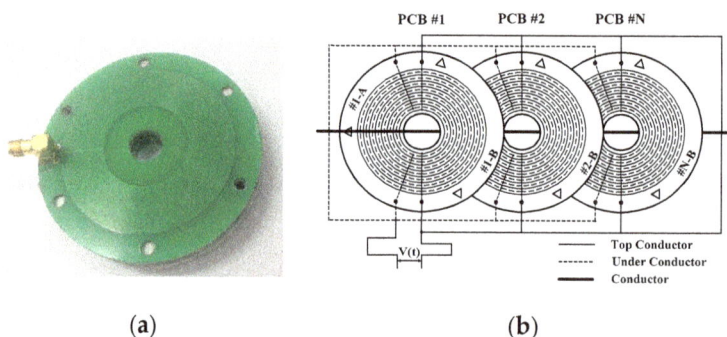

(a)   (b)

Figure 9. Design of non-contact voltage sensor: (**a**) physical design of the sensor; and (**b**) schematic diagram for sensor structure.

4. Simulation

In this paper, Ansoft software is used to realize optimization for the structure design of non-contact voltage sensor. Meanwhile, the influence of sensor introduction on measurement and the distortion situations around the electric field are studied when sensor is applied to voltage detection of three-phase transmission line.

The parameters of voltage sensor were shown in Table 1, and those parameters were used in simulation.

Table 1. Parameters of sensor.

Sensor	Number of Electrode	Width of Electrode (mil)	Spacing of Electrode (mm)	Diameter of Innermost Electrode (mm)
Board front	32	6	0.3	
Board opposite	32	6	0.3	28

The actual sensor adopts 12 PCB boards. According to the measurement results of Agilent 4294 A impedance analyzer, electrode mutual capacitance between the output ends of the PCB sensor is 950 pF, and the maximum fluctuation deviation value within the frequency range from 45 Hz to 2 MHz is 2 pF.

4.1. Optimization Design of Sensor Structure

The simulation experiment model is established according to the electrode structure presented in Figure 10 where 1 represents the electrode on the front side of PCB board; 2 means the electrode on the back side of PCB board; R, D and d refer to the radius, width and electrode spacing of #1 electrode, respectively; and N indicates the number of electrodes. Through post-processing for the simulation result, equivalent distributed capacitances under different motivations and different electrode structures can be gained via calculation, as shown in Figure 11. Figure 11a is the relation curve of equivalent capacitance and electrode radius; Figure 11b presents the relation curve of equivalent capacitance and electrode width; and Figure 11c shows the relation between equivalent capacitance and number of electrodes.

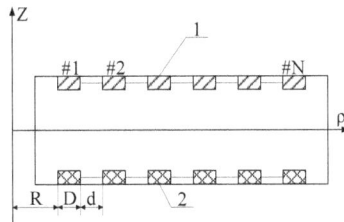

Figure 10. Sensor electrode structure.

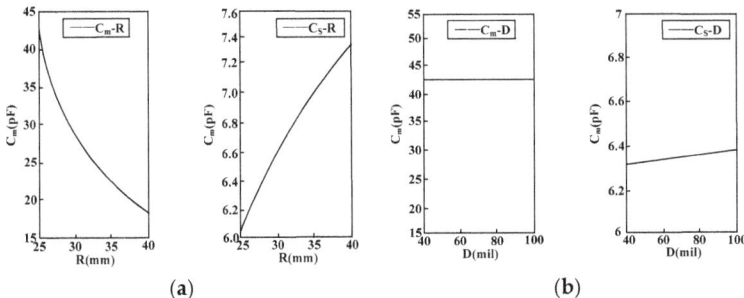

(a)   (b)

Figure 11. *Cont.*

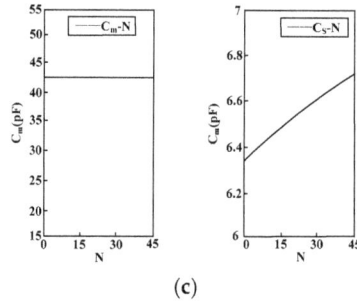

(c)

Figure 11. Simulation result of electrode structure optimization. (**a**) Relation curve of equivalent capacitance and electrode radius; (**b**) Relation curve of equivalent capacitance and electrode width; (**c**) Relation curve of equivalent capacitance and number of electrodes.

The following conclusions can be gained through the simulation result:

(1) With the increase of the transformer electrode radius R, the mutual capacitance C_{ab} between electrodes decreases and the ground stray capacitance C_n of electrode rises.

(2) With the increase of the transformer electrode width D, the ground stray capacitance C_n of electrode rises, while the mutual capacitance C_{ab} between electrodes remains unchanged.

(3) With the increase of the number of electrodes N, the ground stray capacitance C_n of electrode rises, while the mutual capacitance C_{ab} between electrodes remains unchanged.

Therefore, when the induction electrode of sensor is designed, in order to reduce the measurement error disturbance caused by the ground stray capacitance C_n of sensor, the electrode radius should be reduced to the largest extent, and meanwhile the width and distance between electrodes must be decreased. On the premise of comprehensively considering factors like the size of sensor space, more electrodes should be arranged and the mutual capacitance between electrodes must be increased to the largest extent. In this way, the voltage division ratio and measurement accuracy of sensor can be increased, and phase error in measurement will be reduced.

4.2. Insulation Design of Sensor

4.2.1. Insulation Design between Electrodes

In order to increase the mutual capacitance C_{ab} between electrodes to the largest extent, measures to reduce the distance between electrodes and decrease the electrode width are taken in electrode design. These measures have changed the electric field distribution inside the sensor. When the sensor is put in regions with high electric field intensity, the insulating ability of sensor electrode structure should be studied [17–19]. Therefore, simulation calculation should be conducted for the electric field intensity between electrodes, and the simulation result is presented in Figure 12, where 1 represents the diameter of 10 kV electrified body; 2 refers to the air gap between electrode and sensor; and 3 indicates the spacing between sensor support and innermost electrode. According to the analysis, the electric field intensity between electrodes is far lower than the electric field intensity from electrode surface to the innermost side of sensor. In order to clearly reflect the electric field distribution between electrodes, the maximum ruler of the calculation result is 200 V/m, but the actual electric field intensity from electrode surface to the innermost side of sensor is far higher than 200 V/m. Table 2 shows the relative dielectric constant and critical field intensity of medium involved in the calculation. According to the distribution of electric field intensity, the maximum electric field intensity appears at the junction part of the outermost electrode and sensor support (50 V/m) as well as the interface of the sensor support and air (150 V/m). The electric field intensity between electrodes is quite low, almost equal to 0. Therefore, such structure can meet the insulation design between electrodes.

Table 2. Medium parameters.

Medium	Relative Permittivity	Critical Electric Field (kV/cm)
Epoxy (PCB)	3.6	200–300
Air	1	25–30

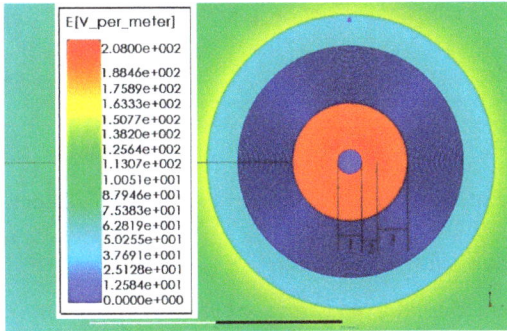

Figure 12. Insulation intensity of sensor electrode and simulation calculation.

4.2.2. Insulation Design of Sensor Support

Figure 13 shows the electric field intensity distribution of the electrified body under the voltage grade of 10 kV in electrified conductor (conductor), conductor and sensor air gap (air), innermost electrode and support of sensor (epxoy), sensor electrode (electrode), outermost electrode and support of sensor (epxoy), and sensor and air (air) successively. According to the simulation result, the maximum value of electric field intensity appears in the sensor support part closest to the electrified body (epoxy resin) and the electric field intensity is about 8000 V/m. As per Table 1, it is far lower than the critical field intensity of epoxy resin. Therefore, this structure can meet the insulation design of sensor support. The electric field intensity between electrodes inside the sensor is quite low, but it is not equal to 0. The electric field distribution after magnifying the coordinate system is presented in Figure 14.

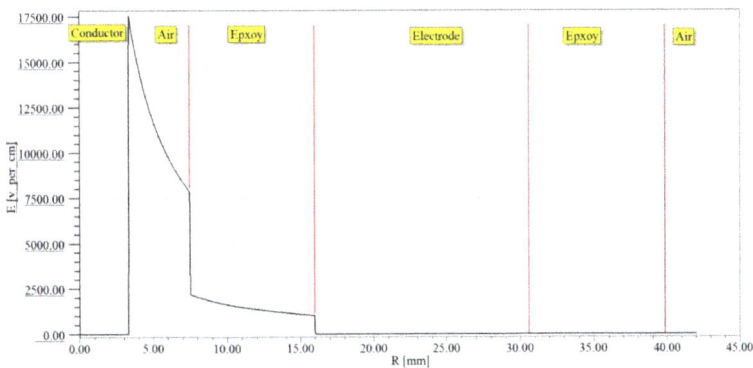

Figure 13. Insulation design simulation result of electrode.

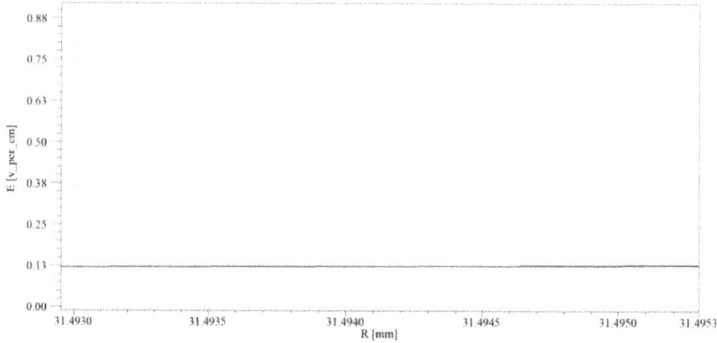

Figure 14. Internal electric field intensity distribution of sensor electrode.

4.3. Influence of Sensor Introduction on Measurement

When non-contact method is adopted to measure the voltage of electrified conductor, the use of sensor will inevitably influence electric field around the conductor and cause electric field distortion. Therefore, the distortion degree of electric field caused by the introduction of sensor should be studied [20–22].

Figure 15 shows the simulation result of electric field intensity in the cross section direction under horizontal arrangement of three-phase conducting wire. Sectional area of the wire is 35 mm², the wire spacing is 1.38 m, and the vertical height from the ground is 1.8 m. Its size and structure are consistent with the test platform actually established. During the simulation, the voltage frequency of three-phase conducting wire is 50 Hz and the voltage amplitude is 12 kV. In Figure 15, A-phase, B-phase and C-phase wires are presented from the left side successively, and the initial phases are 0°, −120° and −240°, respectively. It can be seen that when no sensor is added, electric field distribution around the three-phase conducting wire is uniform. Figure 16 shows electric field distribution around the three-phase conducting wire after sensor is installed on the line. According to the comparison, the introduction of sensor causes distortion to electric field distribution. The electric field intensity between phase and phase decreases obviously. As for the reason, the electric field line sent out by electric charges of the conductor ends in the electrode of voltage sensor. Figure 17a,b are diagrams for local electric field distribution of A-phase wire before and after sensor is used, respectively. The application of sensor changes the electric field distribution. However, according to Figure 17b, electric field distribution between the electrified conductor and voltage sensor is uniform, and no obvious electric field mutation is seen. In this section, the interphase electric field intensity is reduced by using the sensor, but no obvious electric field mutation is seen. The electric field distribution is still uniform.

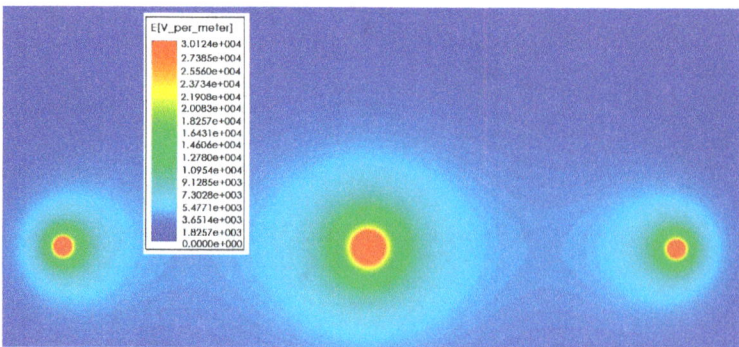

Figure 15. Electric field distribution without sensor.

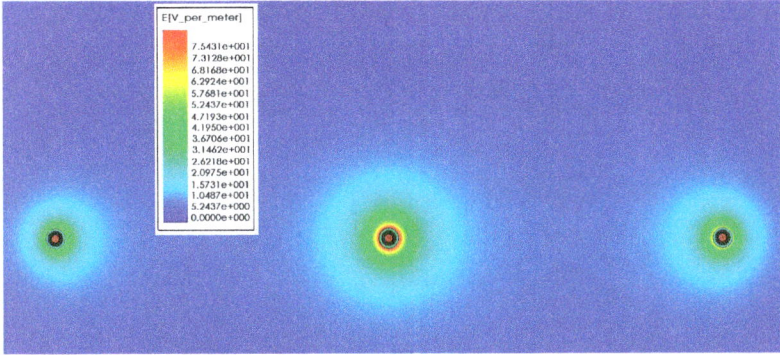

Figure 16. Electric field distribution after introducing sensor.

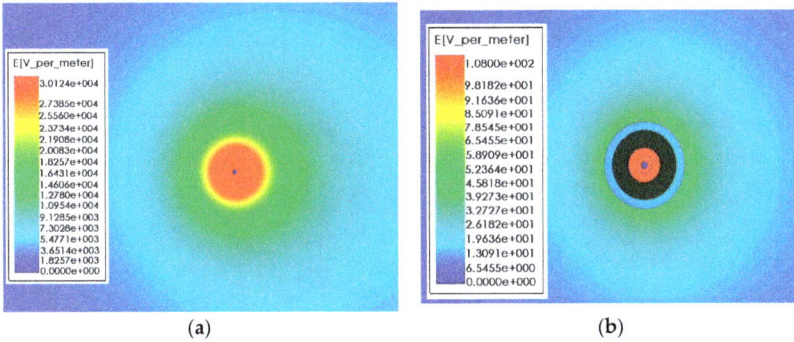

(a)   (b)

Figure 17. Influence of sensor on electric field. (**a**) A-phase electric field distribution without sensor; and (**b**) A-phase electric field distribution after introducing sensor.

Measurement points a and b are put 5 and 45 mm away from the left side of B-phase wire. Electric field intensity at the above two points is calculated at the same moment with the interval of 1 kV (the phase spacing of three phases is 120°) when the line voltage is within the range of 1–12 kV. Table 3 shows the electric field intensity when sensor is not used, and Table 4 presents the electric field intensity after sensor is used.

Table 3. Electric field intensity at A and B when sensor is not used.

Voltage Phase = 35 [deg] (kV)	Electric Field Intensity of Point A (kV/m)	Electric Field Intensity of Point B (kV/m)
1	2.10	0.15
2	4.20	0.30
3	6.30	0.48
4	8.39	0.60
5	10.49	0.75
6	12.59	0.90
7	14.69	1.04
8	16.79	1.19
9	18.89	1.34
10	20.99	1.49
11	23.089	1.64
12	25.189	1.79

Table 4. Electric field intensity at A and B when sensor is introduced.

Voltage Phase = 35 [deg] (kV)	Electric Field Intensity of Point A (kV/m)	Electric Field Intensity of Point B (kV/m)
1	13.35	0.008
2	25.86	0.012
3	39.00	0.016
4	52.00	0.019
5	65.00	0.023
6	78.00	0.028
7	90.99	0.032
8	103.99	0.035
9	116.99	0.040
10	129.98	0.045
11	132.15	0.058
12	155.99	0.062

The linearity fitting curve graph shown in Figure 18 is drawn according to the results in Tables 2 and 3.

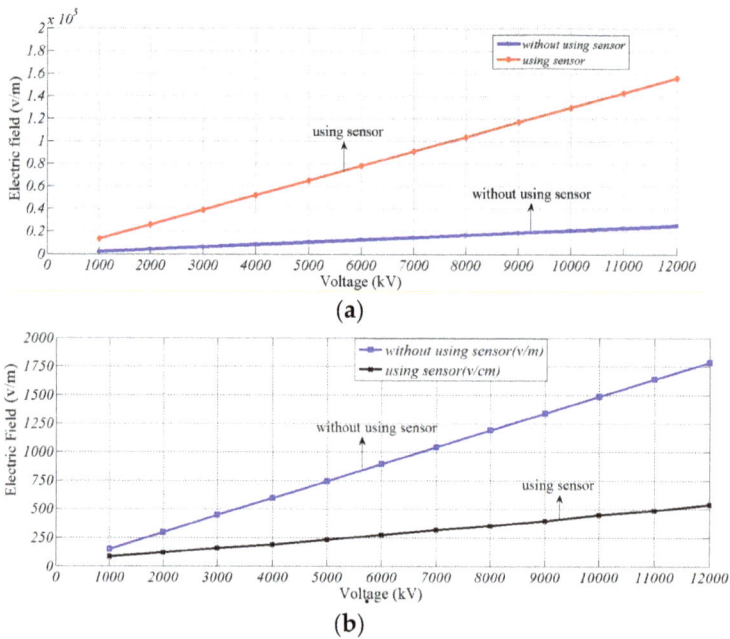

Figure 18. Fitting curve of electric field intensity and voltage grade: (**a**) phase A; and (**b**) phase B.

Least-square fitting is conducted for electric field intensity data at A and B acquired before and after sensor is introduced, and Equations (18) and (19) are obtained, where x means electric field intensity of A and B before using the sensor, and y means electric field intensity of A and B after using the sensor. In the equation the constant term is generated under the combined action of boundary condition setting and fitting method when simulation model is established.

$$y = 6.0291x + 1.06 \tag{18}$$

$$y = 0.0317x + 0.7024 \tag{19}$$

According to the above simulation result, a linear relation is presented between electric field intensity around the circuit and line voltage when sensor is not used. When voltage sensor designed in this paper is used, a linear relation is still presented between electric field intensity around the circuit and line voltage. Therefore, though the use of voltage sensor will influence the electric field to be measured, electric field distribution is still uniform, and electric field intensity and line voltage present linear changes.

5. Experiment

The working performance of non-contact voltage sensor should be verified by experiment. The experiment is composed of sensor linearity test and sensor measurement accuracy test.

The experimental platform is shown in Figure 19. In Figure 19a, 1 refers to the voltage sensor and this sensor is installed on all three-phase lines. Through design of the voltage sensor, its voltage division ratio is set as 1500:1. 2 represents the high-voltage probe of P6051A, its voltage division ratio is set as 1000:1. 3 indicates the wiring terminal of three-phase line. 4 denotes the transformer, its rated capacity is 50 kVA, and the output voltage of transformer can be adjusted through the control cabinet. 5 is the standard voltage transformer. 6 signifies the oscilloscope. 7 means the high-precision digital multimeter. In Figure 19b, 1 refers to the load part composed of resistive load and capacitive load, and the load part is equipped with a large number of fans for heat dissipation and 2 represents the control cabinet which can control the transformation ratio of transformer. Stepless adjustment can be realized for the voltage within the range of 0–35 kV. Meanwhile, the control cabinet can also switch the load, and the line current can change within the range of 0–200 A through the matching relation between voltage and load. Thus voltage and current of three-phase changes can coexist in the line and the purpose of simulating actual three-phase transmission line will be realized. Moreover, the control cabinet can realize real-time display for voltage, current and power factor of different phases. Sectional area of the wire is 35 mm^2, the distance from the wire to the ground is 1.8 m, and the spacing between phase and phase is 1.38 m.

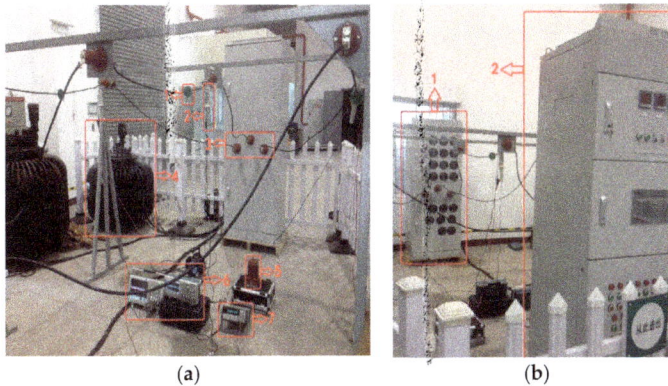

(a)   (b)

Figure 19. Experimental platform structure: (**a**) experimental platform structure 1; and (**b**) experimental platform structure 2.

5.1. Sensor Linearity Test

Sensor linearity test aims to study whether the detection value of sensor and the voltage of conductor to be measured still present linear changes when the use of sensor causes electric field distortion.

The schematic diagram for experimental platform is shown in Figure 20. The high-voltage probe is connected to A-phase, B-phase and C-phase wires, respectively; meanwhile, lines of the three phases all pass through the non-contact voltage sensor. Output connection to oscilloscope is realized between the high-voltage probe and sensor, the voltage of control cabinet is adjusted, and waveform and phase difference data are recorded every 1 kV within the range of 1–12 kV.

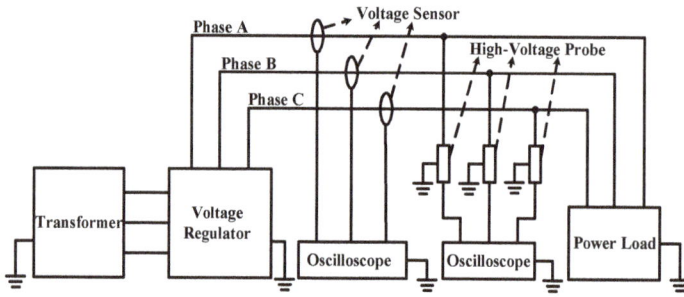

Figure 20. Schematic diagram for the structure of linearity test platform.

Figure 21a shows the line voltage waveform detected by high-voltage probe under the voltage grade of 8 kV, and Figure 21b shows the detection signal waveform of voltage sensor under the voltage grade of 8 kV. According to Figure 21, detection signal of the designed voltage transformer has the same waveform with the actual line voltage signal detected by high-voltage probe, and the voltage division ratio of voltage transformer is 1500:1.

Besides resistive load, capacitive and inductive loads also exist in the part of experimental platform design and platform load implementation, so there are some errors between phase and phase in the three lines. Therefore, the difference is not exactly 120°.

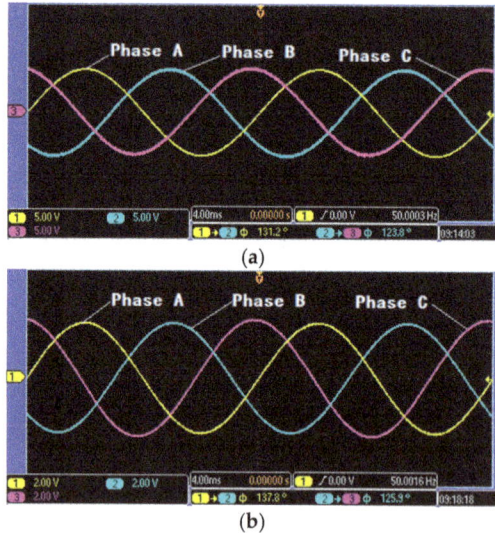

(a)

(b)

Figure 21. Waveform figure: (a) Signal of high-voltage probe under 8 kV; and (b) signal of voltage sensor under 8 kV.

Figure 22a shows the output waveform of A-phase line high-voltage probe and non-contact voltage sensor under the voltage grade of 4 kV, and Figure 22b shows the output waveform of A-phase line high-voltage probe and non-contact voltage sensor under the voltage grade of 6 kV.

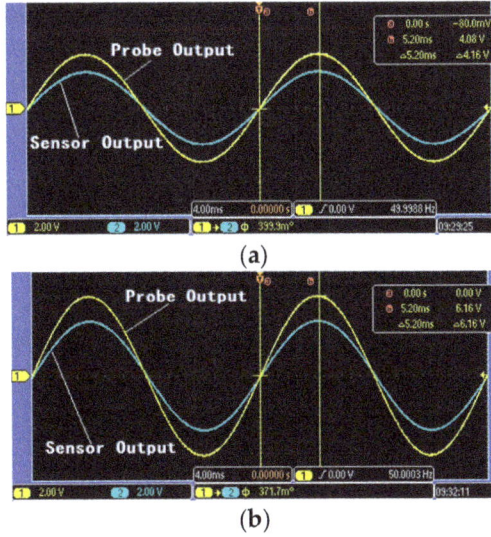

Figure 22. Comparison figure of high-voltage probe and voltage sensor: (**a**) signals comparison under 4 kV; and (**b**) signals comparison under 6 kV.

The phase error of the three-phase line between these two under different voltage grades is gained, as shown in Table 5. In the table, φ_S indicates the output voltage phase of voltage sensor and φ_H signifies the output voltage phase of high-voltage probe. According to Table 5, voltage sensor works under self-integration mode. According to Table 4, the maximum phase error of voltage sensor and high-voltage probe is 33.9′ and the average error is 27.925′. The phase detection accuracy of voltage sensor under steady state meets the accuracy class of 0.5-level voltage sensors.

Table 5. Phase error.

Voltage (kV)	$\varphi_H\text{-}\varphi_S$ (′)
1	33.2
2	30.7
3	25.4
4	23.9
5	25.7
6	22.3
7	31.7
8	32.6
9	25.4
10	22.8
11	33.9
12	27.5

5.2. Sensor Measurement Accuracy Test

Sensor measurement accuracy test aims to verify the measurement accuracy grade of voltage sensor, and it is the important index for the working performance of voltage sensor.

The experimental platform of simulating three-phase transmission line shown in Figure 19 is still adopted, but the verification test is conducted by adopting the method of phase measurement. Figure 23 is the schematic diagram for the structure of sensor measurement accuracy test platform. The standard voltage transformer adopts EVT1-20 electronic voltage transformer. The rated voltage of primary side is 20 kV, the working frequency is 50 Hz, the rated phase shift is 0, the rated output of secondary side is 0–4 V(AC), and the accuracy class is 0.2. 34410A high-performance digital multimeter of Keysight Truevolt series is adopted as high-precision digital multimeter, and the measurement accuracy of measuring the alternating voltage of 50 Hz and 0–50 V is 0.006%. Therefore, the above experimental equipment can guarantee the accuracy of the test conclusion.

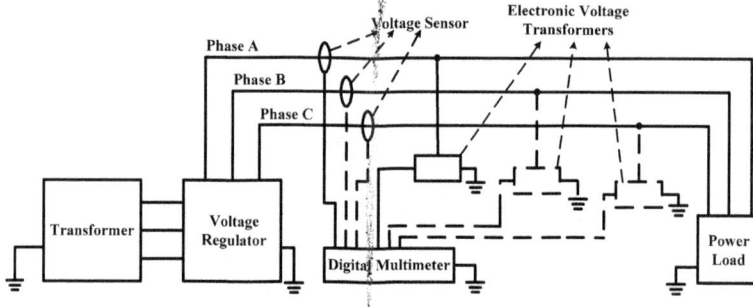

Figure 23. Schematic diagram for the structure of accuracy test platform.

The output ends of standard electronic voltage transformer and non-contact voltage sensor are connected to digital multimeter, the three-phase voltage is adjusted, and the output results of A-phase, B-phase and C-phase standard voltage transformer as well as the output results of non-contact voltage sensor designed in this paper are recorded every 1 kV within the range of 1–12 kV. The ratio error $\varepsilon\%$ between sensor signal and actual voltage signal is gained according to the provisions in IEC60044-7 standard, and its definition is as follows:

$$\varepsilon\% = \frac{KU_S - MU_H}{MU_H} \times 100\% \tag{20}$$

where K refers to the voltage division ratio of voltage sensor designed in this paper and the rated transformation ratio after calibration is 1500:1; U_S means the detection signal of voltage sensor; M signifies the standard electronic voltage transformer and the transformation ratio is 5000:1; and U_H denotes the output voltage of standard electronic voltage transformer. The experimental results are shown in Tables 6–8.

Table 6. Results of Phase A.

Voltage (kV)	U_S (kV)	U_H (kV)	$\varepsilon\%$
1	0.6582	0.1967	0.37
2	1.3245	0.3961	0.31
3	1.9763	0.5912	0.28
4	2.6496	0.7929	0.24
5	3.4023	1.0231	0.21
6	4.0518	1.2138	0.14
7	4.6164	1.3834	−0.11
8	5.3271	1.6010	−0.17
9	6.0846	1.8301	−0.21
10	6.6743	2.0079	−0.23
11	7.3257	2.2027	−0.27
12	8.0710	2.4293	−0.33

Table 7. Results of Phase B.

Voltage (kV)	U_S (kV)	U_H (kV)	$\varepsilon\%$
1	0.6732	0.2011	0.41
2	1.3411	0.4008	0.37
3	2.0177	0.6030	0.39
4	2.6714	0.7994	0.25
5	3.3376	0.9987	0.26
6	4.0365	1.2088	0.18
7	4.6585	1.3943	0.23
8	5.3436	1.6014	0.17
9	6.0094	1.7990	0.21
10	6.6832	2.0106	−0.28
11	7.3431	2.2080	−0.23
12	7.9763	2.4018	−0.37

Table 8. Results of Phase C.

Voltage (kV)	U_S (kV)	U_H (kV)	$\varepsilon\%$
1	0.6641	0.1998	−0.31
2	1.3276	0.3994	−0.27
3	1.9883	0.5985	−0.33
4	2.6475	0.7958	−0.21
5	3.3416	1.0062	−0.31
6	4.0117	1.2010	0.21
7	4.6774	1.3992	0.27
8	5.3169	1.5901	0.31
9	6.0712	1.8154	0.33
10	6.6573	1.9918	0.27
11	7.3448	2.1955	0.36
12	7.9659	2.3829	0.29

According to the above two tests, some errors exist in detection signals of the non-contact voltage sensor designed in this paper and standard electronic voltage transformer, and the errors mainly come from two sources. On the one hand, though the sensor is specially designed, the position of its electrode arrangement has inevitable errors due to the restriction of craftsmanship. As a result, errors between transformer phase position and actual voltage phase are caused. However, such errors can be ignored in actual use. On the other hand, complete coaxial arrangement cannot be realized between the voltage sensor and line, so the electric field lines passing through sensor electrode are imbalanced and the measurement accuracy will be affected. However, such errors can be ignored in actual use. The voltage ratio error of the non-contact voltage sensor designed in this paper within the voltage range of 0–12 kV is $\varepsilon\% < 0.5\%$, and the phase difference is $\varphi < 35'$. Therefore, the measurement accuracy of such non-contact voltage sensor can meet the actual measurement requirement during steady state measurement.

6. Conclusions and Prospect

According to this paper, the following conclusions can be gained:

(1) When non-contact voltage sensor is applied to steady state voltage measurement within 12 kV, the phase error is smaller than 35′, which meets the phase detection accuracy class of 0.5-level voltage transformer used for measurement.

(2) When non-contact voltage sensor is applied to steady state voltage measurement within 12 kV, the ratio error $\varepsilon\%$ is <0.5%, so the accuracy of steady state measurement meets the accuracy class of 0.5-level voltage transformer used for measurement.

(3) Through simulation and experiment, structure design of sensor is optimized, voltage division ratio of voltage sensor is increased, insulation design difficulty of sensor is reduced, and the sensor is able to work under self-integration mode.

In the future, in-depth studies will be made into the aspects of standardization, anti-interference and signal processing circuit of non-contact voltage sensor. In order to meet the electrified installation requirement of sensor, open-type design should be conducted for the sensor electrode; in order to solve the coaxial arrangement problem of sensor and line, a corresponding fixture should be designed.

Acknowledgments: This work was supported by the Funding Program from State Grid Chongqing Electric Power Co. Electric Power Research Institute (SGCQDK00JZJS1400018).

Author Contributions: Wei He is the head of the research group that conducted this study. He contributed to the research through his general guidance and advice. All authors contributed to the writing and revision of the manuscript.

Conflicts of Interest: The authors declare no conflict of interest.

References

1.   John-Paul, H.K.; Cheri, W. An Innovative Approach to Smart Automation Testing at National Grid. In Proceedings of the 2012 IEEE PES Transmission and Distribution Conference and Exposition (T&D), Orlando, FL, USA, 7–10 May 2012; pp. 1–8.

2.   Liu, G.Y.; Kang, L.; Peng, W.N. On-Line Monitoring System for Transformer Partial Discharge. *Appl. Mech. Mater.* **2013**, *303*, 464–467. [CrossRef]

3.   Sarkar, B.; Koley, C.; Roy, N.K.; Kumbhakar, P. Condition monitoring of high voltage transformers using Fiber Bragg Grating Sensor. *Measurement* **2015**, *74*, 255–267. [CrossRef]

4.   Kaveri, B.; Saibal, C. Electric stresses on transformer winding insulation under standard and non-standard impulse voltages. *Electr. Power Syst. Res.* **2015**, *123*, 40–47.

5.   Zhao, S.P.; Li, H.Y.; Peter, C.; Forooz, G. Test and Analysis of Harmonic Responses of High Voltage Instrument Voltage Transformers. *IET* **2014**, *3*, 1–6.

6.   Cecati, C.; Citro, C.; Siano, P. Combined Operations of Renewable Energy Systems and Responsive Demand in a Smart Grid. *IEEE Trans. Sust. Energy* **2011**, *2*, 468–476. [CrossRef]

7.   Browning, C.A.; Vinci, S.J.; Zhu, J.; Hull, D.M.; Noras, M.A. An evaluation of electric-field sensors for projectile detection. In Proceedings of the 2013 IEEE SENSORS, Baltimore, MD, USA, 3–6 November 2013; pp. 1–4.

8.   Farhangi, H. The path of the smart grid. *IEEE Power Energy Mag.* **2010**, *8*, 18–28. [CrossRef]

9.   Pan, F.; Xiao, X.; Xu, Y.; Ren, Y. An optical AC voltage sensor based on the transverse pockels effect. *Sensors* **2011**, *11*, 6593–6602. [CrossRef] [PubMed]

10. Metwally, I.A. Comparative measurement of surge arrester residual voltages by D-dot probes and dividers. *Electr. Power Syst. Res.* **2011**, *81*, 1274–1282. [CrossRef]

11. Metwally, I.A. D-dot probe before fast-front voltage measurement. *IEEE Trans. Instrum. Meas.* **2010**, *59*, 2211–2219. [CrossRef]

12. Harun, T.; Seddik, B.; Daniel, C.; Ahmad, H. Low-Voltage Loss-of-Life Assessments for a High Penetration of Plug-In Hybrid Electric Vehicles. *IEEE Trans. Power Deliv.* **2012**, *27*, 1323–1331.

13. Wang, Y.A.; Xiao, D.M. Prototype design for a high-voltage high-frequency rectifier transformer for high power use. *IET Power Electron.* **2011**, *4*, 615–623. [CrossRef]

14. Metwally, I.A. Coaxial D-dot probe: Design and testing. In Proceedings of the 1995 Annual Report Conference on Electrical Insulation and Dielectric Phenomena, Virginia Beach, VA, USA, 22–25 October 1995; pp. 298–301.

15. Barker, P.P.; Mancao, R.T.; Kvaltine, D.J.; Parrish, D.E. Characteristics of lightning surges measured at metal oxide distribution arresters. *IEEE Trans. Power Deliv.* **1993**, *8*, 301–310. [CrossRef]

16. Zhou, Q.; He, W.; Li, S.N.; Hou, X.Z. Research and Experiment of a Unipolar Capacitive Voltage Sensor. *Sensors* **2015**, *15*, 20678–20697. [CrossRef] [PubMed]

17. Kubo, T.; Furukawa, T.; Itoh, H.; Hisao, W.; Wakuya, H. Numerical electric field analysis of power status sensor observing power distribution system taking into account measurement circuit and apparatus. In Proceedings of the 2011 IEEE SICE Annual Conference (SICE), Tokyo, Japan, 13–18 September 2011; pp. 2741–2746.
18. Koziy, K.; Bei, G.; Aslakson, J. A low-cost power quality meter with series arc-fault detection capability for smart grid. *IEEE Trans. Power Deliv.* **2013**, *18*, 1584–1591. [CrossRef]
19. Bayer, C.F.; Baer, E.; Waltrich, U.; Malipaard, D.; Schletz, A. Simulation of the electric field strength in the vicinity of metallization edges on dielectric substrates. *IEEE Trans. Dielectr. Electr. Insul.* **2015**, *22*, 257–265. [CrossRef]
20. Kubo, T.; Furukawa, T.; Fukumoto, H.; Ohchi, M. Numerical estimation of characteristics of voltage-current sensor of resin molded type for 22 kV power distribution systems. In Proceedings of the ICCAS-SICE 2009, Fukuoka, Japan, 18–21 August 2009; pp. 5050–5054.
21. Richard, C.J.; Michael, W.; Shi, G.Y. Efficient DC Fault Simulation of Nonlinear Analog Circuits: One-Step Relaxation and Adaptive Simulation Continuation. *IEEE Trans. Comput.* **2006**, *25*, 1392–1400.
22. Wang, D.; Zhang, Y.R.; Wang, Y.H.; Lee, Y.S.; Lu, P.J.; Wang, Y. Cutting on Triangle Mesh: Local Model-Based Haptic Display for Dental Preparation Surgery Simulation. *IEEE Trans. Vis. Comput. Graph.* **2005**, *11*, 671–683. [CrossRef] [PubMed]

5

Spatio-Temporal Constrained Human Trajectory Generation from the PIR Motion Detector Sensor Network Data

Zhaoyuan Yu [1,2,3], Linwang Yuan [1,2,3], Wen Luo [1,2,*], Linyao Feng [1] and Guonian Lv [1,2,3]

Academic Editor: Leonhard M. Reindl

[1]   Key Laboratory of VGE (Ministry of Education), Nanjing Normal University, No.1 Wenyuan Road, Nanjing 210023, China; yuzhaoyuan@njnu.edu.cn (Z.Y.); yuanlinwang@njnu.edu.cn (L.Y.); fenglinyao@gmail.com (L.F.); gnlu@njnu.edu.cn (G.L.)

[2]   State Key Laboratory Cultivation Base of Geographical Environment Evolution (Jiangsu Province), No.1 Wenyuan Road, Nanjing 210023, China

[3]   Jiangsu Center for Collaborative Innovation in Geographical Information Resource Development and Application, No.1 Wenyuan Road, Nanjing 210023, China

*   Correspondence: luow1987@163.com

Abstract: Passive infrared (PIR) motion detectors, which can support long-term continuous observation, are widely used for human motion analysis. Extracting all possible trajectories from the PIR sensor networks is important. Because the PIR sensor does not log location and individual information, none of the existing methods can generate all possible human motion trajectories that satisfy various spatio-temporal constraints from the sensor activation log data. In this paper, a geometric algebra (GA)-based approach is developed to generate all possible human trajectories from the PIR sensor network data. Firstly, the representation of the geographical network, sensor activation response sequences and the human motion are represented as algebraic elements using GA. The human motion status of each sensor activation are labeled using the GA-based trajectory tracking. Then, a matrix multiplication approach is developed to dynamically generate the human trajectories according to the sensor activation log and the spatio-temporal constraints. The method is tested with the MERL motion database. Experiments show that our method can flexibly extract the major statistical pattern of the human motion. Compared with direct statistical analysis and tracklet graph method, our method can effectively extract all possible trajectories of the human motion, which makes it more accurate. Our method is also likely to provides a new way to filter other passive sensor log data in sensor networks.

Keywords: sensor networks; trajectory recovering; geometric algebra; spatio-temporal constraints; trajectory filtering; MERL motion sensor

---

1. Introduction

Long-term accurate human motion trajectory analysis is becoming more and more important for indoor navigation [1], smart homes [2], behavior science [3], architectural design for buildings [4] and evacuation [5], *etc*. Continuously obtaining human motion data without violating privacy and at low cost is the main challenge for sensor development and data analysis method construction.

Active sensors (such as RFID tags, WIFI, Bluetooth sensors on mobile devices [6]) and passive sensors (such as cameras [7], passive infrared (PIR) motion detectors [8]) are used widely for human behavior tracking. Active sensors, which send signals on their own initiative, are more accurate in short-time human identification, classification and motion tracking. Active sensors are known to the users themselves and can be used for individual identification. Therefore, they are not suitable for people who are sensitive to privacy intrusions [9]. On the contrary, passive sensors, which only log people's information when there are people in their range, are used more widely in long-term observation of large scale human indoor movements. The continuous use of a large number of passive sensors can be potentially economical, scalable, efficient, and privacy sensitive in human tracking [10]. A lot of available commercial systems for activity monitoring at home (e.g., Quiet Care Systems, e-Neighbor) are based on these passive sensors.

Mitsubishi Electric Research Laboratories (MERL) have deployed a large amount (>200) of cheap, energy-efficient and simple passive infrared (PIR) motion detectors to acquire human motion data continuously for about two years [11]. The sensors are densely placed in large areas and log the human motion data continuously on a large scale [8]. Connolly *et al.* used the MERL dataset to model social behaviors including visiting people, attending meetings and travelling with people with the entropy and graph cuts method [12]. They modelled pairwise statistics over the dataset to extract relationships among the occupancy data. The temporal patterns of the human motion in the MERL dataset were analyzed by T-Pattern algorithm [13,14]. Research trying to recover the social networks from the spatio-temporal patterns of the interactions were also developed [15,16]. Because of the special characteristics of MERL dataset, it has been previously used in the IEEE Information Visualization Challenge, and presents a significant challenge for behavior analysis, search, manipulation and visualization.

Different from other passive motion sensors like cameras, the PIR sensor is a binary passive sensor that cannot classify and locate individual human [11]. The PIR sensor works by sensing light emitted in the far-infrared by warm objects and signal on high-frequency changes. If there is anyone moving into the cognitive range, the sensor will be activated and the logged data will change to 1; otherwise the sensor will output 0. Every sensor works continuously to log the time and the active state data as continuous data stream. Each sensor works independently and the sensor does not distinguish the absolute location and the number of people in the area, *i.e.*, one person and several people in the cognitive range will produce the same active 1 output from the sensor. As has been discussed extensively in the literature, it is not possible to log the complete track of people moving around the space using only motion detectors [17]. Although the sensor is related to the geographical locations, only adjacent relations between different sensors can be revealed from the sensor activation log data [18,19]. Therefore, functional mapping and filtering should be applied to transform the observed response sequences back into the spatio-temporal location relations to make the trajectory complete. In the process of the backward mapping from the adjacent relations to absolute spatial coordinates, various spatio-temporal constraints should be integrated in human motion tracking. Because of the complexity of the data, it is hard to solve such high-dimensional and uncertain problems with classical methods. Due to such reasons, the statistical analysis of the human motion pattern from the PIR motion sensors also has considerable uncertainties [14].

It has already been proved that the PIR sensor network cannot provide enough information to recover the trajectory of an individual. Only the statistical behavior pattern can be extracted from the PIR sensor network log data [8]. Technologies such as tracklet graph models were developed to support the dynamic query and visualization of the possible human motion patterns in the spatio-temporal domain [20]. Other methods, including the Kalman filter [21], hidden Markov chain model [2] and topic models [22], are applied to try to extract human motion patterns in a statistical way. However, since the sensor data logged to the human trajectory mapping are not a unique mapping, *i.e.*, the same sensor logging may be caused by different human motions, the accuracy of these existing methods can still been dubious. For example, human guidance and carefully defined tracklet construction rules are required for accurate trajectory visualization [20]. To overcome these drawbacks, some researches try to use multiple device including cameras [17] to help determine the true trajectories, which makes the problem complicated and costly. Since not all the possible human motion trajectories are completely known as the full set, the accuracy of the statistical models may also be problematic.

From this perspective, generations of all the possible human trajectory patterns from the sensor log data are important for the sensor data analysis. However, to our best knowledge, there is no method that can retrieve the complete possible trajectories from the PIR sensor network data.

Besides the direct analysis of the human motion patterns from the sensor log data, there's another way to analyze the human motions from the generation-refinement paradigm. In the generation-refinement paradigm, all the possible human motion trajectories can be firstly generated and then dynamically refined according to the spatio-temporal constraints and sensor activation logs. Then the people tracking problem can be formulated as how to generate all the possible trajectory with several different spatio-temporal constraints according to the sensor activation log. To summarize, the following advantages can be achieved in the human trajectory analysis under the generation-refinement paradigm: (1) the complete possible human motion patterns can be generated; (2) the human trajectory taking place in the geographic space activates the sensor with unique pattern. Mapping from the trajectory to the sensor activation log is unique; (3) both the spatio-temporal correspondence of the sensor activation and the sensor network topology can be used to reduce the uncertainties of the trajectory analysis [17,23]. With the well-designed trajectory generation algorithm, all the possible human motion trajectories can be better extracted and classified.

The key issue of accurate analysis of the human trajectory from PIR sensor networks is how to reduce the uncertainties of the trajectory reconstruction. Several issues should be carefully studied for the human motion analysis using the PIR sensor network data under the generation-refinement paradigm. First, the human motion trajectory is time-varying (*i.e.*, dynamical). Thus, the complete sets of the possible trajectories should be generated dynamically. However, there are rare methods that can support flexible and dynamical trajectory generation according to the sensor network topologies. Second, both the sensor network topology and the sensor activation log data should be formulated as spatio-temporal constraints, but the sensor activation log data and the spatio-temporal constraints data are significantly different. Few method can support the unified representation of both the sensor activation log data and spatio-temporal constraints. Third, the formulated spatio-temporal constraints should be dynamically integrated into the trajectory generation to refine the trajectories. Yet, hardly is there any method that can support such integration of the complex spatio-temporal constraints with the dynamical trajectory generation.

To overcome the above problems, we developed a new GA-based method to refine the human motion trajectories in this paper. At first, the mathematical definition of the geographical network, sensor activation response sequences and the human motion are defined under the unified GA framework. The relations among the three are analyzed. Then a GA-based dynamical trajectory generation process is defined to generate all the possible human motion trajectories according to the sensor network topology. By integrating both the temporal and spatial constraints, which are extracted from the sensor activation log and predefined rules during the trajectory generation, all the possible human motion patterns that satisfy both the temporal and spatial constraints are dynamically generated. Finally, a complete algorithm to extract all the possible human motion trajectories are proposed. The algorithm is applied to the MERL datasets to evaluate the correctness and performance.

The paper is organized as follows: the problem definition and basic ideas are described in Section 2. The methods, including the human trajectory generation and refinement algorithm, are described in detail in Section 3. The case study and the performance analysis are given in Section 4. Discussion and conclusions are given in Section 5.

2. Problems and Basic Ideas

2.1. GA and GA Representation of PIR Sensor Networks

In the whole trajectory analysis, there are several concepts that should be defined and analyzed. These are geographical network, sensor activation response sequences, and human behavior semantical sequences. Geometric algebra (GA), founded on the dimensional computation, is an ideal tool for the multidimensional algebraical element representation [24–29]. Under the GA framework, any network topology can be mapped into a special Clifford Algebra space $Cl(n)$, and the fundamental elements of the network (*i.e.*, nodes, edges and routes) can be coded as algebraic elements (*i.e.*, blades and multivectors) of this mathematical space [24]. Then the route can be dynamically generated according to the GA products using matrix multiplication [26]. With well-defined computation mechanism, the algebraic network computation can make the route generation and analysis symbolically with low complexities [26,29]. The constraints as well as multi-constrained routing can also be achieved under the GA framework [30,31]. With the GA-based network presentation, the construction of the network expression and calculation model, where there is a unified relationship among the network expression, relation computation and the path search, can be achieved.

Given any positive number $n > 0$, the Clifford algebra/Geometric Algebra system $Cl\,(n)$ can be generated by the vector set $\{f_i\}$, $1 \leqslant i \leqslant n$. The elements of the Clifford algebra space $Cl\,(n)$ are:

$$\begin{cases} scalars : f_0 = 1 \in R \\ vectors : f_1, \ldots, f_n \\ bivectors : f_i f_j = f_{ij}, 0 < i < j < n \\ \vdots \\ n - vectors : f_1 f_2 \ldots f_n = f_{12\ldots n} \end{cases} \qquad (1)$$

Assuming $G(V, E)$ is a graph that have n nodes, we can code each node as an individual algebraical basis of a special Clifford space $Cl\,(n)$. Given $e_i, 1 \leqslant i \leqslant n$ be the basis vectors of the $Cl\,(n)$, the metric matrix which determined by the GA adjacent matrix of the network can be formulated as:

$$A[i, j] = \begin{cases} e_{ij} & \text{if } (v_i, v_j) \in E \\ 0 & \text{otherwise} \end{cases} \qquad (2)$$

where the element $A[i, j] = e_{ij}$, which represents the edge from the node i to the node j, is the i-th row and the j-column of the geometric adjacent matrix. $A[i, j] = 0$ means there is no edge connected from the node i to the node j. According to the definition, we can formally define the geographical sensor network as follows:

Definition 1: The Geographical sensor network. The geographical sensor network is a physical geographical space where the sensors are located. In this sensor network, each sensor represents one node of the network. The sensors are only connected with the adjacent node according to the geographical spatial topology. A typical representation of a geographical sensor network is depicted in Figure 1.

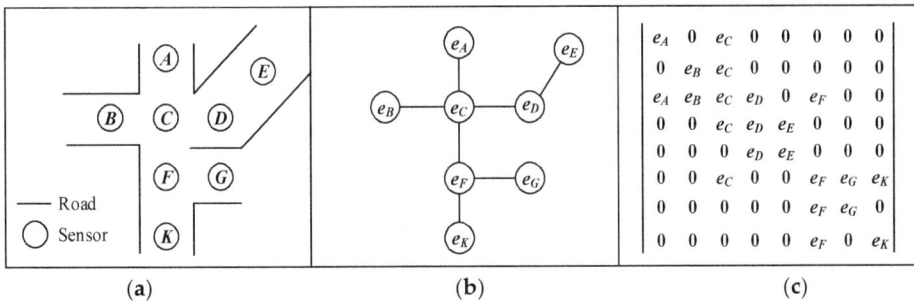

Figure 1. Definition of the geographical sensor network. (a) Geographic Secsor Distribution; (b) Geographic Secsor Network; (c) Adjacent matrix M.

Since the geographical space can be indicated by the network structure of the geographical sensor network (Figure 1a), it is possible to directly represent the geographical space using the GA-based network representation [24,25]. Here, we code each sensor in the geographical sensor network as an individual vector basis (Figure 1b), and then the sensor network can be represented using the GA adjacent matrix M. The adjacent relations between different nodes are inherited in M. To make the computation more efficient, only the network node of the route is logged in the adjacent matrix M. For example, if e_i and e_j are connected, the i-th row and j-th column element of the adjacent matrix M is logged as elements of e_j (Figure 1c). Because the sensor distribution is dense, no route weights are required to be stored in the adjacent matrix. As Figure 1c shows, the route in the sensor network can be represented as blades [26]. According to the construction rule, the adjacent network structure is directly correspondent to the adjacent matrix. All the routes between the sensors are logged in the elements with certain grade of the adjacent matrix. Therefore, the trajectory construction can be seen as the multiplication of the adjacent matrix. The spatio-temporal constraints can also be applied to filter the trajectories during the adjacent matrix multiplication.

The PIR sensor network is a set of PIR sensors installed with the intention to cover the floor area completely with little or no overlap between the sensor viewing fields. Assuming there is a PIR sensor i located in the place with a coordinate of $L(X_i,Y_i)$, the output of this sensor in the time period from 1 to t is a time series with a binary active state $D = \{x_1, x_2, \cdots, x_t\}$, where:

$$x_i = \begin{cases} 1, \ the \ sensor \ is \ active \\ 0, \ otherwise \end{cases}$$

Given a PIR sensor network composed of n sensors with each sensor having its own location L, which can be expressed as $L = \{L_1,L_2,L_i, \ldots ,L_n\}$. The state of the PIR sensor network at time instant t can be seen as an observation, therefore all the observation time can be expressed as $T = \{t_1,t_2,t_j, \ldots ,t_m\}$, and the state series of each time instant can produce a state of the whole sensor network. The observation, notated as X, can thus be encoded as a binary set as $X = \{0,1\}$, where 1 means the corresponding sensor is active, 0 otherwise. For the given sensor located at L_i, the observation can be represented as a feature vector $O(L_i) = \left\{X_{t_1}^{L_i}, X_{t_2}^{L_i}, \cdots, X_{t_j}^{L_i}, \cdots, X_{t_m}^{L_i}\right\}$; Similarly, the feature vector at an instant time t_j can be encoded as $O(t_j) = \left\{X_{t_j}^{L_1}, X_{t_j}^{L_2}, \cdots, X_{t_j}^{L_i}, \cdots, X_{t_j}^{L_n}\right\}$. Therefore, the observation of all the sensors during all the time can be expressed as a feature matrix $O(L, T) = \left\{X_{t_1}^{L_1}, X_{t_1}^{L_2}, \cdots, X_{t_j}^{L_i}, \cdots, X_{t_m}^{L_n}\right\} = \sum_{j=1}^{m} \sum_{i=1}^{n} X_{t_j}^{L_i}$. If there is any person moving in the PIR sensor network area, a trajectory as location series $P = \{P_j\} = \left\{P_{t_1}, P_{t_2}, \cdots, P_{t_j}, \cdots, P_{t_m}\right\}$ can be logged by the sensor network. With this people motion trajectory, a corresponding observation feature vector series $\{X_p\} = \left\{X_{t_1}^{L_{P_1}}, X_{t_2}^{L_{P_2}}, \cdots, X_{t_j}^{L_{P_j}}, \cdots, X_{t_m}^{L_{P_m}}\right\}$ can be outputted from the sensor network. In the feature vector series, there exist $X_{t_j}^{L_{P_j}} = 1$, it means that the person/people are located in the cognitive range of sensor L_{P_j} at time t_j. Since the people are walking in the geographical space time, the observation feature vector series should be constrained by several spatio-temporal constraints $\{C\}$. For example, people's movement should be constrained by the spatial structure (i.e., topology) of the PIR sensor network, i.e., the person/people cannot move between none adjacent sensors. Similarly, there are also temporal constraints that should make the time intervals between different motions acceptable. e.g., the human motion will not exceed the maximum possible velocities.

Since the PIR sensor only logs the binary response of certain locations and the response time of different responses between different sensors, which can be seemed as activated sequences. It is also possible to represent the sensor activation log as blades. With these blades, the sensor activation sequence can further be projected into the network space. In this way, network structures can be extracted from the sensor activation sequence. So, with the sensors coded with the GA space basis, we define the sensor activation response network as follows:

Definition 2: The sensor activation response network. A sensor activation response network is a possible sub-network of the response from the neighbor sensors of activated sensors. For example, in the Figure 2a, if the sensors $\{e_B, e_C\} \rightarrow \{e_C, e_F\}$ is one of the activated sequences within acceptable time constraints, the response network can be defined as the sub-network of the original network from nodes e_B, e_C to e_C, e_F. As shown in Figure 2b, the response adjacent matrix $M_{F\{e_B,e_C\}T\{e_C,e_F\}}$ takes the e_B, e_C rows and e_C, e_F columns of the network matrix M, which means the path begins from the sensor C,F and ends at the sensors B,C. From the matrix the possible paths can be also estimated as shown in the Figure 2c.

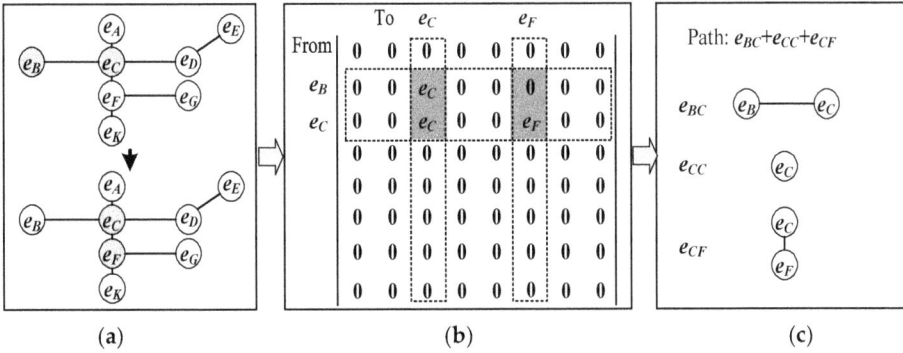

Figure 2. Definition of the activation response network. (a) Activated sequences of sensors: $\{e_B, e_C\} \rightarrow \{e_C, e_F\}$; (b) Response matrix $M_{F\{e_B,e_C\}T\{e_C,e_F\}}$; (c) The possible paths.

The sensor activation response network can be seen as the representation of the spatial constraints of the PIR analysis. It is not possible to directly move a person from one sensor to a non-adjacent sensor in the geographical space. Therefore, it is impossible to construct a real trajectory to represent the human motion, so in this paper, we think it is not suitable to consider the sensor activation sequence between two non-adjacent sensors as a single trajectory, but a combination of several trajectories.

In a sensor activation response sequence, the two adjacent sensor activities must happen in acceptable time intervals [32]. Because the human motion may end at a certain sensor, the sensor itself cannot distinguish the individuals from each other. The sensor activation with a big time difference may be caused by different people. With these assumptions, the determination of the starting and ending nodes of a trajectory depends on the spatial and temporal intervals to process the sensor activation response sequence. The sensor activation response sequence is caused by the human motion trajectory, which is also connected node-by-node in the network. Therefore, we can define the human motion trajectory similarly to the sensor activation sequence. The human motion sequence is defined as follows:

Definition 3: The human motion trajectory sequence. The human motion trajectory sequence is a sequence that represents the human motion structure from any sensor node to another sensor node in the geographic spatio-temporal space. The human motion trajectory sequence is an orderly sequence that can also be represented as a series path matrix, which applied the oriented join product of timely-adjacent response matrix according to the real human motion.

Taking the sensor network in the Figure 1 as an example. Assuming there are activate sensor sequences $\{e_B, e_C\} \rightarrow \{e_C, e_F\} \rightarrow \{e_A, e_G\}$, then two response networks can be constructed by $\{e_B, e_C\} \rightarrow \{e_C, e_F\}$ and $\{e_C, e_F\} \rightarrow \{e_A, e_G\}$. To better represent the path extension using Clifford algebra, we extend the oriented join product to the matrix (notated as oriented join matrix product $\cup$). The '$\cup$' is defined as:

$$
\begin{vmatrix} a_{11} & \cdots & a_{1n} \\ \vdots & \ddots & \vdots \\ a_{m1} & \cdots & a_{mn} \end{vmatrix} \cup \begin{vmatrix} b_{11} & \cdots & b_{1n} \\ \vdots & \ddots & \vdots \\ b_{m1} & \cdots & b_{mn} \end{vmatrix} = \begin{vmatrix} a_{11} \cup b_{11} & \cdots & a_{1n} \cup b_{1n} \\ \vdots & \ddots & \vdots \\ a_{m1} \cup b_{m1} & \cdots & a_{mn} \cup b_{mn} \end{vmatrix} \tag{3}
$$

Therefore, the path matrix can be calculated by the equation:

$$
\begin{aligned}
A_{\{e_B,e_C \to e_C,e_F \to e_A,e_G\}} &= M_{F\{e_B,e_C\}}T_{\{e_C,e_F\}} \cup M_{F\{e_C,e_F\}}T_{\{e_A,e_G\}} \\
&= \begin{array}{c|cc} & e_C & e_F \\ \hline e_B & e_C & 0 \\ e_C & e_C & e_F \end{array} \cup \begin{array}{c|cc} & e_A & e_G \\ \hline e_C & e_A & 0 \\ e_F & 0 & e_G \end{array} = \begin{array}{c|cc} & e_A & e_G \\ \hline e_B & e_{CA} & 0 \\ e_C & e_{CA} & e_{FG} \end{array}
\end{aligned} \tag{4}
$$

Then the possible human motion trajectory sequence can be calculated as: $P_{\{e_B,e_C \to e_C,e_F \to e_A,e_G\}} = e_{BCA} + e_{CCA} + e_{CFG}$. Since each trajectory segment for the real human motion can be seen as a blade and the whole motion trajectory sequences can be seen as linkers between different motion trajectory segments, we can apply the oriented join product to generate all possible connections between trajectory segments [24]. In this way, we can reconstruct and filter the sensor activation response sequences to analyze the human motion trajectory by using algebraic calculation.

2.2. The Problems of Trajectory Generation from PIR Sensor Networks

According to the characteristics of the PIR sensor, the mapping $\{L\} \to \{X\}$ is a unique one-to-one mapping, *i.e.*, the same human trajectory or trajectory combinations will definitely produce the same sensor network activation series. However, the binary sensor activation characteristics of the PIR sensor makes the sensor only able to log the human passing states but not the individual information. If there is only one human motion, the sensor activation sequence (*i.e.*, the feature vector) directly corresponds to the trajectory and activation sequence in the time domain. Both the spatial topology of the sensor network and the human trajectory can be revealed from the sensor activation sequence [27]. However, if there are more than one trajectory made by different people, we cannot classify the spatio-temporal correlations and correspondence between different sensor responses according to the sensor log. In this situation, the inverse mapping $\{X\} \to \{L\}$ may not be unique, *i.e.*, the same sensor activation observation may indicate different human motion trajectories. What's worse, since different trajectories can be intersected, it is not easy to directly determine the starting and the ending nodes of a certain trajectory. A typical example of this uncertainty is illustrated in the Figure 3. With the same sensor activation sequence, different trajectory patterns are likely to be revealed. Then the human motion trajectory analysis problem under the generation-refinement paradigm can be formulated as how we can extract all the possible trajectories $\{L\}$ from the sensor activation data $\{X\}$ according to both the temporal and spatial constraints C.

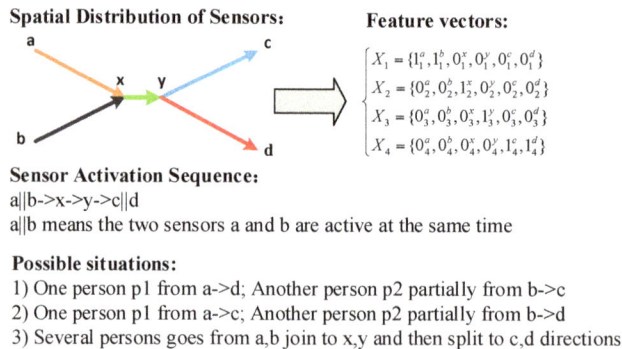

Spatial Distribution of Sensors:

Feature vectors:

$$
\begin{cases}
X_1 = \{1_1^a, 1_1^b, 0_1^x, 0_1^y, 0_1^c, 0_1^d\} \\
X_2 = \{0_2^a, 0_2^b, 1_2^x, 0_2^y, 0_2^c, 0_2^d\} \\
X_3 = \{0_3^a, 0_3^b, 0_3^x, 1_3^y, 0_3^c, 0_3^d\} \\
X_4 = \{0_4^a, 0_4^b, 0_4^x, 0_4^y, 1_4^c, 1_4^d\}
\end{cases}
$$

Sensor Activation Sequence:
a‖b->x->y->c‖d
a‖b means the two sensors a and b are active at the same time

Possible situations:
1) One person p1 from a->d; Another person p2 partially from b->c
2) One person p1 from a->c; Another person p2 partially from b->d
3) Several persons goes from a,b join to x,y and then split to c,d directions

Figure 3. The uncertainty of the sensor response.

2.3. Basic Ideas

According to the above problem definition, the sensor network can be seen as a geographical network with each sensor being one node of such a network. Because the sensors do not overlap, the sensor activation only affects the adjacent sensor. In addition, the sensor node is also a construction node of the human motion trajectories, *i.e.*, the sensor node in an individual trajectory is adjacent to and only affects the adjacent sensor node. Because each individual human motion has unique map to the activation of the sensor logs, the human motion is restricted by the spatial topology of the real sensor distribution. Therefore, the trajectory construction can be seen as the route generation one node by one node. The sensor activation sequences are not only a fusion of several human trajectory motions, but also the spatial and temporal constraints that will limit the possible and impossible trajectories in the geographical spatio-temporal space. Therefore, the human trajectory analysis can be split into two steps: (1) dynamically generate all possible routes for human trajectories node by node; (2) refine/filter the possible human trajectories to make the trajectory consistent with the sensor activation log data and other spatio-temporal constraints. In this way, the problem can be solved in the generation-refinement paradigm.

According to the generation-refinement paradigm, the problem can be further decomposed into the following sub-problems: (1) how to represent and link both the sensor networks and trajectories in the same unified mathematical framework; (2) how to dynamically generate the trajectories according to both the spatial network topology constraints and the temporal constraints from the sensor log data; (3) how to determine the starting and the ending nodes of each individual human motion trajectory. To deal with the sub-problems (1) and (2), we should develop a flexible mathematical expression framework that can represent both the network and human trajectories. Not only the nodes, routes and whole networks should be represented using the same paradigm of mathematics, but also the representation should support the dynamical route generation node by node in the network. In addition, the spatial and temporal constraints should also be represented with the same mathematical tool and integrated in the dynamical route generation. For sub-problem (3), we should determine certain rules that can be used to classify the trajectories to determine the starting and the ending nodes.

With the advantages of the spatio-temporal representation properties of GA, we modeled the whole PIR sensor network by GA network coding. The geographical space is first defined and represented as a geographical sensor network. The sensors are coded as the geometric basis of the high dimensional GA space. Then the interaction and response sequence can be embedded as blades. These blades can then represent both the possible human motion trajectories and the sensor activation sequence. By integrating constrains from the real sensor activation observational data, the possible human motion trajectories can be dynamically generated. The overall framework of our basic idea is illustrated in the Figure 4.

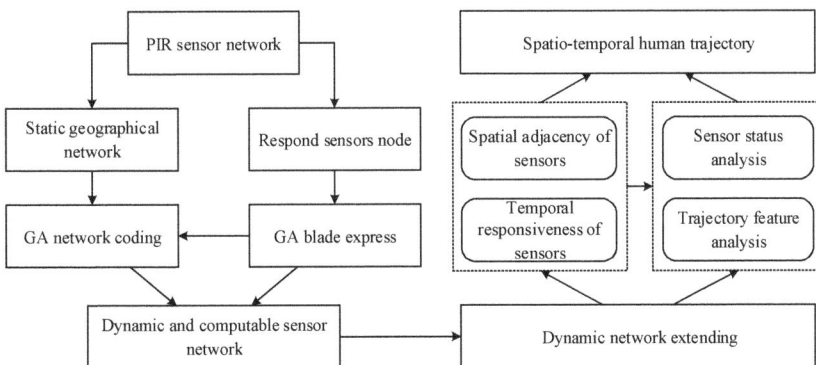

Figure 4. Basic idea.

3. Methods

3.1. Classification of Sensor Status According to the Trajectory

Different from classical network analysis, which has determined the starting and the ending nodes, the ending node during the trajectory tracking is not constant but changed according to the human motion. Since there is more than one trajectory in the geographical space, the classification of different trajectories is important in our analysis. By considering the trajectory generation as a trajectory tracking problem from the dense adjacent matrix, several different situations can be revealed according to the sensor activation log [20] (Figure 5). Both the tracking of the individual trajectory and multiple trajectories are considered. For individual trajectory, the sensors in the trajectory can be classified into three different states: moving, stop and still (Figure 5a–c). The moving status means the trajectory will lead an adjacent sensor to continuously extend the trajectory. The sensor activation sequences will be: the sensor is active in a short time and then the adjacent sensor is activated. The stop states means the trajectory ends at the senor node. In this condition, the sensor is continually active, but after a certain period, there's no adjacent sensor active. The still status means the person stays in the range of certain sensor for a long time, but this sensor is not stopped. Therefore, we cannot make that sensor a stop node of its trajectories. The data sequence should be: the sensor is continuously active for a time period and then after a period, the adjacent sensor is activated, and the trajectory continues.

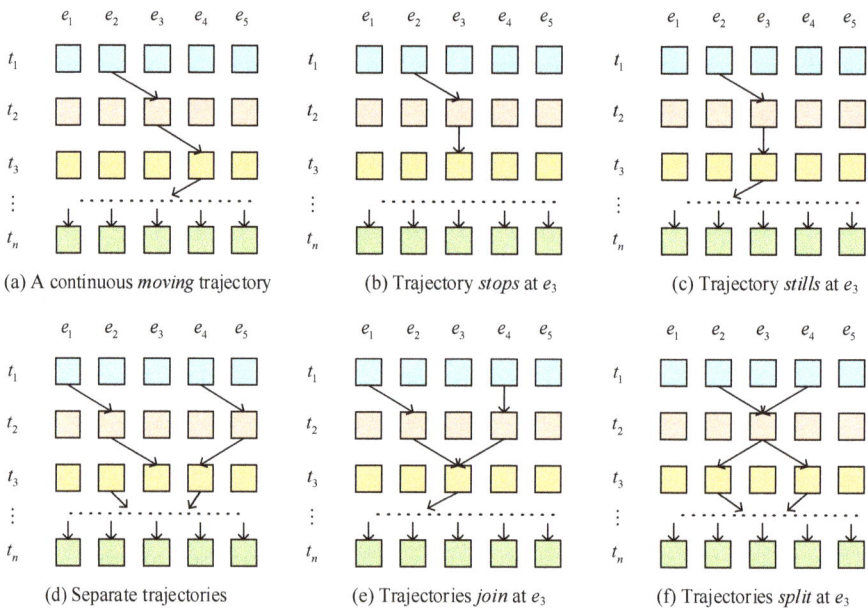

(a) A continuous *moving* trajectory   (b) Trajectory *stops* at e_3 (c) Trajectory *stills* at e_3

(d) Separate trajectories   (e) Trajectories *join* at e_3 (f) Trajectories *split* at e_3

Figure 5. Different situation of the trajectory tracking.

To classify the sensor states of the still and stop nodes is important in the trajectory tracking, because it relates to the starting and ending nodes of the trajectories in the foundation of the trajectory generation/tracking. However, it is not an easy task to simply distinguish the two states directly from the sensor activation data. Since the trajectory stills at a node for a very long time, it can be seen that the trajectory stops at the node and a new trajectory is then started from the same sensor node. A more practical way to classify the two states should be defined at a certain time interval of t to determine the granularity of the trajectory segment. In this way, if the stay time in one sensor is larger than t, we can classify them into the state of stop. And we assume the next trajectory starts from the current sensor. With this configuration, the tracking for individual trajectories can be simply achieved by direct adjacent matrix multiplication [25,26,28,30].

In most common conditions, there should be more than one trajectory in the sensor networks. With more than one trajectory the sensor activation status becomes complex. The simplest condition is that the trajectories are not intersected (Figure 5d). These two non-intersected trajectories can be seen as two separated individual trajectories. However, there will be people from two distinct trajectories joining into one single trajectory (Join, Figure 5e). And there are also people who are originally in a single trajectory but separated into several trajectories (Split, Figure 5f). To the responses of the sensor network, the join condition will lead to several non-adjacent sensor activities, which then will be combined into one common sensor, and only adjacent sensors of the join sensor will be activated afterwards, thus indicating the two different trajectories are combined into a single trajectory. In the split situation, the sensor active trajectory is a single trajectory that only the adjacent sensor is activated, and after a certain sensor, the trajectory splits and two or more non-adjacent sensors are activated, which means the trajectories are split from one.

As the human motion trajectory sequence can be expressed and calculated in the GA space, we can also construct the GA-based classification method of trajectories. Given the sensor activation sequences $\{e_A, e_B\} \rightarrow \{e_C, e_D\} \rightarrow \{e_E, e_F\}$, the associated human motion trajectory can be calculated as:

$$A_{\{e_A, e_B \rightarrow e_C, e_D \rightarrow e_E, e_F\}} = M_{F\{e_A, e_B\}} T_{\{e_C, e_D\}} \cup M_{F\{e_C, e_D\}} T_{\{e_E, e_F\}}$$

$$= \begin{array}{c|cc} & e_C & e_D \\ \hline e_A & \alpha_{11} e_C & \alpha_{12} e_D \\ e_B & \alpha_{21} e_C & \alpha_{22} e_D \end{array} \cup \begin{array}{c|cc} & e_E & e_F \\ \hline e_C & \beta_{11} e_E & \beta_{12} e_F \\ e_D & \beta_{21} e_E & \beta_{22} e_F \end{array} \tag{5}$$

where α and β can take the value 1 or 0, which indicate the connectivity of sensors, for example: if $\alpha_{11} = 0$, sensors A and B are not adjacent. Therefore, the trajectories structures are depending on the values of these two adjacent matrix:

$$m_1 = \begin{vmatrix} \alpha_{11} & \alpha_{12} \\ \alpha_{21} & \alpha_{22} \end{vmatrix}, m_2 = \begin{vmatrix} \beta_{11} & \beta_{12} \\ \beta_{21} & \beta_{22} \end{vmatrix} \tag{6}$$

(1) Continuous *moving* trajectory

As the continuous *moving* trajectory is extended with no branch, the path can only have a single starting node and ending node. So, the response matrixes must be the diagonal matrix that avoids having two or more nonzero value in one row or column. The typical expresses of adjacent matrix are:

$$\begin{cases} m_1 = \begin{vmatrix} 1 & 0 \\ 0 & 1 \end{vmatrix}, m_2 = \begin{vmatrix} 1 & 0 \\ 0 & 1 \end{vmatrix} \\ m_1 = \begin{vmatrix} 0 & 1 \\ 1 & 0 \end{vmatrix}, m_2 = \begin{vmatrix} 0 & 1 \\ 1 & 0 \end{vmatrix} \end{cases} \tag{7}$$

According to the Equation (4), the trajectories can be calculated as $e_{ACE} + e_{BDF}$ or $e_{ADF} + e_{BCE}$. The result shows that the paths are continuous moving trajectories and separate trajectories.

(2) *Join* and *split* trajectories

Unlike the Continuous moving trajectory, the *join* and *split* trajectories need the path that have the two or more starting nodes or ending nodes. Therefore, the typical expresses of adjacent matrix are:

$$
\begin{cases}
m_1/m_2 = \begin{vmatrix} 1 & 1 \\ / & / \end{vmatrix} & \textit{join trajectories} \\[2em]
m_1/m_2 = \begin{vmatrix} 1 & / \\ 1 & / \end{vmatrix} & \textit{split trajectories}
\end{cases}
\tag{8}
$$

where "/" means the element can be 1 or 0, which will not influence the results. As Equation (8) shown, only if m_1 or m_2 has two "to" nodes, the human motion paths will result in *join* trajectories; only if m_1 or m_2 has two "from" nodes, the human motion paths will result in *split* trajectories.

(3) *Still* and *stop* nodes

According to the GA-based trajectory representation methods, the trajectory with *still* nodes can be expressed as $e_{x_1 x_2 \cdots cc \cdots x_n}$, and e_c is the *still* node. The repetitions of e_c suggest the duration time of the *still* node. If the duration time bigger than predefined threshold value, e_c can be also seemed as the *stop* node. Therefore, the most important is to identify the e_{cc} structure in trajectory sequence. According to the definition of human motion trajectory sequence, if there is a diagonal element (which has the same row and column number) in response matrix, this paths must exist a *still* node.

In a common situation of the human motion, the adjacent sensors will be active and respond continuously. With continuous sensor response log, we can track the spatio-temporal correlations between different sensors in the geographical space and then transform them into possible trajectories. In this paper, we assume that a typical human walks at a velocity of 1.2 m/s. Since no gap exists between different sensors, a human passing through a sensor may need a response time of 2–4 s. So we select the median time window of 3 s to do the object tracking. If any object activates a sensor and activates the next adjacent sensor within the next node, it will probably be a trajectory. If there is no adjacent node activated, then we can consider the node the ending node of a trajectory.

3.2. The Generation of All Possible Trajectories

We refer to the generative route construction method to extend the possible path according to the sensor network topology as well as the sensor log data. Since the route generation in the GA space may produce redundant or impossible routes in real geographical space, spatial and temporal constraints to each generation are applied firstly to refine the generated trajectories. Based on the spatial object tracking idea, we propose the following route generation and constraints filtering method.

(1) The trajectory refinement based on spatio-temporal constraints

The spatio-temporal constrains should be applied during the route generation process to filter out the real possible trajectories. At first, we apply a time window as temporal constraints to determine the start and end of the trajectory. For the PIR sensor data, a time window of 3 s is suitable to segment the trajectories as individual trajectory. With this configuration, we can separate the responses of the time constraints during the route extension. The spatial constraints are determined by the indoor topological structure and the relations between the tracked nodes. Here, we query any active neighborhood node in the next time window and construct the spatial constraint matrix C. The spatial constraint matrix C is a diagonal matrix, where e_k are the possible active sensor nodes, and we define $C_{kk} = 1$, otherwise, the $C_{kk} = 0$. Therefore, we can define the whole spatial constraint matrix as follows:

$$
C_{ij}^n = \begin{cases} 1, & i = j, \ e_i \text{ is possible node} \\ 0 \end{cases}
\tag{9}
$$

To extract the trajectory, we apply moving window query to the sensor activation log from the temporal dimension. When there is sensor activation in the next time window, a spatial constraint matrix C is constructed, except when the node queried is considered the ending node of a trajectory. If there are some nodes that are active but not responded to neighborhood nodes, we consider the nodes the new starting nodes of a trajectory. According to the new starting nodes, we can construct a new judgement matrix T, which is also a diagonal matrix, where the element in the new judgement matrix T defines the new starting node e_i. The construction rule of the new judgement matrix T is as follows:

$$\mathbf{T}_{ij}^n = \begin{cases} e_i, & i = j, \ e_i \ is \ the \ response \ node \\ 0 \end{cases} \tag{10}$$

According to the Equations (9) and (10), n is the same with the network adjacent matrix. With the construction of the spatial constraints matrix C and judgement matrix T, we can multiply the adjacent matrix M, the spatial constraint matrix C and the judgement matrix T to construct the new adjacent matrix M' in this matrix. The non-zero element is the real motion trajectory in the n-th order. This can be formulated as:

$$\mathbf{M}' = \mathbf{M} \cup \mathbf{C}^n + \mathbf{T}^n \tag{11}$$

(2) The human motion trajectory sequence based route extension

The route extension is based on the route expanding using the oriented join product based on the GA-representation of the sensor networks. The adjacent relations between different nodes are inherited in M. For the given sensor activation sequence $\{X\} = \{X_1, \cdots, X_2, \cdots, X_n\}$, the n-order adjacent matrix can be extended as:

$$\mathbf{M}'^n = \mathbf{M}'_{F\{x_1\}T\{x_2\}} \cup \mathbf{M}'_{F\{x_2\}T\{x_3\}} \cup \cdots \cup \mathbf{M}'_{F\{x_{n-1}\}T\{x_n\}} \tag{12}$$

To covert the matrix into the n-order routes, the starting node matrix Q is defined here. Q is a diagonal matrix, in which the i-th row and column has the elements of e_i, which means all the routes have the starting node of e_i. The construction of the starting node matrix Q is as follows:

$$\mathbf{Q}_{ij} = \begin{cases} e_i & i = j, \ e_i \ is \ the \ starting \ node \\ 0 \end{cases} \tag{13}$$

According to the above route extension and construction rule, we can then realize the spatial trajectory reconstruction according to the sensor log data. The reconstruction is defined according to the following equations:

$$\begin{cases} A^2 = \mathbf{Q} \cup \mathbf{M}_{F\{x_1\}T\{x_2\}} \\ A^n = \mathbf{Q} \cup A^{n-1} \cup \mathbf{M}_{F\{x_{n-1}\}T\{x_n\}} \end{cases} \tag{14}$$

3.3. Possible Trajectories Generation and Refinement Algorithm

Based on the above definitions, we can then develop a unified algorithm to generate and refine all the possible trajectories according to the sensor activation log. The algorithm uses the oriented join product to realize the route extension and uses the spatial and temporal constraints to refine the generated trajectories. The algorithm starts with the fundamental adjacent matrix M that is constructed according to the topology of the sensor network. Clearly, the multiplication of the matrix M can produce all possible routes that the human can walk in the geographical space. According to the sensor network activation data, the starting nodes of the trajectories are first queried to construct the starting node matrix Q. All feasible trajectories are contained in the matrix M^n. Then the spatial and temporal constraints can be applied to filter the high order matrix M^n to extract the more accurate trajectories. By determining the completeness of a trajectory and repeating the trajectory generation process, all the possible trajectories can be extracted. The overall process of the algorithm is illustrated in the Figure 6.

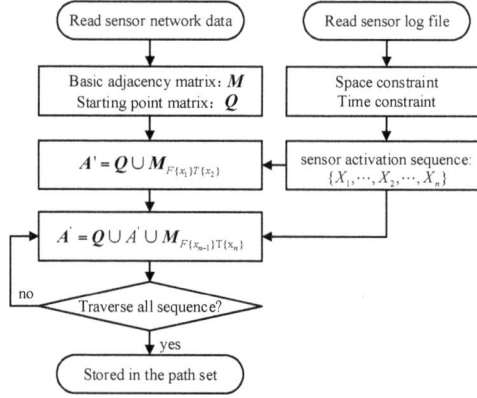

Figure 6. The process of the trajectory generation and refinement algorithm.

The example of the simple sensor network depicted in Figure 1a is used to illustrate the algorithm. In the Figure 1a, every sensor represents the sensor activation of the human motion. Assuming the sensor response sequence is $A(0s \sim 1s)$, $C(2s \sim 4s)$, $D(2s \sim 3s)$, $B(4s \sim 5s)$, $E(4s \sim 6s)$, $F(5s \sim 6s)$, the reconstruction process of the network trajectory is:

(1) According to the time constraint (3s) of human motion trajectory, the sensor activation sequences is $\{A\} \rightarrow \{C, D\} \rightarrow \{B, C, E\} \rightarrow \{E, F\}$. Therefore, the human motion trajectory reconstructing equation is:

$$\begin{cases} A^2 = Q \cup M_{F\{A\}T\{C,D\}} \\ A^3 = A^2 \cup M_{F\{C,D\}T\{B,C,E\}} \\ A^4 = A^3 \cup M_{F\{B,C,E\}T\{E,F\}} \end{cases} \tag{15}$$

(2) Introducing the starting node matrix Q, and adjacent matrix M, Equation (15) can be written as:

$$\begin{cases} A^2 = Q \cup \begin{vmatrix} e_C & 0 \end{vmatrix} \\ A^3 = A^2 \cup \begin{vmatrix} e_B & e_C & 0 \\ 0 & e_C & e_E \end{vmatrix} = Q \cup \begin{vmatrix} e_{CB} & e_{CC} & 0 \end{vmatrix} \\ A^4 = A^3 \cup \begin{vmatrix} 0 & 0 \\ 0 & e_F \\ e_E & 0 \end{vmatrix} = Q \cup \begin{vmatrix} 0 & e_{CCF} \end{vmatrix} \end{cases} \tag{16}$$

Multiply the starting node matrix Q, we can get the n-order routes:

$$\begin{cases} A^2 = e_{AC} \\ A^3 = e_{ACB} + e_{ACC} \\ A^4 = e_{ACCF} \end{cases} \tag{17}$$

(3) The routes in Equation (17) can be classification according to the activated sensor sequences expresses in Equation (16). Firstly, since the response matrix of 2-order routes are diagonal matrix, the results are continuous *moving* trajectory; in the response matrix of 3-order routes, the matrix $\begin{vmatrix} / & e_C \\ / & e_C \end{vmatrix}$ meet the condition of *split* trajectories. So, in the 3-order routes there exist *split* trajectories e_{ACB} and e_{ACC}. At the same time, the e_{CC} is the diagonal element of the adjacency matrix M, e_C is also a *still* node here; in the response matrix of 4-order routes, there are not any special structure, so it only inherited the *still* node e_C of 3-order routes.

Clearly, the algorithm we proposed is automated and is clearer to represent the real and possible trajectories from the sensor log data. The spatial constraints, network topologies and the route uncertainties are largely reduced. The route extension is dynamic and generative, which can also be interactively updated. All the possible trajectories can be extracted algebraically, which can be simply computed and analyzed in the future.

4. Case Studies

The algorithm is applied to the MERL sensor database published by the Mitsubishi Electric Research Labs. There are a total of 213 sensors, which logged the human motion in the building from 21st May 2006 to 24th March 2007 and were distributed on the 7th and 8th floor [8]. The spatial distribution of the sensors is depicted in the Figure 7. The sensors are installed with the intention of covering the floor area completely with little or no overlap between sensor fields of view. Although the minimum inter-detection time varies, the average minimum inter-detection time is around 1.5 s. The event log of the Lab is also provided as reference data for validation. The computation is performed on an Inspur NP 3560 server with two Intel Xeon E5645 (2.4 G) processor, 48 GB DDR-3 ECC Memory and a Raid five disk volume made up of three Seagate ST4000NM0023 SAS hard disk (7200 RPM). The operation system is Windows Server 2008 R2. All the data are imported in a PostgreSQL v9.4 database. The algorithm is implemented as a plug-in of the system CAUSTA [25] with ODBC connection to the PostgreSQL database server. To make better comparison with our method, the trajectory results extracted by the tracklet graph model is also used and imported in the database.

Firstly, the overall active frequency is summarized to get the spatial distribution of the active sensors (Figure 7). The hotspot graph suggests that the hottest area is around sensor 310, which is in the kitchen, because every workday people walk from various directions to the kitchen. Other hotspot regions are mostly meeting rooms (e.g., Belady Meeting Room (452)). The Belady meeting room is the most commonly used, based on the evidence from the activity log. Other frequently active sensors such as 255 and 356 are the ladders people frequently use. Therefore, from the sensor activation data, the spatial distribution of the human walking can be summarized. However, the statistics cannot reveal further trajectory information about people's walking.

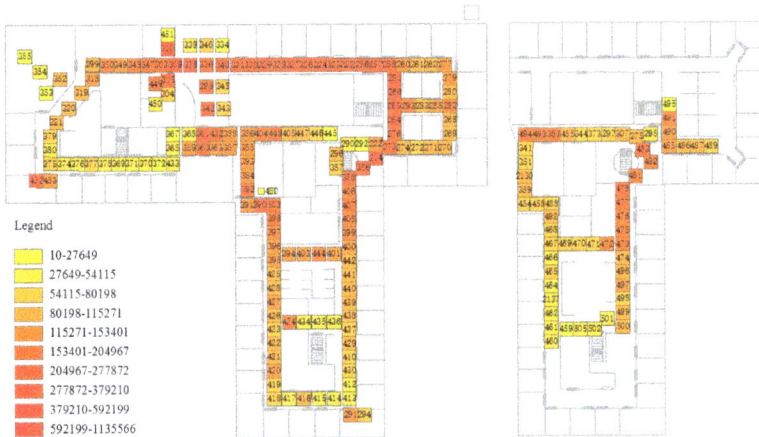

Figure 7. Spatial distribution of the sensor activation.

Since generating all possible trajectories from the whole data set is too complex, we selected one week of data of the 8th floor from 7 August to 13 August to generate all the possible people trajectories from the sensor activation data. There are totally 153 sensors and the total record of the active sensor log is 414,552. Based on the trajectory reconstruction algorithm, we have reconstructed 414,552 possible sensor statues and 563,386 total trajectories. After splitting the trajectories in a one-minute window and removing the duplicated trajectories, which have totally same nodes, we finally have 202,700 complete different trajectories.

The overall time cost of both the computation of the statues of each sensor and generation of all possible trajectories is 3 h and 19 min and the peak memory cost is 3.2 GB. The final result database file (including the original query data and generated path data) occupies a space of 3.07 GB in the NTFS file system. To test the correctness and performance of the method, we use the version 2 tracklets data (which is generated by the tracklet graph model at 22 March 2008, and published together as the reference data along with the MERL sensor data for the IEEE Information Visualization Challenge 2008) as the reference data. Since the original performance of the tracklet graph model are not possible to be accessed, we apply our method onto different numbers of sensor activation logs to evaluate the performance of the method.

The time and memory cost of the generation of the sensor status and trajectories from the start to different end time are logged during the computation (Table 1). The computation time is less than two and a half hours of the whole week's data.

Table 1. The computation performance evaluation (Start from 2006/August/7 0:00).

Time Range (End Time)	No. of Sensor Logs	Time Cost of Sensor Status(s) *	Memory Cost of Sensor Status(MB)	Time Cost of Trajectory Generation(s) *	Memory Cost of Trajectory Generation (MB)
2006/August/7 16:10	53517	1014.89	200.11	47.12	312.31
2006/August/8 07:19	77031	1314.89	213.82	67.96	450.15
2006/August/8 20:00	152031	2919.82	420.32	134.37	888.07
2006/August/9 23:46	229309	3214.71	480.68	202.67	1338.53
2006/August/12 10:30	315130	4231.27	510.24	277.59	1839.47
2006/August/12 23:05	414397	8919.82	718.48	365.85	2419.09
2006/August/13 23:59	414552	8992.71	729.14	365.26	2419.77

* The time cost logged here not include the time of data I/O.

Since in the sensor status determination, the oriented join products are not computed in realistic terms, but only the binary matrix is used to classify the *pass*, *join* and *split* status. The major performance bottleneck in this procedure may be the determination of the *still* and *stop* status. Not so restrictedly speaking, the time efficiency of the sensor status query is nearly increased as a linear relation of the total number of sensor logs. For the performance of the trajectory generation, the computation cost is growing much larger, this is because the generation of the trajectories requires one to compute the real oriented join product to generate the path. The generated path should also be compared with the spatial and temporal constraints of filtering the path. The optimization of the oriented join product and optimized data structure of the computation may largely improve the computation performance. Compared with the tracklet graph models proposed by [11], which extracted 105,757 tracklet graphs, our method can provide more complete possible human motions. This is because the tracklet graph model can only reveal one trajectory from a single sensor network active state [17,20]. The detailed comparison between our result and the tracklet graph method is illustrated in Figure 8. From Figure 8, we can clearly see both the extracted path and the sensor status in one minute of time, which is more semantically meaningful than the tracklet graph model. In addition, our method can provide more detailed trajectories that are not generated by the tracklet graph model, *i.e.*, our method can provide the join, split, still and pass status of each sensor to provide more detailed descriptions about the human motion. In the tracklet graph model result, several possible trajectories are not extracted (Table 2). There are also wrong trajectories extracted (e.g., trajectory from sensor id 257 to 342 in 16:30) by the tracklet graph model. Since the tracklet graph models are a statistical method that extract the start and end node of the trajectory using the graph cut statistics, the internal structures of the trajectory are not fully used to restrict the detailed analysis. Since the single sensor activation log data may lead to several different human motion trajectories, there must be several trajectories lost in the tracklet graph construction. From the result of our algorithm, we can extract all possible trajectories. Although not all the trajectories happened in the real world, there exists a possibility that these trajectories produce the active sensor pattern. Comparing and combining the use of the tracklet graphs and our data may produce interesting and more accurate human motion pattern results. It is also helpful to apply modern statistical methods to analyze the detailed pattern of the human motions [21,22].

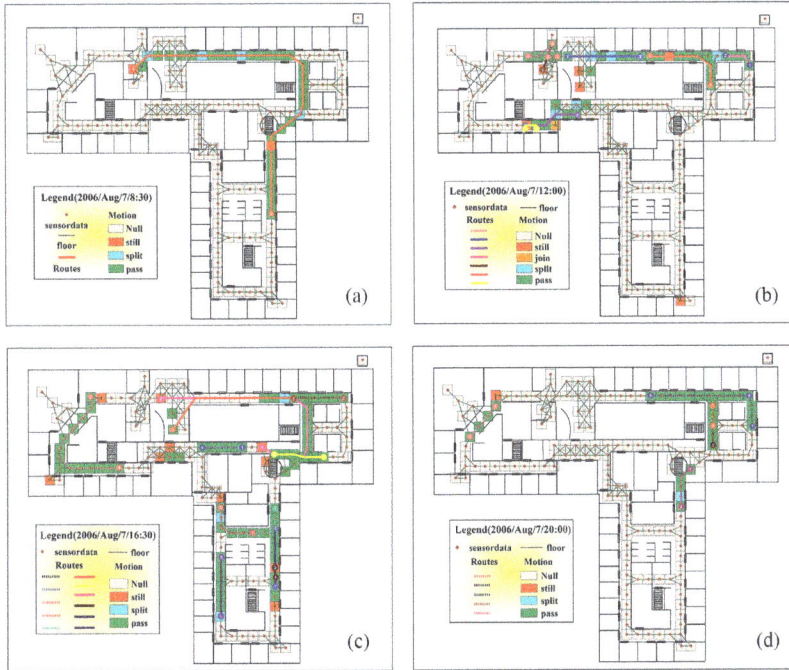

Figure 8. Route path and sensor status generated by our method compared with the path generated by the tracklet graph method in the same one minute. The solid line in the graph is the trajectory extracted by the tracklet graph model. The dashed line in the graph is the missing trajectories that have been extracted by our method but missing by the tracklet graph model. (**a**) Trajectory and sensor status extracted at 8:30; (**b**) Trajectory and sensor status extracted at 12:00; (**c**) Trajectory and sensor status extracted at 16:30; (**d**) Trajectory and sensor status extracted at 20:00.

Table 2. Absent trajectories that are missing compared with the tracklet graph method.

Time	Start Node (Sensor ID)	End Node (Sensor ID)
2006/August/7 12:00	309	348
2006/August/7 16:30	408	444
	444	398
	371	299
	405	386
	256	277
2006/August/7 20:00	265	276
	318	321
	356	408
	282	326
	281	265

To further validate the correctness of our method, we counted all the trajectories by visualizing the start-end node connections using matrix and circular graph (Figure 9a,b). We conclude that most trajectories have similarities, *i.e.*, the people trajectory starting from the same sensor usually have some fixed ending nodes of the sensor. This is also the real case due to the work pattern in the office area. As everyone's work has special tasks that may be connected with people who have working relations. To further reveal the trajectory patterns, we also provide the visualization of the spatial and frequency distribution of the ending nodes of the trajectories started with the same starting node. For example, starting from the sensor ID 272, we have totally 47 sensors that have been selected as ending nodes (Figure 9c). However, the most frequently linked sensors are neighborhood sensors, which suggests that people working in the office located at the sensor 272 mostly have working relations with the neighborhood office. This result can also be supported by similar analysis with the entropy method [15], which suggests the individuals associated with this sensor are frequently concentrated in the large office. From these statistical results, we believe that our method can be used to reveal the human motion patterns directly according to the sensor activation log data.

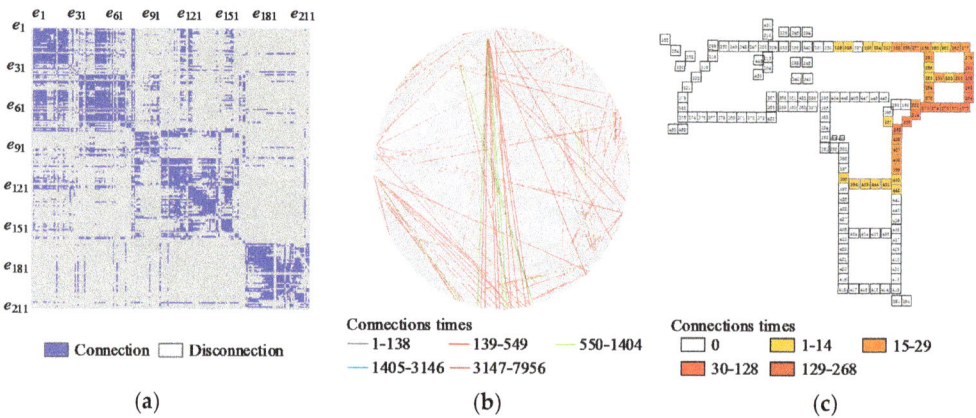

Figure 9. The start and end node analysis. (**a**) The matrix representation; (**b**) The circular representation; (**c**) End nodes for start sensor 272.

With our method, the different states of the sensors according to the trajectories can also be extracted (Figure 10). Since the sensor nodes at different locations have different functions in the working place and the human passing patterns will be affected by the spatial topology of the network, the frequency and spatial distribution of the sensor states can also be evaluated by the sensor locations.

Four different states, including passing, join, split and still, are analyzed. The passing status has the most numbers, and the region of the highest passing frequency is in the line from sensor 348 to sensor 273. This area is the dining room and every noon people walk to this area to have lunch. For the join status, the highest frequency happens at sensors 309, 338, 331, 330, 329, 327, 326, 323, 322, 256, and 257. These places are near the dining room. People from different offices join their trajectories to have lunch. Other individual high frequency join statuses happen at sensors 214, 407, 441, 281, and 264; these sensors are located near the cross where people from different directions join their trajectories. In [15], sensor 442 is classified as the hub that connects different groups of people. In our result, the statements of [15] can also be partially supported. However, we can filter out more such kinds of hubs from the possible trajectories. In addition, the location of the hub is slightly different from the entropy method. Since our result is directly generated according to the spatial topology structure and the sensor activation log, which also integrates different spatio-temporal constraints, it may be more accurate than the statistical segmentation which uses the entropy. For the sensors with status of split, highest frequency sensors are located at the corner or across the work place. The exceptions are sensor 295, 330, 328, 324 and 323. These five sensors are near the dining room, where people walk to their own offices after lunch. The sensors with high frequency of still status include 355, 352, 452, 449, 315, 342, 261, 301 and 435. Among them the sensors 355 and 352 are the Nitta seminar room; sensor 452 is the Belady conference room; sensors 449 and 315 are mail and supply rooms. Sensors 342 and 261 are restrooms; sensor 301 is a lunch room; sensor 435 are stairs to the 7th floor. According to the function of these locations of the sensors, especially for the mail and supply rooms and restrooms, people are definitely more likely to stand still here instead of stopping their trajectories. The above results further validate the correctness of our method.

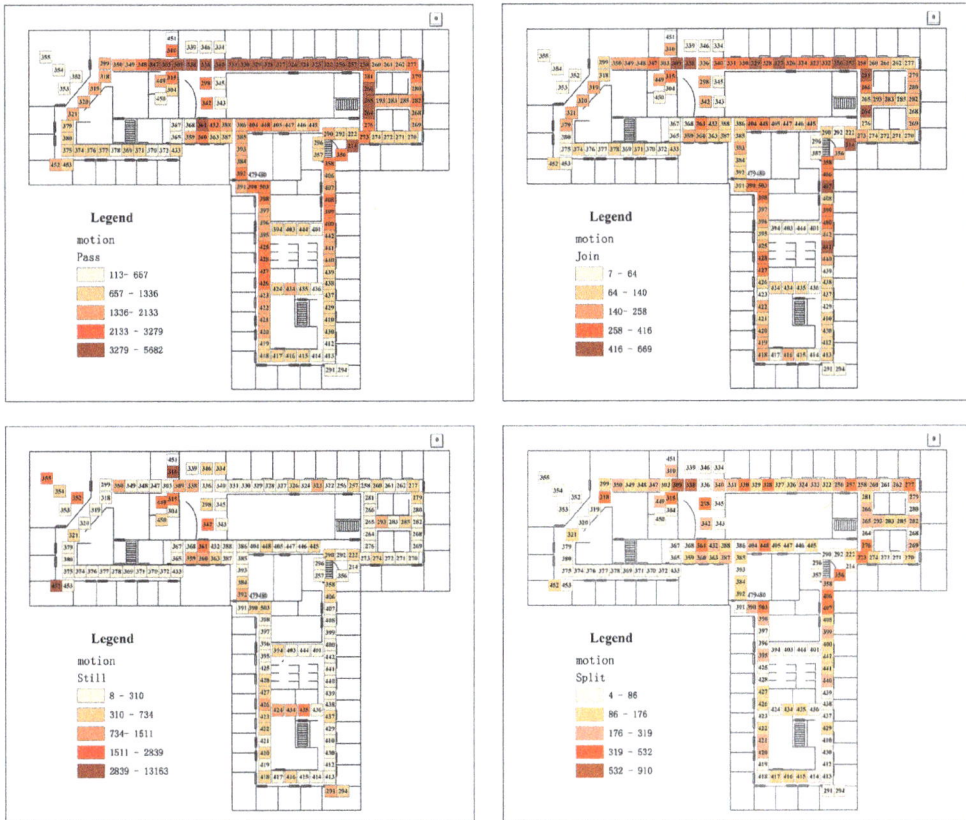

Figure 10. The statistics of the different behaviors.

5. Discussion and Conclusions

In our research, the spatial and temporal constraints play a key role. The major uncertainty of our method is the time interval used to determine the starting and the ending nodes of the trajectories. Future analysis can be applied in the direction of what's the sensitivity of the time windows. Although we have generated all the possible trajectories from the sensor activation data, not all the possible differences between different trajectories are considered. The long-term data observation and the detailed lab event log data make the statistical inference of the possibility difference of each reconstructed trajectory possible. For example, the reconstructed trajectories can be further reanalyzed by the HMM, LDA, Bayesian filtering or temporal segment methods [2,21,22]. The analysis based on the refined trajectories data may largely improve the accuracy and conclusion ability of such researches. Further applications in smart homes or buildings can also benefit from our method.

The GA-based route generation and constraint integration method provide unified and flexible tools for network and trajectory analysis. In the GA-based approach, the route generation is based on the matrix approach according to the topological information of the original network. The approach makes the stepwise route generation of each possible route very flexible. Although the matrix-based storage is memory and computational inefficient, a lot of optimizations can be applied to support the large-scale analysis. Since the route generation is symbolic and independent of other orders of routes, the pre-complier and parallel computation (e.g., Gaalop) technologies may be applied to help improve the efficiency [33–35]. In addition, we can also develop specialized data structures for online data stream computation for the passive sensor networks such as the MERL sensor data sets. Tensors and other compression method may also be helpful for the data representation and analysis [36,37].

In our GA-based trajectory reconstruction method, the route extension is not only the real geographical spatio-temporal sequence, but also an algebraical element that can be directly calculated. The clear geometric and physical meaning of the motion can be directly revealed from the algebraical equations. In this way, the representation and analysis method can make the analysis of the motions much simpler and clearer. Since both the outer product and the oriented join product is asymmetric, the route extension has its own orientations, which can be well classified by the spatial and temporal adjacent relations in the sensor activation log. Since the orientation information is very important for the activity identification, our GA method may also be helpful to improve the activity recognition from the sensor data. According to the orientation information, we can reveal not only the classical activity of walking, entering, joining and splitting, but also the activities like turning left and turning right *etc.* This further detailed analysis may make the GA-based activity analysis a wider potential area. Since the GA representation and analysis is inherently high dimensional, the representation and analysis can be made simple and direct.

In this paper, we developed a human trajectory refinement method to reveal all the possible human motions from the passive MERL sensor activation log data. Our method unifies the representation of the sensor network, sensor activation data and the human moving trajectory under a unified mathematical framework. All the possible human motion trajectories are tracked according to the dense sensor activation log using the matrix approach. The geometric algebra can well express the absolute and relative coordinates of the human motion. The network and trajectory representation can well express the network and trajectory information in a unified multivector structure. The spatio-temporal constraints as well as the sequence information can be unitedly represented using the outer product. With our method, not only can all the possible trajectories of human motion be extracted, but also the spatial and temporal constraints can be flexibly applied during the route extension. The extracted motion trajectories can more accurately reflect the real human motion in the office environment. Our method provides a new solution that can deal with the uncertain problem of the trajectory reconstruction from the sensor network data. Further integration of our method and the statistical inference method may provide new possibility in passive sensor analysis. In addition, our method is also useful for sensor-network-guided indoor navigation and optimal routing.

Acknowledgments: This study was supported by the National Natural Science Foundation of China (Nos. 41571379; 41231173) and the PAPD Program of Jiangsu Higher Education Institutions. We thank for Mingsong Xu, Shuai Zhu, Shuai Yuan, Gege Shi and the anonymous reviewers for their helpful comments.

Author Contributions: All five authors have contributed to the work presented in this paper. Zhaoyuan Yu, Linwang Yuan and Guonian Lü formed the initial idea. Zhaoyuan Yu, Linwang Yuan and Wen Luo developed the overall theoretical structure of the research. Linyao Feng and Wen Luo developed the experiments. Zhaoyuan Yu and Linyao Feng developed the human trajectory generation algorithm. Wen Luo implemented the optimized algorithm, do the performance comparison and produce the graph illustrations. All authors worked collaboratively on writing main text paragraph.

Conflicts of Interest: The authors declare no conflict of interest.

References

1.   Zhu, L.; Wong, K.H. Human Tracking and counting using the KINECT range sensor based on adaboost and kalman filter. In *Advances in Visual Computing*; Bebis, G., Boyle, R., Parvin, B., Koracin, D., Li, B., Porikli, F., Zordan, V., Klosowski, J., Coquillart, S., Luo, X., *et al*, Eds.; Springer: Berlin, Germany, 2013; pp. 582–591.

2. Fahad, L.G.; Khan, A.; Rajarajan, M. Activity recognition in smart homes with self verification of assignments. *Neurocomputing* **2015**, *149*, 1286–1298. [CrossRef]

3. Yin, J.Q.; Tian, G.H.; Feng, Z.Q.; Li, J.P. Human activity recognition based on multiple order temporal information. *Comput. Electr. Eng.* **2014**, *40*, 1538–1551. [CrossRef]

4. Howard, J.; Hoff, W. Forecasting building occupancy using sensor network data. In Proceedings of the 2nd International Workshop on Big Data, Streams and Heterogeneous Source Mining: Algorithms, Systems, Programming Models and Applications, Chicago, IL, USA, 11 Auguest 2013; pp. 87–94.

5. Liu, T.; Chi, T.; Li, H.; Rui, X.; Lin, H. A GIS-oriented location model for supporting indoor evacuation. *Int. J. Geogr. Inf. Sci.* **2014**, *29*, 305–326. [CrossRef]

6. Alexander, A.A.; Taylor, R.; Vairavanathan, V.; Fu, Y.; Hossain, E.; Noghanian, S. Solar-powered ZigBee-based wireless motion surveillance: A prototype development and experimental results. *Wirel. Commun. Mob. Comput.* **2008**, *8*, 1255–1276. [CrossRef]

7. Creusere, C.D.; Mecimore, I. Bitstream-based overlap analysis for multi-view distributed video coding. In Proceeding of the IEEE Southwest Symposium on Image Analysis and Interpretation, Santa Fe, NM, USA, 24–26 March 2008; pp. 93–96.

8. Wren, C.R.; Ivanov, Y.A.; Leigh, D.; Westhues, J. The MERL motion detector dataset. In Proceedings of the 2007 Workshop on Massive Datasets, Nagoya, Japan, 12 November 2007; pp. 10–14.

9. Gong, N.W.; Laibowitz, M.; Paradiso, J.A. Dynamic privacy management in pervasive sensor networks. In *Ambient Intelligence*; de Ruyter, B., Wichert, R., Keyson, D.V., Markopoulos, P., Streitz, N., Divitini, M., Georgantas, N., Gomez, A.M., Eds.; Springer: Berlin, Germany, 2010; pp. 96–106.

10. Wren, C.R.; Tapia, E.M. Toward scalable activity recognition for sensor networks. In *Location- and Context-Awareness*; Hazas, M., Krumm, J., Strang, T., Eds.; Springer: Berlin, Germany, 2006; pp. 168–185.

11. Wong, K.B.Y.; Zhang, T.D.; Aghajan, H. Extracting patterns of behavior from a network of binary sensors. *J. Ambient. Intell. Humaniz. Comput.* **2015**, *6*, 83–105. [CrossRef]

12. Connolly, C.I.; Burns, J.B.; Bui, H.H. Sampling stable properties of massive track datasets. In Proceedings of the 2007 Workshop on Massive Datasets, Nagoya, Japan, 12 November 2007; pp. 2–4.

13. Magnusson, M.S. Discovering hidden time patterns in behavior: T-patterns and their detection. *Behav. Res. Meth. Instrum. Comput.* **2000**, *32*, 93–110. [CrossRef]

14. Salah, A.A.; Pauwels, E.; Tevenard, R.; Gevers, T. T-Patterns revisited: Mining for temporal patterns in sensor data. *Sensors* **2010**, *10*, 7496–7513. [CrossRef] [PubMed]

15. Connolly, C.I.; Burns, J.B.; Bui, H.H. Recovering social networks from massive track datasets. In Proceedings of the 2008 IEEE Workshop on Applications of Computer Vision (WACV), Copper Mountain, CO, USA, 7–9 January 2008; pp. 1–8.

16. Salah, A.A.; Lepri, B.; Pianesi, F.; Pentland, A.S. Human behavior understanding for inducing behavioral change: application perspectives. In *Human Behavior Understanding*; Salah, A.A., Lepri, B., Eds.; Springer: Berlin, Germany, 2011; pp. 1–15.

17. Ivanov, Y.; Sorokin, A.; Wren, C.; Kaur, I. Tracking people in mixed modality systems. In Proceedings of the SPIE Electronic Imaging 2007, San Jose, CA, USA, 28 January 2007; pp. 65080L-1–65080L-11.

18. Anker, T.; Dolev, D.; Hod, B. Belief propagation in wireless sensor networks—A practical approach. In *Wireless Algorithms, Systems, and Applications*; Li, Y., Huynh, D.T., Das, S.K., Du, D.Z., Eds.; Springer: Berlin, Germany, 2008; pp. 466–479.

19. Wren, C.R.; Ivanov, Y.A. Ambient intelligence as the bridge to the future of pervasive computing. In Proceedings of the 8th IEEE Int'l Conference on Automatic Face and Gesture Recognition, Amsterdam, The Netherlands, 17–19 September 2008; pp. 1–6.

20. Ivanov, Y.A.; Wren, C.R.; Sorokin, A.; Kaur, I. Visualizing the history of living spaces. *IEEE Trans. Visual. Comput. Graph.* **2007**, *13*, 1153–1160. [CrossRef]

21. Cerveri, P.; Pedotti, A.; Ferrigno, G. Robust recovery of human motion from video using kalman filters and virtual humans. *Human Move. Sci.* **2003**, *22*, 377–404. [CrossRef]

22. Castanedo, F.; de-Ipiña, D.L.; Aghajan, H.K.; Kleihorst, R. Learning routines over long-term sensor data using topic models. *Expert Syst.* **2014**, *31*, 365–377. [CrossRef]

23. Friedler, S.A.; Mount, D.M. Spatio-temporal range searching over compressed kinetic sensor data. In *Algorithms—ESA 2010*; de Berg, M., Meyer, U., Eds.; Springer: Berlin, Germany, 2010; pp. 386–397.

24. Yuan, L.; Yu, Z.; Luo, W.; Zhang, J.; Hu, Y. Clifford algebra method for network expression, computation, and algorithm construction. *Math. Meth. Appl. Sci.* **2014**, *37*, 1428–1435. [CrossRef]
25. Yuan, L.; Yu, Z.; Luo, W.; Yi, L.; Lü, G. Geometric algebra for multidimension-unified geographical information system. *Adv. Appl. Clifford Algebras* **2013**, *23*, 497–518. [CrossRef]
26. Schott, R.; Staples, G.S. Generalized Zeon Algebras: Theory and application to multi-constrained path problems. *Adv. Appl. Clifford Algebras* **2015**. [CrossRef]
27. Wren, C.R.; Minnen, D.C.; Rao, S.G. Similarity-based analysis for large networks of ultra-low resolution sensors. *Pattern Recognit.* **2006**, *39*, 1918–1931. [CrossRef]
28. Hitzer, E.; Nitta, T.; Kuroe, Y. Applications of clifford's geometric algebra. *Adv. Appl. Clifford Algebras* **2013**, *23*, 377–404. [CrossRef]
29. Schott, R.; Staples, G.S. Dynamic geometric graph processes: Adjacency operator Approach. *Adv. Appl. Clifford Algebras* **2010**, *20*, 893–921. [CrossRef]
30. Schott, R.; Staples, G.S. Complexity of counting cycles using zeons. *Comput. Math. Appl.* **2011**, *62*, 1828–1837. [CrossRef]
31. Nefzi, B.; Schott, R.; Song, Y.Q.; Staples, G.S.; Tsiontsiou, E. An operator calculus approach for multi-constrained routing in wireless sensor networks. In Proceedings of the 16th ACM International Symposium on Mobile Ad Hoc Networking and Computing, Hangzhou, China, 22–25 June 2015; pp. 367–376.
32. Elman, J.L. Finding structure in time. *Cogn. Sci.* **1990**, *14*, 179–211. [CrossRef]
33. Hildenbrand, D. *Foundations of Geometric Algebra Computing*; Springer: Berlin, Germany, 2013.
34. Hildenbrand, D. Geometric computing in computer graphics using conformal geometric algebra. *Comput. Graph-UK* **2005**, *29*, 795–803. [CrossRef]
35. Cheng, C.; Song, X.; Zhou, C. Generic cumulative annular bucket histogram for spatial selectivity estimation of spatial database management system. *Int. J. Geogr. Inf. Sci.* **2013**, *27*, 339–362. [CrossRef]
36. Yu, Z.; Yuan, L.W.; Lv, G.N.; Luo, W.; Xie, Z.-R. Coupling characteristics of zonal and meridional sea level change and their response to different ENSO events. *Chin. J. Geophys.* **2011**, *54*, 1972–1982.
37. Yuan, L.; Yu, Z.; Luo, W.; Hu, Y.; Feng, L.; Zhu, A.-X. A hierarchical tensor-based approach to compressing, updating and querying geospatial data. *IEEE Trans. Knowl. Data Eng.* **2015**, *27*, 312–325. [CrossRef]

6

Classifier Subset Selection for the Stacked Generalization Method Applied to Emotion Recognition in Speech

Aitor Álvarez [1,*], Basilio Sierra [2], Andoni Arruti [2], Juan-Miguel López-Gil [2] and Nestor Garay-Vitoria [2]

Academic Editor: Vittorio M. N. Passaro

[1] Vicomtech-IK4. Human Speech and Language Technologies Department, Paseo Mikeletegi 57,
     Parque Científico y Tecnológico de Gipuzkoa, 20009 Donostia-San Sebastián, Spain
[2] University of the Basque Country (UPV/EHU), Paseo de Manuel Lardizabal 1,
     20018 Donostia-San Sebastián, Spain; b.sierra@ehu.eus (B.S.); andoni.arruti@ehu.eus (A.A.);
     juanmiguel.lopez@ehu.eus (J.-M.L.-G.); nestor.garay@ehu.eus (N.G.-V.)
*   Correspondence: aalvarez@vicomtech.org

Abstract: In this paper, a new supervised classification paradigm, called classifier subset selection for stacked generalization (CSS stacking), is presented to deal with speech emotion recognition. The new approach consists of an improvement of a bi-level multi-classifier system known as stacking generalization by means of an integration of an estimation of distribution algorithm (EDA) in the first layer to select the optimal subset from the standard base classifiers. The good performance of the proposed new paradigm was demonstrated over different configurations and datasets. First, several CSS stacking classifiers were constructed on the RekEmozio dataset, using some specific standard base classifiers and a total of 123 spectral, quality and prosodic features computed using in-house feature extraction algorithms. These initial CSS stacking classifiers were compared to other multi-classifier systems and the employed standard classifiers built on the same set of speech features. Then, new CSS stacking classifiers were built on RekEmozio using a different set of both acoustic parameters (extended version of the Geneva Minimalistic Acoustic Parameter Set (eGeMAPS)) and standard classifiers and employing the best meta-classifier of the initial experiments. The performance of these two CSS stacking classifiers was evaluated and compared. Finally, the new paradigm was tested on the well-known Berlin Emotional Speech database. We compared the performance of single, standard stacking and CSS stacking systems using the same parametrization of the second phase. All of the classifications were performed at the categorical level, including the six primary emotions plus the neutral one.

Keywords: affective computing; machine learning; speech emotion recognition

1. Introduction

Affective computing is an emerging area that tries to make human-computer interaction (HCI) more natural to humans. This area covers topics, such as affect or emotion recognition, understanding and synthesis. Computing systems can better adapt to human behavior taking non-verbal information into account. As Mehrabian suggested [1], verbal information comprises around 10% of the information transmitted between humans, while around 90% is non-verbal. This is why the inclusion of emotion-related knowledge in HCI applications improves the interaction by increasing the level of understanding and decreasing the ambiguity of the messages.

The expression of emotions by humans is multimodal [2]. Apart from verbal information (written or spoken text), emotions are expressed through speech [3–5], facial expressions [6], gestures [7] and other nonverbal clues (mainly psycho-physiological). With regard to the speech communication modality, the literature shows that several parameters (e.g., volume, pitch and speed) are appropriate to generate or recognize emotions [8]. This knowledge is important either to emulate diverse moods reflecting the user's affective states or, in the case of a recognizer, to create patterns for classifying the emotions transmitted by the user.

Affective speech analysis refers to the analysis of spoken behavior as a marker of emotion, with a focus on the nonverbal aspects of speech [9]. Speech emotion recognition is particularly useful for applications that require natural human-machine interaction, in which the response to the user may depend on the detected emotion. Furthermore, it has been demonstrated that emotion recognition through speech can also be helpful in a wide range of other several scenarios, such as e-learning, in-card board safety systems, medical diagnostic tools, call centers for frustration detection, robotics, mobile communication or psychotherapy, among others.

Nevertheless, recognizing emotions from a human's voice is a challenging task due to multiple issues. First, it must be considered that emotions' expression is highly speaker, culture and language dependent. In addition, one spoken utterance can include more than one emotion, either as a combination of different underlying emotions in the same portion or as an individual expression of each emotion in different speech segments. Another interesting aspect is that there is no definitive consensus among the research community regarding which are the most useful speech features for emotion recognition. One possible cause may be the high impact of the variability introduced by the different speakers in commonly-used prosodic features. Finally, selecting the set of emotions to classify is an important decision, which can affect the performance of the speech emotion recognizer. Many works on the topic agree that any emotion is a combination of primary emotions. The primary six emotions include anger, disgust, fear, joy, sadness and surprise [10].

In this paper, we present a study on emotion recognition based on two different sets of speech features extracted from emotional audio signals recorded by professional actors. The analysis was performed on two datasets called RekEmozioand the Berlin Emotional Speech (Emo-DB) database. RekEmozio contains bilingual utterances in Basque and Spanish languages [11], whilst Emo-DB [12] includes sentences recorded in German. Both databases were designed to cover the six primary emotions plus the neutral one, and each recording contained one acted emotion. The classification approach was focused on the categorical recognition of the seven emotions included in the open Emo-DB and the RekEmozio dataset, which is currently in the process of becoming publicly available to the community.

The experiments were divided into three main phases. The first phase corresponded to the construction and evaluation of 10 base supervised classifiers, multi-classifier systems (bagging, boosting and standard stacking generalization) and bi-level multi-classifiers based on the classifier subset selection for stacked generalization (CSS stacking) method on the RekEmozio dataset. For this end, local and global speech parameters containing prosodic, quality and spectral information were computed from each recording through in-house feature extraction algorithms. The selected supervised classifiers for this phase were the following: Bayesian Network (BN), C4.5, k-Nearest Neighbors (kNN), KStar, Naive Bayes Tree (NBT), Naive Bayes (NB), One Rule (OneR), Repeated Incremental Pruning to Produce Error Reduction (RIPPER), Random Forest (RandomF) and Support Vector Machines (SVM). These classifiers were also used to build the CSS stacking classifiers in this first phase.

The aim of the second phase was to verify the efficiency of the CSS stacking classification paradigm on the RekEmozio dataset using: (1) a well-known set of acoustic parameters (extended version of the Geneva Minimalistic Acoustic Parameter Set (eGeMAPS)); and (2) different base classifiers in the first layer. For this purpose, CSS stacking classifiers were built using the best meta-classifier of the first phase. In the second phase, we applied the following base classifiers in the first layer: MultiLayer Perceptron (MLP), Radial Basis Function network (RBF), Logistic Regression (LR), C4.5, kNN, NB, OneR, RIPPER, RandomF and SVM. Hence, the MLP, RBF and LR classifiers were added with respect to the first phase, and the BN, KStar and NBT were discarded.

The third phase consisted of testing the CSS stacking paradigm over the well-known and open Emo-DB. To this end, the same standard classifiers and acoustic features (eGeMAPS) of the second phase were used to build single, standard stacking and CSS stacking classifiers. We decided to leave out the bagging and boosting classifiers because of their poor performance in the first phase.

The paper presents the results from three phases. Regarding the first phase, the results obtained when applying each classification method to each actor were presented, providing a comparison and discussion between each of the several classification paradigms proposed. Concerning the second phase, the results obtained with the CSS stacking classifier are given for each actor, and a comparison with the CSS stacking classifiers from the first phase is also provided. Finally, in the third phase, only the three classifiers with a better score have been presented for each of the constructed systems (single, standard stacking and CSS stacking). The performances of these classifier systems have been compared to each other and to other results obtained in related works in the literature over the same Emo-DB.

In addition, this paper aims to serve as a forum to announce that the RekEmozio dataset will be publicly available soon for research purposes. The aim is to provide the scientific community a new resource to make experiments in the speech emotion recognition field over audio and video acted recordings, several made by actors, others by amateurs, in the Spanish and Basque languages.

The rest of the paper is structured as follows. Section 2 introduces related work. Section 3 details the RekEmozio and Emo-DB datasets, in addition to the two sets of speech features used in this work. In Section 4, how EDA was applied for the stacking classification method is explained. Section 5 describes how the experiments that have been carried out were performed, specifying which techniques have been used in each step of the process. Section 6 explains the obtained results and provides a discussion. Section 7 concludes the paper and presents future work.

2. Related Work

Many studies in psychology have examined vocal expressions of emotions. Eyben *et al.* [8], Schuller *et al.* [3,9], Scherer [4] and Scherer *et al.* [5] provide reviews of these works. Besides, during recent years, the field of emotional content analysis of speech signals has been gaining growing attention. Scherer [4] described the state of research on emotion effects on voice and speech and discussed issues for future research efforts. The analyses performed by Sundberg *et al.* [13] suggested that the emotional samples could be better described by three physiological mechanisms, namely the parameters that quantified subglottal pressure, glottal adduction and vocal fold length and tension. Ntalampiras and Fakotakis [14] presented a framework for speech emotion recognition based on feature sets from diverse domains, as well as on modeling their evolution in time. Wu *et al.* [15] proposed modulation spectral features (MSFs) for the automatic recognition of human affective information from speech. More recently, [16] proposed a novel feature extraction based on multi-resolution texture image information (MRTII), including a BS-entropy-based acoustic activity detection (AAD) module and using an SVM classifier. They improve the performance of other systems based on Mel-frequency cepstral coefficients (MFCC), prosodic and low-level descriptor (LLD) features for three artificial corpora (Emo-DB, eNTERFACE, KHUSC-EmoDB) and a mixed database. There have been several challenges on emotion and paralinguistics in INTERSPEECH, as shown in [3,9].

An important issue to be considered in the evaluation of an emotional speech recognizer is the quality of the data used to assess its performance. The proper design of emotional speech databases is critical to the classification task. Work in this area has made use of material that was recorded during naturally-occurring emotional states of various sorts, that recorded speech samples of experimentally-induced specific emotional states in groups of speakers and that recorded professional or lay actors asked to produce vocal expressions of emotion as based on emotion labels and/or typical scenarios [4].

Several reviews on emotional speech databases have been published. Douglas-Cowie *et al.* [17] provided a list of 19 data collections, while El Ayadi *et al.* [18] and Ververidis and Kotropoulos [19] provided a record of an overview of 17 and 64 emotional speech data collections, respectively. Most of these references of affective databases are related to English, while fewer resources have been developed for other languages. This is particularly true to languages with a relatively low number of speakers, such as the Basque language. To the authors' knowledge, the first affective database in Basque is the one presented by Navas *et al.* [20]. Concerning Spanish, the work of Iriondo *et al.* [21] stands out; and relating to Mexican Spanish, the work of Caballero-Morales [22] can be highlighted. On the other hand, the RekEmozio dataset is a multimodal bilingual database for Spanish and Basque [11], which also stores information that came from processes of some global speech feature extractions for each audio recording.

Popular classification models used for emotional speech classification include, among others, different decision trees [23], SVM [8,24–26], neural networks [27] and hidden Markov models (HMM) [28,29]. Which one is the best classifier often depends on the application and corpus [30]. El Ayadi *et al.*[18] and Ververidis and Kotropoulos [19] provide a review of appropriate techniques in order to classify speech into emotional states.

In order to combine the benefits of different classifiers, classifier fusion is starting to become common, and several different examples can be found in the literature [31]. Pfister and Robinson [30] proposed an emotion classification framework that consists of n(n-1)/2 pairwise SVMs for n labels, each with a differing set of features selected by the correlation-based feature selection algorithm. Arruti *et al.* [32] used four machine learning paradigms (IB, ID3, C4.5, NB) and evolutionary algorithms to select feature subsets that noticeably optimize the automatic emotion recognition success rate. Schuller *et al.* [24] combined SVMs, decision trees and Bayesian classifiers to yield higher classification accuracy. Scherer *et al.* [33] combined three different KNN classifiers to improve the results. Chen *et al.* [34] proposed a three-level speech emotion recognition model combining Fisher rate, SVM and artificial NN in comparative experiments. Attabi and Dumouchel [35] proved that, in the context of highly unbalanced data classes, back-end systems, such as SVMs or a multilayer perceptron (MLP), can improve the emotion recognition performance achieved by using generative models, such as Gaussian mixture models (GMMs), as front-end systems, provided that an appropriate sampling or importance weighting technique is applied. Morrison *et al.* [36] explored two classification methods that had not previously been applied in affective recognition in speech: stacked generalization and unweighted vote. They showed how these techniques can yield an improvement over traditional classification methods. Huang *et al.* [37] developed an emotion recognition system for a robot pet using stacked generalization ensemble neural networks as the classifier for determining human affective state in the speech signal. Wu and Liang [38] presented an approach to emotion recognition of affective speech based on multi-classifiers using acoustic-prosodic information (AP) and semantic labels. Three types of models, GMMs, SVMs and MLPs, are adopted as the base-level classifiers. A meta decision tree (MDT) is then employed for classifier fusion to obtain the AP-based emotion recognition confidence. Several methods have been used for decision fusion in speech emotion recognition. Kuang and Li [39] proposed the Dempster–Shafer evidence theory to execute decision fusion among the three kinds of emotion classifiers to improve the accuracy of the speech emotion recognition. Huang *et al.* [40] used FoCalfusion, AdaBoost fusion and simple fusion on their studies of the effects of acoustic features, speaker normalization methods and statistical modeling techniques on speaker state classification.

3. Case Study

In this section, the main characteristics of the RekEmozio and Emo-DB datasets used for the experiments are presented first. In addition, the speech features used to train and test classifiers are described.

3.1. RekEmozio Dataset

The RekEmozio dataset was created with the aim of serving as an information repository to perform research on user emotions. The RekEmozio dataset is based on data acquired through user interaction and metadata used to describe and label each interaction and provides access to the data stored and the faculty of performing transactions over them, so new information can be added to the dataset by analyzing the data included in it. When building the RekEmozio dataset, the aim was adding descriptive information about the performed recordings, so processes, such as extracting speech parameters and video features, may be done currently on them.

The RekEmozio dataset is composed of audio and video acted recordings, several made by professional actors, while others are by amateurs. In this study, we use the audio recordings made by professional actors. Those recordings are either in the Basque or Spanish languages.

The classification of emotions was performed at the categorical level. For this purpose, seven emotions were used: the six basic emotions described by [6], that is sadness, fear, joy, anger, surprise and disgust, and a neutral emotion. The selection of these specific emotions was based on the work by Ekman and Friesen [6], which suggested that these emotions are universal for all cultures. This is interesting considering the bilingualism of the RekEmozio dataset.

There are 88 different sentences with 154 recordings over them for each actor. Seven actors recorded sentences for Basque, while 10 recorded for Spanish. The total length of the audio recordings was $130'41''$ for Basque and $166'17''$ for Spanish.

A validation for normative study was performed by experimental subjects in order to obtain affective values for each recording and to see what the validity of the recorded material and the affective values for each recording are [41]. Achieved results show that the material recorded in the RekEmozio database was correctly identified by 57 experimental subjects, with a mean accuracy of 66.5% for audio recordings. In Table 1, audio recognition accuracy percentages for the different types of utterances (depending on the language) are presented. It has also to be noted that several automatic emotion recognition systems have used the RekEmozio dataset in previous works, such as [32,42].

Table 1. Human recognition accuracy percentages for utterances as a function of language and emotions (taken from [41]).

	Sadness	Fear	Joy	Anger	Surprise	Disgust	Neutral
Spanish	75%	51%	78%	71%	66%	52%	80%
Basque	77%	52%	68%	74%	59%	51%	77%

The RekEmozio dataset is currently in the process of being made publicly available (until the process is completed and as the RekEmozio dataset remains unavailable from a public repository, anyone interested can contact Karmele López de Ipiña or the co-author Nestor Garay-Vitoria with the aim of the community having access to the dataset for research purposes.

A complete description of the RekEmozio dataset characteristics can be seen in [11].

3.2. Emo-DB

The widely extended German Emo-DB [12] is composed of recordings of 10 actors (five female and five male), which simulated the six primary emotions defined by [6] plus the neutral one. The complete database was evaluated through a perception test with 20 subjects, achieving a human performance of 84% accuracy [43]. The Emo-DB is publicy available via the Internet.

3.3. Speech Features

The selection of suitable features to extract from the voice signal is one of the most difficult and important decisions to be made in the speech emotion recognition task. It is even more critical when pattern recognition techniques are involved, since they are highly dependent on the domain and training material. The voice characteristics most commonly employed in the literature involve the computation of prosodic and continuous features, qualitative features, spectral features and Teager energy operator (TEO)-based features. A deep description of these categories is given in the survey on speech emotion recognition presented in [18]. With the aim of creating a common baseline and agreed set of speech features to use by the speech emotion recognition community, a minimalistic set of voice parameters were recently compiled and presented in [8].

The feature extraction method is also a regular topic of discussion within the speech emotion recognition field. Because of the non-stationary nature of speech signals, the features are usually extracted from overlapped small frames, which consist of a few milliseconds portions of signal. The features extracted at the frame level are known as local features. Using these local features and computing statistics among them, global features are also usually calculated at the utterance level. Even if the best results were obtained in many works [44–46] using global features instead of local features, it is not clear whether global features performed better for any emotion classification. In fact, in the work presented in [28], they proved that global features do not perform correctly when recognizing emotions with similar arousal, e.g., happiness and anger.

In this work, two sets of speech features were computed along the three phases. In the first phase, local and global features containing prosodic, spectral and quality information were extracted using in-house algorithms, considering a total set of 123 features for each spoken utterance. The extraction of local features was done at both the frame and region levels. In the first case, a 20-millisecond frame-based analysis window was used, with an overlapping of 10 milliseconds. Concerning the feature extraction at the region level, the work presented in Tato *et al.* [47] was followed. They defined a technique for signal treatment and information extraction from emotional speech, not only extracting information by frames, but also by regions consisting of more than three consecutive speech frames. With regard to global features, statistics containing measures, such as the mean, variance, standard deviation and the maximum and minimum values and their positions, were computed, among others. The full set of the 123 features we used in the first phase of this work, including local characteristics and their correlated global statistics, were described in more detail in [32].

With regard to the second and third phases, the extended version (eGeMAPS) of the Geneva Minimalistic Acoustic Parameter Set (GeMAPS) was used to extract a different set of speech features. The complete description of the parameters involved in the eGeMAPS set is given in [8]. The extraction of this set of features was done through the OpenSMILE toolkit presented in [48].

4. Classifier Subset Selection to Improve the Stacked Generalization Method

One of the main goals of this work was the construction of a multi-classifier system with optimal selection of the base classifiers in the speech emotion recognition domain. For this purpose, a method proposed in [49] was applied to select an optimal classifier subset by means of the estimation of distribution algorithms (EDAs).

In order to combine the results of the base classifiers, we employed stacked generalization (SG) as a multi-classifier system. Stacked generalization is a well-known ensemble approach, and it is also called stacking [50,51]. While ensemble strategies, such as bagging or boosting, obtain the final decision after a vote among the predictions of the individual classifiers, SG applies another individual classifier to the predictions in order to detect patterns and improve the performance of the vote.

As can be seen in Figure 1, SG is divided into two levels: for Level 0, each individual classifier makes a prediction independently, and for Level 1, these predictions are treated as the input values of another classifier, known as the meta-classifier, which returns the final decision.

The data for training the meta-classifier is obtained after a validation process, where the outputs of the Level 0 classifiers are taken as attributes, and the class is the real class of the example. This implies that a new dataset is created in which the number of predictor variables corresponds to the number of classifiers of the bottom layer, and all of the variables have the same value range as the class variable.

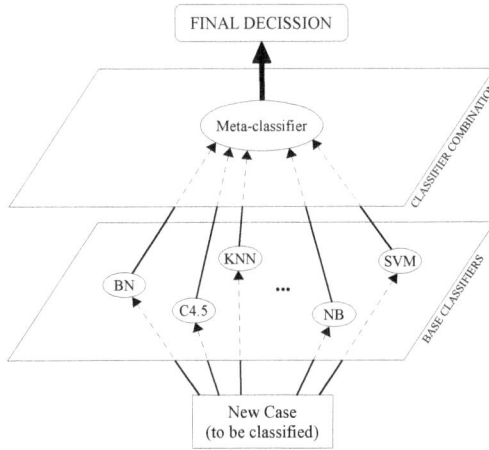

Figure 1. Stacked generalization schemata.

Within this approach, using many classifiers can be very effective, but selecting a subset of them can reduce the computational cost and improve the accuracy, assuming that the selected classifiers are diverse and independent. It is worth mentioning that a set of accurate and diverse classifiers is needed in order to be able to improve the classification results obtained by each of the individual classifiers that are to be combined. This fact has been taken into account to select the classifiers that take part in the first layer of the stacked generalization multi-classifier used.

In [49], an extension of the staking generalization approach is proposed, reducing the number of classifiers to be used in the final model. This new approach is called classifier subset selection (CSS), and a graphical example is illustrated in Figure 2. As can be seen, an intermediate phase is added to the multi-classifier to select a subset of Level 0 classifiers. The classification accuracy is the main criterion to make this selection. As can be seen in Figure 2, discarded classifiers, those with an X, are not used in the multi-classifier.

The method used to select the classifiers could be any, but in this type of scenario, evolutionary approaches are often used. Currently, some of the best known evolutionary algorithms for feature subset selection (FSS) are based on EDAs [52]. EDA combines statistical learning with population-based search in order to automatically identify and exploit certain structural properties of optimization problems. Inza *et al.* [53] proposed an approach that used an EDA called the estimation of Bayesian network algorithm (EBNA) [54] for an FSS problem. Seeing that in [55], EBNA shows better behavior than genetic and sequential search algorithms for FSS problems (and hence, for CSS in this approach), we decided to use EBNA. Moreover, EBNA has been selected as the model in the recent work that analyses the behavior of the EDAs [56].

In our approach, an individual in the EDA algorithm is defined as an n-tuple with $0, 1$ binary values, so-called binary encoding. Each position in the tuple refers to a concrete base classifier, and the value indicates whether this classifier is used (1 value) or not (0 value). An example with 10 classifiers (the value used in this paper) can be seen in Figure 3. In this example, Classifiers 1, 4 and 7 (Cl1, Cl4 and Cl7) are the selected classifiers, and the remaining seven are not used.

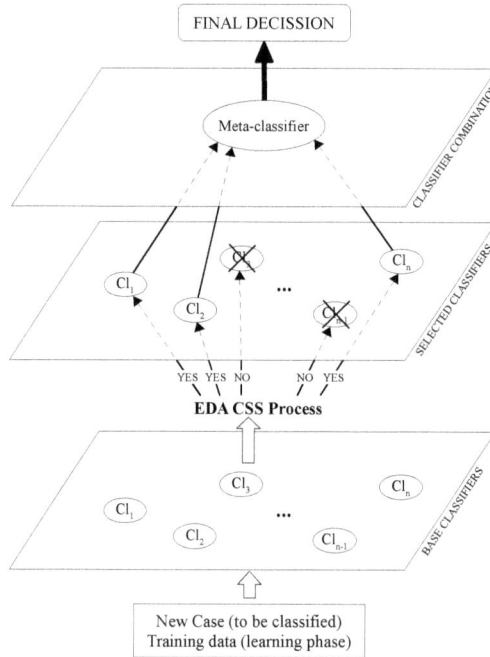

Figure 2. Classifier subset selection stacked generalization.

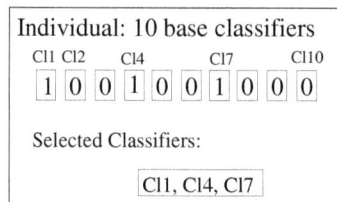

Figure 3. The combinations of base classifiers as the estimation of distribution algorithm (EDA) individuals.

Once an individual has been sampled, it has to be evaluated. The aim is to consider the predictive power of each subset of base classifiers. To this end, a multi-classifier is built for each individual using the corresponding subset of classifiers, and the obtained validated accuracy is used as the fitness function. Thus, when looking for the individual that maximizes the fitness function, the EDA algorithm is also searching the optimal subset of base classifiers.

5. Experiments

In this section, the whole experimental design is described. Firstly, the single classifiers employed in all phases are presented, followed by the definition of the experiment steps. In the end, the experimental setup and the main measure used for the analysis of the obtained results are detailed.

5.1. Base Classifiers

5.1.1. First Phase

The experiments of the first phase were carried out over 10 well-known machine-learning (ML) supervised classification algorithms through the Weka software package [57], which includes a collection of machine learning algorithms for data mining tasks. A brief description of the classifiers of the first phase is presented below.

- Bayesian Networks (BN): A Bayesian network [58], belief network or directed acyclic graphical model is a probabilistic graphical model that represents a set of random variables and their conditional independencies via a directed acyclic graph (DAG).
- C4.5: C4.5 [59] represents a classification model by a decision tree. The tree is constructed in a top-down way, dividing the training set and beginning with the selection of the best variable in the root of the tree.
- k-Nearest Neighbors (KNN): This algorithm is a case-based, nearest-neighbor classifier [60]. To classify a new test sample, a simple distance measure is used to find the training instance closest to the given test instance, and then, it predicts the same class as this nearest training instance.
- KStar: This classifier is an instance-based algorithm that uses an entropy-based distance function [61].
- Naive Bayes Tree (NBT): This classification method uses a decision tree with naive Bayes classifiers at the leaves [62].
- Naive Bayes (NB): The naive Bayes rule [63] uses the Bayes theorem to predict the class for each case, assuming that the predictive genes are independent given the category. To classify a new sample characterized by d genes $\mathbf{X} = (X_1, X_2, ..., X_d)$, the NB classifier applies the following rule:

$$c_{NB} = \arg\max_{c_j \in C} p(c_j) \prod_{i=1}^{d} p(x_i|c_j)$$

where c_{NB} denotes the class label predicted by the naive Bayes classifier, and the possible classes of the problem are grouped in $C = \{c_1, ..., c_l\}$.

- One Rule (OneR): This simple classification algorithm is a one-level decision tree, which tests just one attribute [64]. The chosen attribute is the one that produces the minimum error.
- Repeated Incremental Pruning to Produce Error Reduction (RIPPER): The rule-based learner presented in [65] forms rules through a process of repeated growing (to fit training data) and pruning (to avoid overfitting). RIPPER handles multiple classes by ordering them from least to most prevalent and then treating each in order as a distinct two-class problem.
- Random Forest (RandomF): This constructs a combination of many unpruned decision trees [66]. The output class is the mode of the classes output by individual trees.
- Support Vector Machines (SVM): These are a set of related supervised learning methods used for classification and regression [67]. Viewing input data as two sets of vectors in a n-dimensional space, an SVM will construct a separating hyperplane in that space, one that maximizes the margin between the two datasets.

5.1.2. Second and Third Phases

For the second and third phases, the BN, K-Star and NBT classifiers of the first phase were discarded, the rest of the base classifiers were kept, and three new classifiers were included for experimentation, including multilayer perceptron, radial basis function networks and logistic regression, as they are described below.

- Multilayer Perceptron (MLP): A multilayer perceptron is a feedforward artificial neural network model to map sets of input data onto a set of appropriate outputs [68]. An MLP consists of multiple layers of nodes in a directed graph, with each layer fully connected to the next one. Except for the input nodes, each node is a processing element with a nonlinear activation function.

- Radial Basis Function (RBF) network: A radial basis function network is an artificial neural network using radial basis functions as activation functions [69]. The output of the network is a linear combination of radial basis functions of the inputs and neuron parameters.
- Logistic Regression: A logistic regression (also known as logit regression or logit model) [70] is considered in statistics a regression model where the dependent variable is categorical.

As it can be seen from the three phases, classifiers with different approaches for learning and widely used in different classification tasks were selected. The goal was to combine them in a multi-classifier to maximize the benefits of each modality by intelligently fusing their information and by overcoming the limitations of each modality alone.

5.2. Experimental Steps

As described above, the experiments were organized in three phases. In the first phase, single classifiers, standard multi-classifier systems and CSS stacking classifiers were built over the RekEmozio dataset and compared. During the second phase, new CSS stacking classifiers were built for each of the 17 actors in the same dataset, using new parametrization and configuration of the base classifiers in the first layer. These CSS stacking systems were compared to the CSS stacking classifiers of the first phase. Finally, new single, standard stacking and CSS stacking classifiers were built on the Emo-DB, employing the same acoustic features and standard classifiers of the second phase.

5.2.1. First Phase

1. Single classifiers: build 10 classifiers on the RekEmozio dataset, applying the 10 base machine learning algorithms of the first phase to the training dataset and get validated classification accuracies.
2. Standard multi-classifiers: build one classifier on the RekEmozio dataset applying bagging, another one applying boosting and ten more applying stacking generalization, one for each base classifier at Level 1, and get validated classification accuracies.
3. Classifier subset selection for stacked generalization: build 10 stacking generalization classifiers on the RekEmozio dataset, one for each base classifier acting as a meta-classifier at Level 1, and select, by means of an evolutionary algorithm, a subset of the ten classifiers to participate in the Level 0 layer.

It is worth mentioning that in all of the experiments, a 10-fold cross-validation technique was used. In the case of the classifier subset selection method, this validation was also employed to select the classifier configuration that performed better on average.

5.2.2. Second Phase

1. Classifier subset selection for stacked generalization: build one stacking generalization classifier on the RekEmozio dataset, using the best meta-classifiers from the first phase, and select, by means of an evolutionary algorithm, a subset of the ten classifiers to participate in the Level 0 layer.

5.2.3. Third Phase

1. Single classifiers: build 10 classifiers on the Emo-DB, applying the 10 base machine learning algorithms of the second phase to the training dataset, and get validated classification accuracies.
2. Standard multi-classifiers: build ten classifiers applying stacking generalization, one for each base classifier at Level 1, and get validated classification accuracies.
3. Classifier subset selection for stacked generalization: build 10 stacking generalization classifiers on the Emo-DB, one for each base classifier acting as the meta-classifier at Level 1, and select, by means of an evolutionary algorithm, a subset of the classifiers to participate in the Level 0 layer.

5.3. Experimental Setup

In all of the experiments, 10-fold cross-validation [71] was applied to get a validated classification accuracy (well-classified rate), and this accuracy has been the criterion to define the fitness of an individual, inside the evolutionary algorithm.

For classifier subset selection, the selected EDA algorithm was EBNA, with Algorithm B [72] for structural learning of the Bayesian network. Population size N was set to 50 individuals, representing 50 combinations of classifiers; the number S of selected individuals at each generation was 20 (40% of the population size); and the maximum number of generations of new individuals was set to 10.

5.4. Obtained Results Analysis

The main measure that has been used in this study to evaluate classification methods was the accuracy. The accuracy reflects how many times the emotions are recognized, comparing this to the metadata stored in the RekEmozio and Emo-DB datasets. Accuracy is expressed as a percentage with respect to the total of the recordings.

6. Results and Discussion

6.1. First Phase

Table 2 presents the results obtained for each of the 17 actors in the first phase when a single classification is applied for categorical emotion recognition, in addition to the mean values and standard deviation (SD) of each classifier in the last two rows. The best accuracies obtained per actor are highlighted in bold. The results suggest that SVM is the classifier that performs better when the single classifier method is applied, as for 13 of the 17 actors, the SVM classifier obtains the best results compared to the rest of the single classifiers, and its mean value is 6.43 percentage points higher than the second mean value. Only BN and RandomF get the better accuracies than SVM in the single classification, in the case of two actors for each one. The best accuracy (73.79%) is achieved for the actor P1. The rest of best accuracies for each actor range from 41.82% (P8) to 68.93% (P6).

Table 2. First phase. Accuracy percentages for each person using single classifiers. Mean and SD rows denote the average and standard deviation for each classifier considering all of the actors. BN, Bayesian network; NBT, naive Bayes tree; OneR, one rule; RIPPER, repeated incremental pruning to produce error reduction; RandomF, random forest.

	BN	C4.5	KNN	KStar	NBT	NB	OneR	RIPPER	RandomF	SVM
P1	69.90%	64.08%	65.05%	54.37%	55.34%	61.17%	53.40%	48.54%	64.08%	**73.79%**
P2	58.25%	45.63%	49.51%	36.89%	44.66%	39.81%	34.95%	46.60%	53.40%	**66.02%**
P3	46.60%	45.63%	49.51%	36.89%	44.66%	39.81%	34.95%	46.60%	34.95%	**53.40%**
P4	**59.22%**	43.69%	39.81%	35.92%	43.69%	34.95%	45.63%	33.01%	54.37%	**59.22%**
P5	52.43%	42.72%	41.75%	36.89%	54.37%	49.51%	42.72%	48.54%	52.43%	**68.93%**
P6	53.40%	48.54%	46.60%	38.83%	56.31%	38.83%	43.69%	46.60%	63.11%	**66.99%**
P7	42.72%	33.98%	32.04%	25.24%	41.75%	37.86%	38.83%	38.83%	**50.49%**	45.63%
P8	18.18%	29.09%	29.09%	20.91%	29.09%	19.09%	12.73%	12.73%	26.36%	**41.82%**
P9	44.55%	43.64%	37.27%	33.64%	40.00%	38.18%	30.00%	36.36%	43.64%	**52.73%**
P10	54.55%	43.64%	50.91%	26.36%	52.73%	55.45%	30.00%	41.82%	58.18%	**64.55%**
P11	56.36%	42.73%	37.27%	28.18%	50.00%	43.64%	38.18%	42.73%	55.45%	54.55%
P12	44.55%	32.73%	37.27%	27.27%	31.82%	27.27%	38.18%	39.09%	37.27%	45.45%
P13	44.55%	40.91%	33.64%	35.45%	44.55%	27.27%	45.45%	42.73%	50.91%	**61.82%**
P14	**64.55%**	51.82%	37.27%	31.82%	54.55%	35.45%	40.00%	45.45%	60.00%	56.36%
P15	51.82%	53.64%	53.64%	37.27%	52.73%	40.91%	40.00%	57.27%	**62.73%**	62.73%
P16	58.18%	48.18%	47.27%	36.36%	50.00%	40.91%	43.64%	54.55%	53.64%	**59.09%**
P17	50.91%	46.36%	40.91%	23.64%	45.45%	37.27%	40.00%	40.91%	**53.64%**	50.91%
Mean	51.22%	44.53%	42.87%	33.29%	46.57%	39.26%	38.37%	42.49%	51.45%	**57.88%**
SD	10.98	7.91	8.83	7.57	7.69	9.72	8.52	9.51	10.13	8.69

Table 3. First phase. Accuracy percentages for each person using stacking and bagging and boosting multi-classifiers. Mean and SD rows denote the average and standard deviation for each standard multi-classifier considering all of the actors.

	BN	C4.5	KNN	KStar	NBT	NB	OneR	RIPPER	RandomF	SVM	Bagging	Boosting
P1	64.08%	59.22%	72.82%	57.28%	67.96%	61.17%	43.69%	62.14%	66.99%	**73.79%**	65.05%	34.95%
P2	47.57%	54.37%	42.72%	36.89%	**60.19%**	47.57%	36.89%	49.51%	49.51%	52.43%	57.28%	30.10%
P3	49.51%	44.66%	48.54%	41.75%	46.60%	**58.25%**	45.63%	46.60%	52.43%	49.51%	49.51%	35.92%
P4	49.51%	48.54%	41.75%	33.98%	53.40%	45.63%	39.81%	42.72%	50.49%	44.66%	**55.34%**	33.01%
P5	51.46%	45.63%	48.54%	32.04%	**55.34%**	50.49%	41.75%	52.43%	**55.34%**	**55.34%**	54.37%	32.04%
P6	**59.22%**	56.31%	53.40%	42.72%	**59.22%**	54.37%	55.34%	54.37%	**59.22%**	55.34%	52.43%	32.04%
P7	46.60%	33.98%	41.75%	39.81%	46.60%	44.66%	37.86%	41.75%	46.60%	40.78%	**48.54%**	37.86%
P8	**31.82%**	23.64%	30.91%	24.55%	29.09%	25.45%	19.09%	20.91%	23.64%	24.55%	30.91%	17.27%
P9	45.45%	41.82%	47.27%	36.36%	49.09%	45.45%	41.82%	36.36%	**50.91%**	46.36%	**50.91%**	30.91%
P10	55.45%	55.45%	53.64%	40.91%	56.36%	48.18%	37.27%	52.73%	**60.91%**	55.45%	54.55%	35.45%
P11	49.09%	48.18%	43.64%	39.09%	49.09%	40.91%	39.09%	50.00%	55.45%	52.73%	**58.18%**	35.45%
P12	39.09%	35.45%	36.36%	20.91%	38.18%	48.18%	30.91%	35.45%	35.45%	43.64%	**47.27%**	35.45%
P13	39.09%	41.82%	34.55%	30.91%	42.73%	43.64%	40.91%	**46.36%**	40.91%	40.91%	44.55%	34.55%
P14	50.00%	56.36%	57.27%	40.91%	63.64%	62.73%	37.27%	56.36%	**60.00%**	**60.00%**	44.55%	34.55%
P15	44.55%	57.27%	50.91%	40.91%	57.27%	52.73%	37.27%	50.91%	56.36%	59.09%	**60.91%**	36.36%
P16	55.45%	45.45%	49.09%	40.91%	50.00%	49.09%	40.00%	48.18%	54.55%	48.18%	**56.36%**	32.73%
P17	41.82%	44.55%	38.18%	38.18%	46.36%	44.55%	36.36%	43.64%	43.64%	**50.00%**	43.64%	30.91%
Mean	48.22%	46.63%	46.55%	37.54%	51.24%	48.41%	38.88%	46.50%	50.73%	50.16%	**51.43%**	32.91%
SD	7.44	9.05	9.33	7.53	9.04	8.08	6.82	8.96	9.86	9.86	7.53	4.32

The performance of the standard multi-classifiers systems for all of the actors in the first phase is presented in Table 3, with mean and SD values in the last two rows. In the first 10 columns, the results obtained by the stacked generalization method with the single classifiers as meta classifiers are presented. In addition, the accuracy achieved by the bagging and boosting multi-classifiers are shown in the last two columns. The best results per actor are marked in bold. In contrast to single classifiers, there is no meta classifier that performs much better than the others. This is evident looking at their mean values, with four classifiers in the range from 50.16% to 51.43%, showing low differences between them. For seven actors, the best accuracies are reached using the bagging multi-classifier; RandomF gets the best accuracies for five actors, SVM for four actors, NBT for three actors, BN for two actors, and NB and RIPPER get the best accuracy for one actor each. This happens because in some cases, there are several meta-classifiers that get the best accuracies for a given actor (for example, P5). On the other hand, for 14 actors, the worst results are obtained with the boosting multi-classifier. Compared to the results from the single classifiers in Table 2, only for three of the 17 actors (P3, P11 and P12) are improvements achieved on their best classification results using multi-classifiers. For the rest of the actors, the accuracies are lower when compared to single classifiers.

The results reached by CSS stacking classifiers in the first phase are shown in Table 4, including their mean and SD values. If we focus on the highlighted values, which correspond to the best accuracies for each of the actors, the SVM classifier achieves the best scores, on average, when it is used as a meta classifier (an increase of 2.22 percentage points over the second one) and for 13 actors. The other actors obtained best accuracies with C4.5, NBT, NB and RIPPER meta-classifiers. In general, the best accuracies are improved using the CSS stacking classification method against the standard multi-classifiers. Besides, if we compared the results from CSS stacking with the best accuracies achieved by the single classifiers, 13 of the 17 actors obtain higher classification results. This point is clearly demonstrated in Table 5, where the best accuracies obtained per actor are presented for each of the classification methods, including single classifiers, multi-classifiers (boosting, bagging and stacking) and CSS stacking classifiers. In addition, two columns are presented that show the differences obtained when comparing the best accuracies achieved by multi-classifiers against the single classifiers (Differences_1), and the ones obtained by the CSS stacking classifiers against the best between the single and multi-classifiers (Differences_2).

Table 4. First phase. Accuracy percentages for each person applying CSS stacking with the EDA classification method. Mean and SD rows denote the average and standard deviation for each classifier working as meta classifiers and considering all of the actors.

	BN	C4.5	KNN	KStar	NBT	NB	OneR	RIPPER	RandomF	SVM
P1	71.84%	70.87%	**73.79%**	72.82%	68.93%	70.87%	44.66%	68.93%	71.84%	**73.79%**
P2	66.02%	70.87%	**73.79%**	72.82%	60.19%	70.87%	36.89%	68.93%	71.84%	**73.79%**
P3	57.28%	70.87%	**73.79%**	72.82%	65.05%	70.87%	45.63%	68.93%	71.84%	**73.79%**
P4	55.34%	60.19%	51.46%	53.40%	51.46%	**62.14%**	39.81%	51.46%	59.22%	61.17%
P5	55.34%	63.11%	57.28%	49.51%	51.46%	62.14%	42.72%	58.25%	62.14%	**65.05%**
P6	64.08%	**66.99%**	59.22%	58.25%	50.49%	54.37%	57.28%	60.19%	66.02%	59.22%
P7	51.46%	47.57%	51.46%	41.75%	39.81%	51.46%	37.86%	47.57%	50.49%	**52.43%**
P8	30.91%	34.55%	35.45%	30.00%	**43.64%**	34.55%	26.36%	33.64%	33.64%	32.73%
P9	48.18%	50.91%	48.18%	47.27%	46.36%	45.45%	42.73%	47.27%	50.00%	**54.55%**
P10	58.18%	60.91%	58.18%	56.36%	45.45%	60.91%	37.27%	**61.82%**	60.91%	60.91%
P11	53.64%	53.64%	54.55%	54.55%	46.36%	55.45%	41.82%	50.91%	56.36%	**60.00%**
P12	34.55%	49.09%	45.45%	42.73%	32.73%	49.09%	40.91%	41.82%	49.09%	**51.82%**
P13	47.27%	**59.09%**	51.82%	54.55%	52.73%	52.73%	40.91%	52.73%	50.91%	**59.09%**
P14	50.91%	62.73%	64.55%	59.09%	55.45%	64.55%	38.18%	64.55%	63.64%	**66.36%**
P15	58.18%	62.73%	59.09%	59.09%	50.00%	60.91%	37.27%	59.09%	59.09%	**63.64%**
P16	55.45%	50.91%	59.09%	53.64%	44.55%	60.00%	40.91%	52.73%	56.36%	**60.91%**
P17	48.18%	50.91%	52.73%	45.45%	45.45%	49.09%	37.27%	44.55%	48.18%	**54.55%**
Mean	53.34%	58.00%	57.05%	54.36%	50.01%	57.38%	40.50%	54.90%	57.74%	**60.22%**
SD	9.57	9.34	9.73	10.87	8.40	9.26	5.75	9.63	9.55	9.37

The results from Table 5 show that using multi-classifiers does not outperform the classification accuracies in this classification problem. Nevertheless, when applying the CSS stacking classification method, the improvements are noticeable for many of the actors. As is detailed in the last column Differences_2, 11 actors outperform the best accuracies when compared to the ones obtained with the single and multi-classifiers, giving a mean increase of 1.48 percentage points. The highest improvement is achieved by the actor P3, which increases the accuracy by 15.54 percentage points. The rest of the improvements are in the range from 0.91 to 7.77 points. In addition, two of the actors (P1 and P6) reached the same best accuracy with no significant improvements, and there are four cases (P5, P8, P10 and P13) where the single classifiers reach the best accuracies. A comparison of the best accuracies obtained per actor for each of the classification methods is presented in Figure 4.

Table 5. First phase. Best accuracy per person by using each classification method. Improvements comparing the best accuracy from multi-classifiers (bagging, boosting and stacking) against single classifiers are presented in the Differences_1 column. In addition, the improvements between the CSS stacking with EDA and the best accuracy from both single and standard multi-classifiers are shown in the Differences_2 column. Mean and SD rows denote the average and standard deviation for each classification method and the type of differences considering all of the actors. Differences are expressed in percentage points.

	Single	Bagging	Boosting	Stacking	Differences_1	CSS Stacking	Differences_2
P1	**73.79%**	65.05%	34.95%	**73.79%**	0.00	**73.79%**	0.00
P2	66.02%	57.28%	30.10%	60.19%	−5.83	**73.79%**	+7.77
P3	53.40%	49.51%	35.92%	58.25%	+4.85	**73.79%**	+15.54
P4	59.22%	55.34%	33.01%	53.40%	−3.88	**62.14%**	+2.92
P5	**68.93%**	54.37%	32.04%	55.34%	−13.59	65.05%	−3.88
P6	66.99%	52.43%	32.04%	59.22%	−7.77	**66.99%**	0.00
P7	50.49%	48.54%	37.86%	46.60%	−1.95	**52.43%**	+1.94
P8	**41.82%**	30.91%	17.27%	31.82%	−10.00	35.45%	−6.37
P9	52.73%	50.91%	30.91%	50.91%	−1.82	**54.55%**	+1.82
P10	**64.55%**	54.55%	35.45%	60.91%	−3.64	61.82%	−2.73
P11	56.36%	58.18%	35.45%	55.45%	+1.82	**60.00%**	+1.82
P12	45.45%	47.27%	35.45%	48.18%	+2.73	**51.82%**	+3.64
P13	**61.82%**	44.55%	34.55%	46.36%	−15.45	59.09%	−2.73
P14	64.55%	44.55%	34.55%	63.64%	−0.91	**66.36%**	+1.82
P15	62.73%	60.91%	36.36%	59.09%	−1.82	**63.64%**	+0.91
P16	59.09%	56.36%	32.73%	55.45%	−2.73	**60.91%**	+1.82
P17	53.64%	43.64%	30.91%	50.00%	−3.64	**54.55%**	+0.91
Mean	58.92%	51.43%	32.91%	54.62%	−3.74	**60.95%**	+1.48
SD	8.30%	7.75%	4.45%	8.77%	5.28	9.33%	4.70

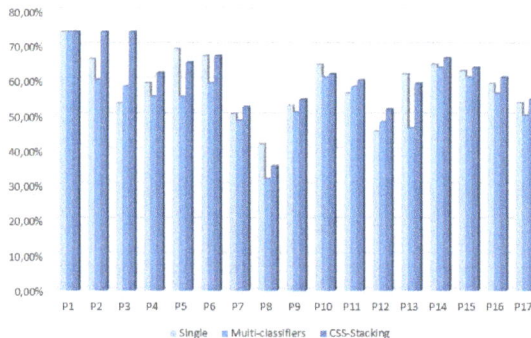

Figure 4. First phase. Best accuracies per person considering single, multi-classifiers and CSS stacking with EDA classification methods.

Finally, we selected one of the classifiers as the meta classifier (SVM) for both stacking and CSS stacking classification methods and presented the results obtained per actor in Table 6 and the mean and SD values at the end. The results prove that using the CSS stacking classification method, the recognition accuracy is outperformed for all of the actors, except for actor P1, in which no improvements are appreciated. The improvements using the CSS stacking classification method range from 3.88 to 24.27 percentage points, with an average improvement of 10.06 points.

Table 6. First phase. Accuracies and improvements per person in percentage points comparing stacking and CSS stacking with EDA classification methods using SVM as the meta classifier. Mean and SD rows denote the average and standard deviation for each classification method and improvements considering all of the actors.

	Stacking	CSS Stacking	Improvements
P1	73.79%	73.79%	0.00
P2	52.43%	73.79%	+21.36
P3	49.51%	73.79%	+24.27
P4	44.66%	61.17%	+16.50
P5	55.34%	65.05%	+9.71
P6	55.34%	59.22%	+3.88
P7	40.78%	52.43%	+11.65
P8	24.55%	32.73%	+8.18
P9	46.36%	54.55%	+8.18
P10	55.45%	60.91%	+5.45
P11	52.73%	60.00%	+7.27
P12	43.64%	51.82%	+8.18
P13	40.91%	59.09%	+18.18
P14	60.00%	66.36%	+6.36
P15	59.09%	63.64%	+4.55
P16	48.18%	60.91%	+12.73
P17	50.00%	54.55%	+4,55
Mean	50.16%	60.22%	+10.06
SD	10.14	9.64	6.42

Statistical Tests

According to [73], we employed the Iman and Davenport test to detect statistical differences among the different classification paradigms. This test rejects the null hypothesis of equivalence between algorithms, since the p-value (0.000216) is lower than the α-value (0.1). Thus, Shaffer *post hoc* test is applied in order to find out which algorithms are distinctive among them. Table 7 shows the statistical differences obtained. As can be seen, the new approach statistically outperforms the results obtained with the standard multi-classifier systems (p-value <0.01). It is worth mentioning that there were no significant differences between CSS stacking and the best single paradigm. This is indeed due to the selection phase of the best approach among all of the single approaches used, before applying meta-classification, as explained in Section 4 of this paper.

Table 7. First phase. *p*-values of the pair-wise comparison between CSS stacking and the other multi-classifiers.

Hypothesis	Adjusted *p*
CSS Stacking vs. Boosting	**1.2094622076166072E-10**
CSS Stacking vs. Bagging	**2.2567292727265824E-4**
CSS Stacking vs. Stacking	**0.004635715398394891**

If the comparison is done pair-wise, the new approach shows better accuracy than each of the single classifiers used. For instance, comparing the SVM single classifier (the best one) with the new approach obtained using SVM as the meta classifier, the new paradigm outperforms the single one in 11 up to 17 actors.

6.2. Second Phase

Table 8 presents the results obtained by the CSS stacking classification method during the second phase, in which eGeMAPS parameters and a new combination of base classifiers in the first layer were employed for classification. Besides, a comparison with the CSS stacking built in the first phase and the corresponding improvements achieved are also presented. Both CSS stacking classifiers were constructed using the SVM as the meta-classifier, as it was the best meta-classifier in the first phase. As can be seen, the integration in the first layer of new base classifiers that performed well as single classifiers (especially the MLP classifier) and the employment of the eGeMAPS acoustic parameters, which also demonstrated their efficiency when comparing the results of single classifiers in both phases, helped improve the results for most actors. The most appreciable improvements are given by the actors P13, P12 and P8, which outperformed the previous results in the first phase by 20.70, 20.26 and 16.62 percentage points, respectively. In global terms, the average accuracy of the CSS stacking classifiers of the second phase outperformed the mean accuracy of the first phase by 4.56 percentage points, which demonstrated the effectiveness of the eGeMAPS parameters and the new classifiers included in the first layer of the CSS stacking classifiers of the second phase.

Table 8. Second phase. Accuracy percentages per actor for the CSS stacking classifier systems of the second phase (CSS stacking 2nd_Phase) and the comparison with the CSS stacking classifiers of the first phase (CSS stacking 1st_Phase). Mean and SD rows denote the average and standard deviation for each classifier for all of the actors.

	CSS Stacking 2nd_Phase	CSS Stacking 1st_Phase	Differences
P1	85.06%	73.79%	+11.27
P2	69.48%	73.79%	−4.31
P3	75.32%	73.79%	+1.53
P4	70.78%	61.17%	+9.61
P5	77.27%	65.05%	+12.22
P6	64.29%	59.22%	+5.07
P7	47.4%	52.43%	−5.03
P8	49.35%	32.73%	+16.62
P9	46.01%	54.55%	−8.54
P10	73.38%	60.91%	+12.47
P11	66.88%	60.00%	+6.88
P12	72.08%	51.82%	+20.26
P13	79.87%	59.09%	+20.78
P14	61.69%	66.36%	−4.67
P15	59.09%	63.64%	−4.55
P16	46.1%	60.91%	−14.81
P17	57.14%	54.55%	+2.59
Mean	64.78%	60.22%	+4.56
SD	12.34	9.94	10.46

In Appendix A, the confusion matrices scored by the CSS stacking classifiers in the second phase are presented for all of the actors.

6.3. Third Phase

In the third phase, ten classifiers were built for each of the classification systems (single, standard stacking and CSS stacking) employed on the Emo-DB. In Table 9, the results of the three best classifiers of each system are shown. The best result of the three classification systems is highlighted in bold per actor. Interestingly, MLP, RandomF and SVM are the best three classifiers for each of the classification systems.

Looking at the results, only for the A5 and A9 actors, the single classifier (RandomF) system scored the best accuracies; 80.00% and 82.14%, respectively, whilst the standard stacking classifiers achieved the worst results. However, the CSS stacking systems outperformed the results of single and standard stacking classifiers for the rest of the actors. The best result is achieved by the A2 actor, which scored an accuracy of 96.55% when the SVM acted as the meta-classifier. On average, the CSS stacking classifier with the SVM acting as the meta-classifier reached higher results, obtaining a mean of 82.45% accuracy for all of the actors. Considering that the human perception rate for the Emo-DB was set to 84% [43], this mean value of 82.45% can be seen as a promising result. Moreover, this score outperforms the results of other works in the literature over the Emo-DB, like the scores obtained in [43,74], which reached accuracies of 79% and 77%, respectively, although these works analyzed the whole database and used different machine learning algorithms and audio features. The overall results demonstrate the good performance of the CSS stacking classification paradigm and confirms the robustness of this classification system to deal with the emotion recognition in speech over several conditions and datasets.

Table 9. Third phase. Accuracy percentages per actor for the best three classifiers of each system built on the Berlin Emotional Speech database (Emo-DB). Mean and SD rows represent the average and standard deviation considering all of the actors.

	Single			Standard Stacking			CSS Stacking		
	MLP	RandomF	SVM	MLP	RandomF	SVM	MLP	RandomF	SVM
A1	79.59%	73.46%	77.55%	63.26%	71.42%	61.22%	79.59%	**81.63%**	79.59%
A2	94.82%	87.93%	86.20%	79.31%	89.65%	72.41%	93.10%	94.83%	**96.55%**
A3	74.41%	62.79%	67.44%	62.79%	67.44%	62.79%	74.42%	74.42%	**76.74%**
A4	84.21%	84.21%	81.57%	68.42%	71.05%	68.42%	**89.47%**	84.21%	86.84%
A5	63.63%	**80.00%**	72.72%	56.36%	65.45%	54.54%	67.27%	72.73%	78.18%
A6	77.14%	74.28%	80.00%	71.42%	68.57%	68.57%	**82.86%**	**82.86%**	**82.86%**
A7	78.68%	75.40%	72.13%	67.21%	70.49%	65.57%	77.05%	**80.33%**	78.69%
A8	78.26%	75.36%	78.26%	73.91%	76.81%	78.26%	82.61%	**86.96%**	85.51%
A9	67.85%	**82.14%**	66.07%	69.64%	71.42%	64.28%	76.79%	75.00%	75.00%
A10	74.64%	83.09%	76.05%	73.23%	71.83%	76.05%	83.10%	80.28%	**84.51%**
Mean	77.32%	77.87%	75.80%	68.55%	72.41%	67.21%	80.63%	81.32%	**82.45%**
SD	8.52	7.17	6.28	6.56	6.76	7.13	7.41	6.57	6.33

In Appendix A, the confusion matrices scored by the CSS stacking system with the SVM classifier acting as the meta-classifier are presented for all of the actors.

7. Conclusions and Future Work

Enabling computers the ability to recognize human emotions is an emergent research area. Continuing the authors' previous work on the topic, in this article, different classification approaches have been presented and compared for the speech emotion recognition task. The experimentation was divided into three main phases, which differ from each other in: (1) the speech parametrization; (2) the base classifiers used to construct the classification systems; and (3) the dataset employed.

The experiments were performed over the RekEmozio and Emo-DB datasets, which contain audio recordings in Basque, Spanish and German from several actors. As the emotional annotation in both datasets was performed using categories, the statistical approach was also turned into a categorical classification problem.

In the first phase, 10 single classifiers, 12 multi-classifiers (bagging, boosting and standard stacking generalization) and 10 final CSS stacking classifiers with the EDA classification method were built, evaluated and compared to each other. For single classifiers, the SVM became the best classifier among the ten algorithms employed, as it obtained the best accuracy for 13 of the 17 actors. If we focus on the performance of multi-classifiers, in most cases, they did not achieve better results compared to single classifiers. In addition, it is noticeable that although bagging was the classifier that reached the best results in most cases, it performed better only for seven of the 17 actors. The best accuracies for multi-classifiers ranged between 31.82% and 73.79%.

In comparison, the CSS stacking multi-classifier with EDA achieved higher accuracies than the single and multi-classifiers in most cases. Table 5 shows that, except for four out of 17 actors, CSS stacking with EDA outperformed the results of all of the other single and multi-classifiers tested in the first phase of this work. Furthermore, these results were statistically significant when comparing pair-wise with the other multi-classifiers. Therefore, it can be concluded from this first phase that multi-classifiers based on the CSS stacking method with EDA are a promising approach for emotion recognition in speech.

With regard to the second phase, a new parametrization based on the eGeMAPS acoustic parameters in addition to new base classifiers was employed to construct new CSS stacking classifiers using the best meta-classifier of the first phase. These new CSS stacking classifiers were compared to the CSS stacking classifiers from the first phase, in order to evaluate the impact of the new parameters and base classifiers included. The results from Table 8 concluded that the new configuration of the CSS stacking classifiers of the second phase outperformed the results obtained in the first phase in most cases. This demonstrated the good performance of the acoustic parameters and the new base classifiers employed in the second phase.

Finally, the third phase was focused on constructing single, standard stacking and CSS stacking classifiers for each of the actors in the well-known and freely-available Emo-DB. The results confirmed the good performance of the CSS stacking classifier system, which improved the accuracies obtained by the other classification systems for all actors, except two.

A future work for this research will be to perform new experiments on different databases, such as the Belfast naturalistic emotion database [10], the Vera am Mittag German audio-visual emotional speech database [75] and the FAUAibo Emotion Corpus [76], which include spontaneous speech, and the Berlin Database of Emotional Speech [12] and EMOVO[77] databases, in order to test out the efficiency of the presented new classification paradigm in other dataset conditions and domains. Besides, new standard classifiers will be explored, and a combination of data from several databases will be used with the aim of building speaker- and language-independent classification systems.

Acknowledgments: This research work was partially funded by the Spanish Ministry of Economy and Competitiveness (Project TIN2014-52665-C2-1-R) and by the Department of Education, Universities and Research of the Basque Government (Grants IT395-10 and IT313-10). Egokituz Laboratory of HCI for Special Needs, Galan research group and Robotika eta Sistema Autonomoen Ikerketa Taldea (RSAIT) are part of the Basque Advanced Informatics Laboratory (BAILab) unit for research and teaching supported by the University of the Basque Country (UFI11/45). The authors would like to thank Karmele López de Ipiña and Innovae Vision S.L. for giving permission to use RekEmozio database for this research.

Author Contributions: The current research was completed through the collaboration of all of the authors. Aitor Álvarez was the team leader and responsible for the speech processing part, selecting and extracting the features to be classified from the speech utterances. Basilio Sierra managed the machine learning part, training and evaluating the classifiers used for the project. Andoni Arruti helped with the audio analysis and with designing the new classification paradigm. Juan-Miguel López-Gil worked preparing the RekEmozio and Emo-DB and provided the state of the art to the team. Nestor Garay-Vitoria completed the state of the art, helped in the data and results interpretation and guided the focus of the article writing.

Conflicts of Interest: The authors declare no conflict of interest.

Appendix

Confusion Matrices for the CSS Stacking Classification Method of the Second and Third Phases

In this Appendix, one table per actor is presented, in which the confusion matrices obtained by the CSS stacking classifiers of the second and third phases are detailed. First, the confusion matrices from the RekEmozio database are shown from Tables A1 to A17. Results of the Emo-DB are then presented from Tables A18 to A27.

Table A1. P1 actor confusion matrix from the RekEmozio dataset in the second phase.

	Sadness	Fear	Joy	Anger	Surprise	Disgust	Neutral
Sadness	19	2	0	0	0	1	0
Fear	0	20	0	0	0	1	1
Joy	0	0	18	3	1	0	0
Anger	1	0	2	17	1	1	0
Surprise	0	0	3	2	17	0	0
Disgust	0	0	0	2	0	20	0
Neutral	1	0	0	0	0	1	20

Table A2. P2 actor confusion matrix from the RekEmozio dataset in the second phase.

	Sadness	Fear	Joy	Anger	Surprise	Disgust	Neutral
Sadness	16	2	0	0	0	1	3
Fear	1	20	0	0	0	1	0
Joy	0	0	17	2	3	0	0
Anger	0	0	3	19	0	0	0
Surprise	0	2	7	0	13	0	0
Disgust	2	1	2	0	3	5	9
Neutral	0	1	0	0	1	8	12

Table A3. P3 actor confusion matrix from the RekEmozio dataset in the second phase.

	Sadness	Fear	Joy	Anger	Surprise	Disgust	Neutral
Sadness	18	2	0	0	0	1	1
Fear	1	18	0	0	2	1	0
Joy	0	0	10	8	4	0	0
Anger	0	0	8	14	0	0	0
Surprise	0	1	2	2	16	1	0
Disgust	1	2	0	0	0	16	3
Neutral	3	0	0	0	0	1	18

Table A4. P4 actor confusion matrix from the RekEmozio dataset in the second phase.

	Sadness	Fear	Joy	Anger	Surprise	Disgust	Neutral
Sadness	21	0	0	0	0	0	1
Fear	0	16	5	0	0	1	0
Joy	0	4	12	4	2	0	0
Anger	0	1	4	13	2	0	2
Surprise	0	2	2	1	14	2	1
Disgust	0	2	1	2	0	15	2
Neutral	2	0	0	1	0	1	18

Table A5. P5 actor confusion matrix from the RekEmozio dataset in the second phase.

	Sadness	Fear	Joy	Anger	Surprise	Disgust	Neutral
Sadness	21	0	0	0	0	0	1
Fear	0	15	2	2	1	1	1
Joy	0	1	13	5	1	2	0
Anger	0	2	2	17	1	0	0
Surprise	0	0	1	5	16	0	0
Disgust	1	2	2	0	0	17	0
Neutral	1	0	0	0	0	1	20

Table A6. P6 actor confusion matrix from the RekEmozio dataset in the second phase.

	Sadness	Fear	Joy	Anger	Surprise	Disgust	Neutral
Sadness	20	0	1	0	1	0	0
Fear	0	16	0	1	3	2	0
Joy	2	0	16	1	0	1	2
Anger	2	5	3	8	3	1	0
Surprise	0	6	0	0	14	2	0
Disgust	3	2	2	3	0	9	3
Neutral	5	0	3	0	1	0	13

Table A7. P7 actor confusion matrix from the RekEmozio dataset in the second phase.

	Sadness	Fear	Joy	Anger	Surprise	Disgust	Neutral
Sadness	10	3	0	0	3	4	2
Fear	4	8	0	3	3	4	0
Joy	1	5	10	1	2	1	2
Anger	1	2	7	8	1	2	1
Surprise	2	4	1	0	12	2	1
Disgust	6	0	0	2	3	11	0
Neutral	4	1	1	2	0	0	14

Table A8. P8 actor confusion matrix from the RekEmozio dataset in the second phase.

	Sadness	Fear	Joy	Anger	Surprise	Disgust	Neutral
Sadness	12	5	0	0	0	2	3
Fear	5	8	1	0	0	6	2
Joy	1	1	10	5	4	0	1
Anger	0	1	5	11	5	0	0
Surprise	0	1	8	4	9	0	0
Disgust	3	7	0	0	0	11	1
Neutral	2	3	0	0	0	2	15

Table A9. P9 actor confusion matrix from the RekEmozio dataset in the second phase.

	Sadness	Fear	Joy	Anger	Surprise	Disgust	Neutral
Sadness	12	5	1	0	0	4	0
Fear	5	5	1	0	0	10	1
Joy	0	3	6	7	4	1	1
Anger	0	1	4	7	8	0	2
Surprise	1	1	4	5	11	0	0
Disgust	2	8	1	0	0	11	0
Neutral	0	2	0	2	0	2	16

Table A10. P10 actor confusion matrix from the RekEmozio dataset in the second phase.

	Sadness	Fear	Joy	Anger	Surprise	Disgust	Neutral
Sadness	19	0	0	0	0	1	2
Fear	0	16	1	0	1	4	0
Joy	0	1	12	4	1	4	0
Anger	0	1	3	16	0	2	0
Surprise	1	1	1	0	19	0	0
Disgust	0	4	3	4	1	10	0
Neutral	0	0	0	0	0	1	21

Table A11. P11 actor confusion matrix from the RekEmozio dataset in the second phase.

	Sadness	Fear	Joy	Anger	Surprise	Disgust	Neutral
Sadness	20	0	0	1	0	0	1
Fear	0	14	3	1	2	2	0
Joy	0	2	14	4	2	0	0
Anger	2	0	4	13	1	2	0
Surprise	0	3	3	0	11	4	1
Disgust	0	1	2	5	2	12	0
Neutral	2	0	0	0	0	1	19

Table A12. P12 actor confusion matrix from the RekEmozio dataset in the second phase.

	Sadness	Fear	Joy	Anger	Surprise	Disgust	Neutral
Sadness	20	0	0	0	0	1	1
Fear	1	13	1	0	0	6	1
Joy	0	1	18	2	1	0	0
Anger	0	1	2	16	2	1	0
Surprise	0	0	2	3	17	0	0
Disgust	2	3	0	1	1	11	4
Neutral	1	0	0	0	1	4	16

Table A13. P13 actor confusion matrix from the RekEmozio dataset in the second phase.

	Sadness	Fear	Joy	Anger	Surprise	Disgust	Neutral
Sadness	20	1	0	0	0	1	0
Fear	2	17	1	1	1	0	0
Joy	0	1	18	1	0	0	2
Anger	0	0	1	20	0	0	1
Surprise	0	3	1	1	17	0	0
Disgust	1	0	1	0	1	15	4
Neutral	0	1	1	0	1	3	16

Table A14. P14 actor confusion matrix from the RekEmozio dataset in the second phase.

	Sadness	Fear	Joy	Anger	Surprise	Disgust	Neutral
Sadness	7	1	0	0	0	7	7
Fear	2	14	1	0	2	3	0
Joy	0	1	14	2	5	0	0
Anger	0	0	2	17	2	0	1
Surprise	0	2	10	3	7	0	0
Disgust	5	2	0	0	0	14	1
Neutral	2	0	0	0	0	1	19

Table A15. P15 actor confusion matrix from the RekEmozio dataset in the second phase.

	Sadness	Fear	Joy	Anger	Surprise	Disgust	Neutral
Sadness	10	0	0	0	0	5	7
Fear	3	14	1	1	1	1	1
Joy	0	0	16	3	3	0	0
Anger	0	1	4	8	6	2	1
Surprise	0	0	5	3	13	1	0
Disgust	4	4	0	0	0	13	1
Neutral	4	0	0	0	0	1	17

Table A16. P16 actor confusion matrix from the RekEmozio dataset in the second phase.

	Sadness	Fear	Joy	Anger	Surprise	Disgust	Neutral
Sadness	16	0	0	0	0	4	2
Fear	0	9	5	4	2	0	2
Joy	1	3	7	3	4	3	1
Anger	2	4	3	8	4	1	0
Surprise	0	4	2	3	13	0	0
Disgust	4	0	0	0	2	12	4
Neutral	1	3	2	0	0	3	13

Table A17. P17 actor confusion matrix from the RekEmozio dataset in the second phase.

	Sadness	Fear	Joy	Anger	Surprise	Disgust	Neutral
Sadness	7	1	0	0	0	8	6
Fear	1	9	2	3	4	3	0
Joy	0	2	7	9	3	1	0
Anger	0	6	7	4	5	0	0
Surprise	0	2	3	2	10	5	0
Disgust	3	3	0	1	5	9	1
Neutral	8	0	1	0	0	2	11

Table A18. A1 actor confusion matrix from the Emo-DB in the third phase.

	Sadness	Fear	Joy	Anger	Surprise	Disgust	Neutral
Sadness	13	0	0	1	0	0	0
Fear	0	2	0	0	0	0	3
Joy	0	0	0	0	1	0	0
Anger	3	0	0	1	0	0	0
Surprise	0	0	0	0	7	0	0
Disgust	0	0	0	0	0	7	0
Neutral	0	2	0	0	0	0	9

Table A19. A2 actor confusion matrix from the Emo-DB in the third phase.

	Sadness	Fear	Joy	Anger	Surprise	Disgust	Neutral
Sadness	11	0	0	0	1	0	0
Fear	0	10	0	0	0	0	0
Joy	0	0	0	0	0	0	0
Anger	0	0	0	6	0	0	0
Surprise	0	0	0	0	11	0	0
Disgust	0	0	0	0	0	9	0
Neutral	0	1	0	0	0	0	9

Table A20. A3 actor confusion matrix from the Emo-DB in the third phase.

	Sadness	Fear	Joy	Anger	Surprise	Disgust	Neutral
Sadness	12	0	1	0	0	0	0
Fear	0	0	0	0	0	0	4
Joy	1	0	7	0	0	0	0
Anger	1	0	0	0	0	0	0
Surprise	2	0	0	0	2	0	0
Disgust	0	0	0	0	0	4	0
Neutral	0	1	0	0	0	0	8

Table A21. A4 actor confusion matrix from the Emo-DB in the third phase.

	Sadness	Fear	Joy	Anger	Surprise	Disgust	Neutral
Sadness	10	0	0	0	0	0	0
Fear	0	7	0	0	0	1	0
Joy	0	0	0	1	0	0	0
Anger	0	0	0	7	1	0	0
Surprise	0	0	0	0	4	0	0
Disgust	0	0	0	0	0	3	0
Neutral	0	2	0	0	0	0	2

Table A22. A5 actor confusion matrix from the Emo-DB in the third phase.

	Sadness	Fear	Joy	Anger	Surprise	Disgust	Neutral
Sadness	8	0	0	1	2	0	0
Fear	0	4	0	0	0	3	1
Joy	0	0	0	2	0	0	0
Anger	0	0	0	10	0	0	0
Surprise	2	0	0	1	5	0	0
Disgust	0	0	0	0	0	7	0
Neutral	0	0	0	0	0	0	9

Table A23. A6 actor confusion matrix from the Emo-DB in the third phase.

	Sadness	Fear	Joy	Anger	Surprise	Disgust	Neutral
Sadness	12	0	0	0	0	0	0
Fear	0	3	0	0	2	0	0
Joy	0	0	1	1	0	0	0
Anger	0	1	0	5	0	0	0
Surprise	1	1	0	0	0	0	0
Disgust	0	0	0	0	0	4	0
Neutral	0	0	0	0	0	0	4

Table A24. A7 actor confusion matrix from the Emo-DB in the third phase.

	Sadness	Fear	Joy	Anger	Surprise	Disgust	Neutral
Sadness	11	0	0	0	1	0	0
Fear	0	9	0	0	0	0	1
Joy	0	1	6	0	0	0	1
Anger	1	0	1	5	0	0	0
Surprise	1	0	0	0	9	0	0
Disgust	0	0	0	0	0	5	0
Neutral	0	6	0	0	0	0	3

Table A25. A8 actor confusion matrix from the Emo-DB in the third phase.

	Sadness	Fear	Joy	Anger	Surprise	Disgust	Neutral
Sadness	16	0	0	0	0	0	0
Fear	0	7	0	0	0	0	1
Joy	0	0	7	1	0	0	0
Anger	0	0	1	10	1	0	0
Surprise	6	0	0	0	2	0	0
Disgust	0	0	0	0	0	10	0
Neutral	0	0	0	0	0	0	7

Table A26. A9 actor confusion matrix from the Emo-DB in the third phase.

	Sadness	Fear	Joy	Anger	Surprise	Disgust	Neutral
Sadness	11	0	0	1	1	0	0
Fear	0	7	0	0	0	1	1
Joy	0	0	4	0	0	0	1
Anger	2	0	0	6	0	0	0
Surprise	2	0	0	2	2	0	0
Disgust	0	0	0	0	0	4	0
Neutral	0	2	0	0	0	1	8

Table A27. A10 actor confusion matrix from the Emo-DB in the third phase.

	Sadness	Fear	Joy	Anger	Surprise	Disgust	Neutral
Sadness	12	0	0	1	1	0	0
Fear	0	12	1	0	0	0	1
Joy	0	1	10	0	0	0	0
Anger	1	0	1	4	1	0	0
Surprise	1	0	0	0	10	0	0
Disgust	0	0	0	0	0	9	0
Neutral	0	2	0	0	0	0	3

References

1. Albert, M. *Silent Messages*; Wadsworth: Belmont, CA, USA, 1971.
2. Lang, P.J. The emotion probe: Studies of motivation and attention. *Am. Psychol.* **1995**, *50*, 372.
3. Schuller, B.; Batliner, A.; Steidl, S.; Seppi, D. Recognising realistic emotions and affect in speech: State of the art and lessons learnt from the first challenge. *Speech Commun.* **2011**, *53*, 1062–1087.
4. Scherer, K.R. Vocal communication of emotion: A review of research paradigms. *Speech Commun.* **2003**, *40*, 227–256.
5. Scherer, K.R.; Johnstone, T.; Klasmeyer, G. Vocal expression of emotion. In *Handbook of Affective Sciences*; Oxford University Press: London, UK, 2003; pp. 433–456.
6. Ekman, P.; Friesen, W.V.; Press, C.P. *Pictures of Facial Affect*; Consulting Psychologists Press: Palo Alto, CA, USA, 1975.
7. Lefter, I.; Burghouts, G.B.; Rothkrantz, L.J. Recognizing stress using semantics and modulation of speech and gestures. *IEEE Trans. Affect. Comput.* **2015**, in press.
8. Eyben, F.; Scherer, K.; Schuller, B.; Sundberg, J.; André, E.; Busso, C.; Devillers, L.; Epps, J.; Laukka, P.; Narayanan, S.; *et al.* The Geneva minimalistic acoustic parameter set (GeMAPS) for voice research and affective computing. *IEEE Trans. Affect. Comput.* **2015**, in press.
9. Schuller, B.; Steidl, S.; Batliner, A.; Burkhardt, F.; Devillers, L.; Müller, C.; Narayanan, S. Paralinguistics in speech and language—State-of-the-art and the challenge. *Comput. Speech Lang.* **2013**, *27*, 4–39.
10. Cowie, R.; Douglas-Cowie, E.; Tsapatsoulis, N.; Votsis, G.; Kollias, S.; Fellenz, W.; Taylor, J.G. Emotion recognition in human-computer interaction. *IEEE Signal Process. Mag.* **2001**, *18*, 32–80.

11. López, J.M.; Cearreta, I.; Garay-Vitoria, N.; de Ipiña, K.L.; Beristain, A. A methodological approach for building multimodal acted affective databases. In *Engineering the User Interface*; Springer: London, UK, 2009; pp. 1–17.

12. Burkhardt, F.; Paeschke, A.; Rolfes, M.; Sendlmeier, W.; Weiss, B. A database of German emotional speech. In Proceedings of the Interspeech 2005, Lissabon, Portugal, 4–8 September 2005; pp. 1517–1520.

13. Sundberg, J.; Patel, S.; Bjorkner, E.; Scherer, K.R. Interdependencies among voice source parameters in emotional speech. *IEEE Trans. Affect. Comput.* **2011**, *2*, 162–174.

14. Ntalampiras, S.; Fakotakis, N. Modeling the temporal evolution of acoustic parameters for speech emotion recognition. *IEEE Trans. Affect. Comput.* **2012**, *3*, 116–125.

15. Wu, S.; Falk, T.H.; Chan, W.Y. Automatic speech emotion recognition using modulation spectral features. *Speech Commun.* **2011**, *53*, 768–785.

16. Wang, K.C. Time-Frequency Feature Representation Using Multi-Resolution Texture Analysis and Acoustic Activity Detector for Real-Life Speech Emotion Recognition. *Sensors* **2015**, *15*, 1458–1478.

17. Douglas-Cowie, E.; Campbell, N.; Cowie, R.; Roach, P. Emotional speech: Towards a new generation of databases. *Speech Commun.* **2003**, *40*, 33–60.

18. El Ayadi, M.; Kamel, M.S.; Karray, F. Survey on speech emotion recognition: Features, classification schemes, and databases. *Pattern Recognit.* **2011**, *44*, 572–587.

19. Ververidis, D.; Kotropoulos, C. Emotional speech recognition: Resources, features, and methods. *Speech Commun.* **2006**, *48*, 1162–1181.

20. Navas, E.; Hernáez, I.; Castelruiz, A.; Luengo, I. Obtaining and evaluating an emotional database for prosody modelling in standard Basque. In *Text, Speech and Dialogue*; Springer: Berlin/Heidelberg, Germany, 2004; pp. 393–400.

21. Iriondo, I.; Guaus, R.; Rodríguez, A.; Lázaro, P.; Montoya, N.; Blanco, J.M.; Bernadas, D.; Oliver, J.M.; Tena, D.; Longhi, L. Validation of an acoustical modelling of emotional expression in Spanish using speech synthesis techniques. In Proceedings of the ISCA Tutorial and Research Workshop (ITRW) on Speech and Emotion, Newcastle, Northern Ireland, UK, 5–7 September 2000.

22. Caballero-Morales, S.O. Recognition of emotions in Mexican Spanish speech: An approach based on acoustic modelling of emotion-specific vowels. *Sci. World J.* **2013**, *2013*, 162093.

23. Sobol-Shikler, T.; Robinson, P. Classification of complex information: Inference of co-occurring affective states from their expressions in speech. *IEEE Trans. Pattern Anal. Mach. Intell.* **2010**, *32*, 1284–1297.

24. Schuller, B.; Reiter, S.; Muller, R.; Al-Hames, M.; Lang, M.; Rigoll, G. Speaker independent speech emotion recognition by ensemble classification. In Proceedings of the IEEE International Conference on Multimedia and Expo (ICME 2005), Amsterdam, The Netherland, 6 July 2005; pp. 864–867.

25. Lee, C.C.; Mower, E.; Busso, C.; Lee, S.; Narayanan, S. Emotion recognition using a hierarchical binary decision tree approach. *Speech Commun.* **2011**, *53*, 1162–1171.

26. Pan, Y.; Shen, P.; Shen, L. Speech emotion recognition using support vector machine. *Int. J. Smart Home* **2012**, *6*, 101–107.

27. Batliner, A.; Fischer, K.; Huber, R.; Spilker, J.; Nöth, E. Desperately seeking emotions or: Actors, wizards, and human beings. In Proceedings of the ISCA Tutorial and Research Workshop (ITRW) on Speech and Emotion, Newcastle, Northern Ireland, UK, 5–7 September 2000.

28. Nwe, T.L.; Foo, S.W.; De Silva, L.C. Speech emotion recognition using hidden Markov models. *Speech Commun.* **2003**, *41*, 603–623.

29. Shahin, I. Speaker identification in emotional talking environments based on CSPHMM2s. *Eng. Appl. Artif. Intell.* **2013**, *26*, 1652–1659.

30. Pfister, T.; Robinson, P. Real-time recognition of affective states from nonverbal features of speech and its application for public speaking skill analysis. *IEEE Trans. Affect. Comput.* **2011**, *2*, 66–78.

31. Alhamdoosh, M.; Wang, D. Fast decorrelated neural network ensembles with random weights. *Inf. Sci.* **2014**, *264*, 104–117.

32. Arruti, A.; Cearreta, I.; Álvarez, A.; Lazkano, E.; Sierra, B. Feature Selection for Speech Emotion Recognition in Spanish and Basque: On the Use of Machine Learning to Improve Human-Computer Interaction. *PLoS ONE* **2014**, *9*, e108975.

33. Scherer, S.; Schwenker, F.; Palm, G. Classifier fusion for emotion recognition from speech. In *Advanced Intelligent Environments*; Springer: Berlin/Heidelberg, Germany, 2009; pp. 95–117.

34. Chen, L.; Mao, X.; Xue, Y.; Cheng, L.L. Speech emotion recognition: Features and classification models. *Digit. Signal Process.* **2012**, *22*, 1154–1160.

35. Attabi, Y.; Dumouchel, P. Anchor models for emotion recognition from speech. *IEEE Trans. Affect. Comput.* **2013**, *4*, 280–290.

36. Morrison, D.; Wang, R.; de Silva, L.C. Ensemble methods for spoken emotion recognition in call-centres. *Speech Commun.* **2007**, *49*, 98–112.

37. Huang, Y.; Zhang, G.; Xu, X. Speech Emotion Recognition Research Based on the Stacked Generalization Ensemble Neural Network for Robot Pet. In Proceedings of the Chinese Conference on Pattern Recognition, 2009, CCPR 2009, Nanjing, China, 4–6 November 2009; pp. 1–5.

38. Wu, C.H.; Liang, W.B. Emotion recognition of affective speech based on multiple classifiers using acoustic-prosodic information and semantic labels. *IEEE Trans. Affect. Comput.* **2011**, *2*, 10–21.

39. Kuang, Y.; Li, L. Speech emotion recognition of decision fusion based on DS evidence theory. In Proceedings of the 2013 4th IEEE International Conference on Software Engineering and Service Science (ICSESS), Beijing, China, 23–25 May 2013; pp. 795–798.

40. Huang, D.Y.; Zhang, Z.; Ge, S.S. Speaker state classification based on fusion of asymmetric simple partial least squares (SIMPLS) and support vector machines. *Comput. Speech Lang.* **2014**, *28*, 392–419.

41. López, J.M.; Cearreta, I.; Fajardo, I.; Garay, N. Validating a multilingual and multimodal affective database. In *Usability and Internationalization. Global and Local User Interfaces*; Springer: Berlin/Heidelberg, Germany, 2007; pp. 422–431.

42. Álvarez, A.; Cearreta, I.; López, J.M.; Arruti, A.; Lazkano, E.; Sierra, B.; Garay, N. A comparison using different speech parameters in the automatic emotion recognition using Feature Subset Selection based on Evolutionary Algorithms. In *Text, Speech and Dialogue*; Springer: Berlin/ Heidelberg, Germany, 2007; pp. 423–430.

43. Esparza, J.; Scherer, S.; Brechmann, A.; Schwenker, F. Automatic emotion classification vs. human perception: Comparing machine performance to the human benchmark. In Proceedings of the 2012 11th International Conference on Information Science, Signal Processing and their Applications (ISSPA), Montreal, QC, Canada, 2–5 July 2012; pp. 1253–1258.

44. Ververidis, D.; Kotropoulos, C. Emotional speech classification using Gaussian mixture models and the sequential floating forward selection algorithm. In Proceedings of the IEEE International Conference on Multimedia and Expo, 2005, ICME 2005, Amsterdam, The Netherland, 6 July 2005; pp. 1500–1503.

45. Hu, H.; Xu, M.X.; Wu, W. Fusion of global statistical and segmental spectral features for speech emotion recognition. In Proceedings of the INTERSPEECH, Antwerp, Belgium, 27–31 August 2007; pp. 2269–2272.

46. Shami, M.T.; Kamel, M.S. Segment-based approach to the recognition of emotions in speech. In Proceedings of the IEEE International Conference on Multimedia and Expo, 2005, ICME 2005, Amsterdam, The Netherlands, 6–8 July 2005; pp. 366-369.

47. Tato, R.; Santos, R.; Kompe, R.; Pardo, J.M. Emotional space improves emotion recognition. In Proceedings of the INTERSPEECH, Denver, CO, USA, 16–20 September 2002; pp. 2029–2032.

48. Eyben, F.; Weninger, F.; Gross, F.; Schuller, B. Recent developments in opensmile, the munich open-source multimedia feature extractor. In Proceedings of the 21st ACM international conference on Multimedia, Barcelona, Catalunya, Spain, 21–25 October 2013; pp. 835–838.

49. Mendialdua, I.; Arruti, A.; Jauregi, E.; Lazkano, E.; Sierra, B. Classifier Subset Selection to construct multi-classifiers by means of estimation of distribution algorithms. *Neurocomputing* **2015**, *157*, 46–60.

50. Wolpert, D.H. Stacked generalization. *Neural Netw.* **1992**, *5*, 241–259.

51. Sierra, B.; Serrano, N.; LarrañAga, P.; Plasencia, E.J.; Inza, I.; JiméNez, J.J.; Revuelta, P.; Mora, M.L. Using Bayesian networks in the construction of a bi-level multi-classifier. A case study using intensive care unit patients data. *Artif. Intell. Med.* **2001**, *22*, 233–248.

52. Larrañaga, P.; Lozano, J.A. *Estimation of Distribution Algorithms: A New Tool for Evolutionary Computation*; Springer Science & Business Media: New York, NY, USA, 2002; Volume 2.

53. Inza, I.; Larrañaga, P.; Etxeberria, R.; Sierra, B. Feature subset selection by Bayesian network-based optimization. *Artif. Intell.* **2000**, *123*, 157–184.

54. Etxeberria, R.; Larranaga, P. Global optimization using Bayesian networks. In Proceedings of the Second Symposium on Artificial Intelligence (CIMAF-99), Habana, Cuba, March 1999; pp. 332–339.

55. Inza, I.; Larrañaga, P.; Sierra, B. Feature subset selection by Bayesian networks: A comparison with genetic and sequential algorithms. *Int. J. Approx. Reason.* **2001**, *27*, 143–164.

56. Echegoyen, C.; Mendiburu, A.; Santana, R.; Lozano, J.A. Toward understanding EDAs based on Bayesian networks through a quantitative analysis. *IEEE Trans. Evolut. Comput.* **2012**, *16*, 173–189.

57. Hall, M.; Frank, E.; Holmes, G.; Pfahringer, B.; Reutemann, P.; Witten, I.H. The WEKA data mining software: An update. *ACM SIGKDD Explor. Newsl.* **2009**, *11*, 10–18.

58. Sierra, B.; Lazkano, E.; Jauregi, E.; Irigoien, I. Histogram distance-based Bayesian Network structure learning: A supervised classification specific approach. *Decis. Support Syst.* **2009**, *48*, 180–190.

59. Quinlan, J.R. *C4.5: Programs for Machine Learning*; Elsevier: San Francisco, CA, USA, 1993.

60. Aha, D.W.; Kibler, D.; Albert, M.K. Instance-based learning algorithms. *Mach. Learn.* **1991**, *6*, 37–66.

61. Cleary, J.G.; Trigg, L.E. K*: An instance-based learner using an entropic distance measure. In Proceedings of the 12th International Conference on Machine Learning, Tahoe City, CA, USA, 9–12 July 1995; Volume 5, pp. 108–114.

62. Kohavi, R. Scaling Up the Accuracy of Naive-Bayes Classifiers: A Decision-Tree Hybrid. In Proceedings of the Second International Conference on Knowledge Discovery and Data Mining, Portland, Oregon, 1996; pp. 202–207.

63. Cestnik, B. Estimating probabilities: A crucial task in machine learning. In Proceedings of the 9th European Conference on Artificial Intelligence (ECAI-90), Stockholm, Sweden, 6 August 1990, Volume 90, pp. 147–149.

64. Holte, R.C. Very simple classification rules perform well on most commonly used datasets. *Mach. Learn.* **1993**, *11*, 63–90.

65. Cohen, W.W. Fast effective rule induction. In Proceedings of the Twelfth International Conference on Machine Learning, Tahoe City, CA, USA, 9-12 July 1995; pp. 115–123.

66. Breiman, L. Random forests. *Mach. Learn.* **2001**, *45*, 5–32.

67. Meyer, D.; Leisch, F.; Hornik, K. The support vector machine under test. *Neurocomputing* **2003**, *55*, 169–186.

68. Rosenblatt, F. *Principles oF Neurodynamics: Perceptrons and the Theory of Brain Mechanisms*; Spartan Books: Washington, DC, USA, 1961.

69. Broomhead, D.; Lowe, D. Multivariable functional interpolation and adaptive networks. *Complex Syst.* **1988**, *2*, 321–355.

70. Freedman, D.A. *Statistical Models: Theory and Practice*; Cambridge University Press: New York, NY, USA, 2009.

71. Stone, M. Cross-validatory choice and assessment of statistical predictions. *J. R. Stat. Soc. Ser. B Methodol.* **1974**, *36*, 111–147.

72. Buntine, W. Theory refinement on Bayesian networks. In Proceedings of the Seventh conference on Uncertainty in Artificial Intelligence, Los Angeles, CA, USA, 13–15 July 1991; pp. 52–60.

73. García, S.; Fernández, A.; Luengo, J.; Herrera, F. Advanced nonparametric tests for multiple comparisons in the design of experiments in computational intelligence and data mining: Experimental analysis of power. *Inf. Sci.* **2010**, *180*, 2044–2064.

74. Schwenker, F.; Scherer, S.; Magdi, Y.M.; Palm, G. The GMM-SVM supervector approach for the recognition of the emotional status from speech. In *Artificial Neural Networks–ICANN 2009*; Springer: Berlin/Heidelberg, Germany, 14–17 September 2009; pp. 894–903.

75. Grimm, M.; Kroschel, K.; Narayanan, S. The Vera am Mittag German audio-visual emotional speech database. In Proceedings of the 2008 IEEE International Conference on Multimedia and Expo, Hannover, Germany, 23 June 2008.

76. Batliner, A.; Steidl, S.; Nöth, E. Releasing a thoroughly annotated and processed spontaneous emotional database: The FAU Aibo Emotion Corpus. In Proceedings of the Satellite Workshop of LREC, Marrakesh, Morocco, 26 May 2008; pp. 28–31.

77. Costantini, G.; Iaderola, I.; Paoloni, A.; Todisco, M. EMOVO Corpus: An Italian Emotional Speech Database. In Proceedings of Ninth International Conference on Language Resources and Evaluation (LREC 2014), Reykjavik, Iceland, 26–31 May 2014; pp. 3501–3504.

7

A Novel Method for Speech Acquisition and Enhancement by 94 GHz Millimeter-Wave Sensor

Fuming Chen [1,†], Sheng Li [2,†], Chuantao Li [1,†], Miao Liu [1], Zhao Li [1], Huijun Xue [1], Xijing Jing [1] and Jianqi Wang [1,3,*]

Academic Editor: Vittorio M. N. Passaro

[1] Department of Biomedical Engineering, Fourth Military Medical University, Xi'an 710032, China; cfm5762@126.com (F.C.); lichuantao614@126.com (C.L.); lium90@163.com (M.L); lizhaofmmu@fmmu.edu.cn (Z.L.); xinyin20130419@163.com (H.X.); fmmujxj@fmmu.edu.cn (X.J.)
[2] College of Control Engineering, Xijing University, Xi'an 710123, China; shengli@fmmu.edu.cn
[3] Shaanxi University of Technology, Hanzhong 723001, China
[*] Correspondence: sheng@mail.xjtu.edu.cn
[†] These authors contributed equally to this work.

Abstract: In order to improve the speech acquisition ability of a non-contact method, a 94 GHz millimeter wave (MMW) radar sensor was employed to detect speech signals. This novel non-contact speech acquisition method was shown to have high directional sensitivity, and to be immune to strong acoustical disturbance. However, MMW radar speech is often degraded by combined sources of noise, which mainly include harmonic, electrical circuit and channel noise. In this paper, an algorithm combining empirical mode decomposition (EMD) and mutual information entropy (MIE) was proposed for enhancing the perceptibility and intelligibility of radar speech. Firstly, the radar speech signal was adaptively decomposed into oscillatory components called intrinsic mode functions (IMFs) by EMD. Secondly, MIE was used to determine the number of reconstructive components, and then an adaptive threshold was employed to remove the noise from the radar speech. The experimental results show that human speech can be effectively acquired by a 94 GHz MMW radar sensor when the detection distance is 20 m. Moreover, the noise of the radar speech is greatly suppressed and the speech sounds become more pleasant to human listeners after being enhanced by the proposed algorithm, suggesting that this novel speech acquisition and enhancement method will provide a promising alternative for various applications associated with speech detection.

Keywords: radar speech; 94 GHz MMW; speech enhancement; empirical mode decomposition; mutual information entropy

1. Introduction

Speech is one of the most important and effective means for human communication, thus, speech acquisition is particularly important. There are some methods which can be used to acquire speech signals, such as traditional air-borne microphones and non-air-borne contact detection. However, traditional microphones are easily disturbed by background noise and their propagation distance is very short, while other methods using non-air-borne contact detection such as electroglottography and the bone conduction microphone constrain people's free movement and make users feel uncomfortable.

Thus, non-contact speech detection methods have been studied and developed. Optical speech detection technology, as one such approach, had been used to listen for messages. For example, Avargel *et al.* presented a remote speech-measurement system that utilizes an auxiliary laser Doppler vibrometer sensor, and proposed a speech enhancement algorithm to enhance speech quality [1].

Recently, radar sensor speech detection technology has also been investigated by many researchers. In 1998, Holzrichter's group developed a micro-power impulse radar which was used to measure the movement of the vocal organs [2]. In order to improve the performance synthetic speech and speech pathology as well as allow silent speech recognition, Eid *et al.* explored a novel application of Ultra Wide Band (UWB) radar speech sensing [3]. Chang's group presented a Doppler radar system and successfully extracted speech information from the vocal vibration signals of a human subject [4]. Although these results verified the effectiveness of the radar sensor in speech, they mainly concentrated on measuring the vibration of the speech organs, instead of examining the performance of the radar speech detection.

Millimeter wave (MMW) radars were developed in previous research for speech detection. Li's group used MMW radar to detect speech signals, which were successfully acquired with a 40 GHz MMW radar. He also demonstrated that the 60 GHz or 90 GHz radars performed better than the 40 GHz one in this new application [5]. In addition, a MMW radar was examined in our laboratory [6,7]. Li *et al.* successfully used a 34 GHz MMW radar to acquire speech signals in free space [8,9], however, the quality of the 34 GHz MMW radar speech was found to be unsatisfactory. In our previous research, we found that the high operation frequency demonstrated excellent sensitivity for the acquisition of speech signals [10–12]. Compared with the Ka-band range, MMW frequency in the W-band range (75–110 GHz) provides a good tradeoff between range and sensitivity for the detection of biosignals [12–14].

To further improve sensitivity and achieve high quality speech detection, in this paper a 94 GHz microwave radar sensor with a superheterodyne receiver was employed to acquire speech signals. In addition, in order to avoid the null point, in-phase and quadrature demodulation technology was adopted in this radar. A superheterodyne receiver was employed to reduce the DC offsets and $1/f$ noise. However, the combined sources of noise, which include ambient, harmonic and electrical circuit noise, were combined in the acquired speech signals. These types of noise greatly degrade the quality of radar speech, and seriously affect the applications of the MMW radar speech. Therefore, how to enhance the quality of radar speech is an important question in radar speech acquisition. Many noise reduction methods have been proposed for enhancing the quality of traditional microphone speech; these include mainly the spectral subtraction, Wiener filtering and wavelet shrinkage methods. However, these methods have several shortcomings which limit their further development. The spectral subtraction method [15] can reduce global noise in speech, but introduces some musical noise. The Wiener filtering method is a linear method which is easy to implement and design [16], but since speech signals are always nonlinear, this results in severe speech distortion. The wavelet shrinkage method relies on the threshold of the wavelet coefficient, and has been applied to denoise signals [17,18]. The application of this method is limited because the basis functions of the algorithm are fixed, and it will not entirely fit real signals. Therefore, it is important for the development of speech enhancement systems to find an adaptive method aimed at improving intelligibility and reducing speech distortion.

Recently, empirical mode decomposition (EMD) has been proposed by Huang *et al.* for analyzing signals from nonlinear and nonstationary processes [19]. Unlike other nonlinear methods, the basis functions in this case are derived from the signal itself, so the major advantage of the EMD algorithm is its adaptability. Several authors have studied EMD-based signal noise filtering and successfully reduced the noise of signals [20–22]. Boudraa *et al.* introduced a new signal denoising approach based on the EMD framework. The approach assumes that the noise of the signal is spread across the intrinsic mode functions (IMFs), and it sets a threshold to remove the noise of the signal; the results show that the EMD-soft method can effectively reduce the signal noise [23]. However, for radar speech, the method should also ensure the intelligibility of the speech when reducing noise. If each IMF is filtered, we find that the noise is suppressed, but the intelligibility of the radar speech is poor. In order to find the best tradeoff between the intelligibility of radar speech and noise reduction, an algorithm combining empirical mode decomposition (EMD) and mutual information entropy (MIE) is proposed for enhancing the perceptibility and intelligibility of radar speech. Mutual information entropy (MIE) is a measure of independence between two variables, a theory proposed by Shannon [24]. In this paper, MIE is used to determine the number of reconstructive components.

This paper demonstrates a potential radar sensor for acquiring high quality speech, and we find that the quality of the acquired speech was enhanced by our proposed method. The radar sensor can therefore be used for non-contact speech signal detection over long distances. This will provide a promising alternative for various applications associated with speech detection.

2. The 94 GHz MMW Radar Sensor

2.1. Quadrature Doppler Radar Theory

The 94 GHz MMW radar system typically transmits a single-tone signal by the transmitting antenna, and the signal can be described as below:

$$P_T(t) = A\cos(2\pi f_0 t + \theta_1) \tag{1}$$

where A is the oscillation amplitude, and f_0 is the oscillation frequency of the transmitting signal. θ_1 is the initial phase of the oscillator. When the signal is reflected by the human throat with a distance change $x(t)$, the received signal may be expressed as [4]:

$$P_R(t) = KA\cos(2\pi f_0 t + \theta_2 - \frac{4\pi x(t)}{\lambda}) \tag{2}$$

where λ_0 is the carrier wavelength of the 94-GHz radar sensor, and $x(t)$ is the time-varying displacement by a target. K is the decay factor of the oscillation amplitude. θ_2 is phase modulated by the nominal distance. Then the received signal and local oscillator signal are mixed, and the mixer signal is filtered by a low-pass filtering. Thus, the signal can be expressed as [25,26]:

$$P_M(t) = \frac{KA^2}{2}\cos(\Delta\theta + \frac{4\pi x(t)}{\lambda_0}) + N(t) \tag{3}$$

where $\Delta\theta$ is the constant phase shift dependent on the nominal distance to the target. $N(t)$ is the phase noise and ambient noise.

It is known that there is a null detection point problem for a single channel radar. This null detection point occurs with a target distance every $\lambda/4$ from the radar. In order to avoid the null point of the single-channel radar, a quadrature receiver with I/Q channel was designed [27]. The quadrature receiver with local oscillator phases $\pi/2$ apart, insuring that there is always at least one output not in the null point. The output of the radar quadrature mixer can be expressed as follows [25,27]:

$$W_I(t) = A_I\cos(\Delta\theta + \frac{4\pi x(t)}{\lambda_0}) + N_I(t) \tag{4}$$

and:

$$W_Q(t) = A_Q\sin(\frac{4\pi x(t)}{\lambda_0} + \Delta\theta) + N_Q(t) \tag{5}$$

where, A_I and A_Q are the amplitudes of the quadrature channel I and channel Q, N_I and N_Q are added sources of noise which include ambient noise and electrical-circuit noise for the I-branch and Q-branch. Therefore, if $A_I = A_Q$, the associated phase $\omega(t)$ can be extracted by the following equation:

$$\omega(t) = \arctan\left[\frac{W_Q(t) - N_Q(t)}{W_I(t) - N_I(t)}\right] = \frac{4\pi x(t)}{\lambda_0} + \Delta\theta \tag{6}$$

2.2. The 94 GHz MMW Radar System

Figure 1 shows a schematic diagram of the 94 GHz MMW radar sensor system. The system is composed of an oscillator, transmitter module and receiver module. The W-band double resonant oscillator operates at a local frequency at 7.23 GHz and the power of the reference frequency is 20 mW. The transmitting and receiving antennas of the radar sensor are both Cassegrain antennas, with a diameter of 200 mm, a gain of 41.7 dBi, and a beam width of 1° at –3 dB levels. The output radio frequency (RF) power of the transmitting antenna is 100 mW and the equivalent isotropic radiated power (EIRP) is 61.7 dBm. To begin with, the Dielectric Resonator Oscillator (DRO) of 7.23 GHz emits a continuous wave signal, and then the frequency of the signal is amplified and feeds into both the transmitter module and receiver module. In the transmitter module, the local frequency is multiplied 13 times by the frequency multiplier, first it passes through a band-pass filter of 94 GHz, and then generates a high-stability 94 GHz RF signal, with the beams radiated by the transmitting antenna. In the receiver module, the noise figure is 7.6 dB. The total gain of RF-IF is 65 dB and the I/Q phase balance is +/−1 deg. Firstly, the local frequency is multiplied 12 times by the frequency multiplier, and passes through a band-pass filter of 86.7 GHz, and is then balance-mixed with received signal from receiving antenna. Finally, a signal is amplified with a low-noise amplifier (LNA) and is then mixed with two quadrature local signal for the in-phase and quadrature (I/Q) receiver chains. After I/Q quadrature demodulation, the final signal is sampled by an A/D converter to be transferred to a computer, and then the speech signal is recorded by the computer.

A superheterodyne receiver is employed to avoid the severe DC offsets and the associated 1/f noise at the baseband. Moreover, the transmitting and receiving circuits employ two antennas, and they are separated, which can increase the detection range and reduce interference. The distance and the angle between the two antennas can be easily adjusted. Furthermore, the I/Q quadrature demodulation technology can not only effectively avoid the null detection point problem, but also enhance the signal-to-noise ratio (SNR) by 3 dB compared with the one-signal channel [28].

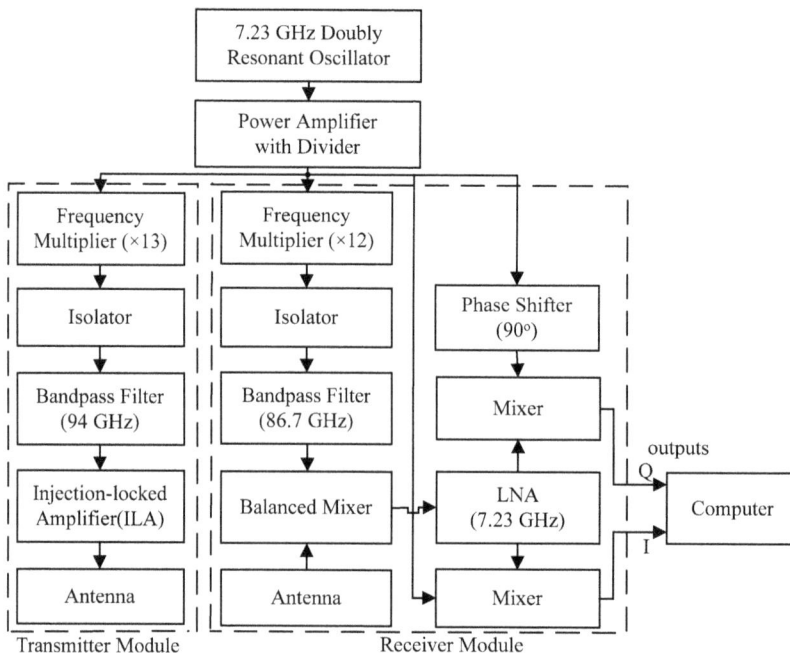

Figure 1. Schematic diagram of the 94 GHz millimeter wave radar sensor.

2.3. Safety

To begin with, the safety issue regarding human exposure to radar electromagnetic fields should be taken into account. Thus, the maximum allowed density which exposed to the human should be computed. In this paper, the radiating power of the radar sensor is 100 mW, the antenna gain is 41.7 dBi. The maximum accepted density exposed S to the human can be computed as [29]:

$$S\left(\frac{W}{m^2}\right) = \frac{\text{raiating power} \times \text{antenna gain}}{4\pi(\text{distance})^2} \tag{7}$$

where the distance represents the minimum distance between the human subject and the radar. Here, the distance is 1 m. Therefore, the maximum acceptable density S is about 0.3318 W/m^2.

The maximum allowed density level accepted safe power density level of 10 W/m^2 [30] for human exposure at frequencies from 10 to 300 GHz. The maximum acceptable power density is much lower than the maximum allowed density level accepted safe power density level. Therefore, the radar sensor poses no risk to the human health.

3. Experimental Section

3.1. Subjects and the Experiment

Ten healthy volunteers (five males and five females) participated in the radar speech experiment. Their ages varied from 20 to 35, and all of them were Chinese native speakers. In the experiment, one of the volunteers sat in front of the radar sensor with his throat kept at the same height as the radar sensor. The radar speech sensor was positioned ranging from 2 m to 20 m away from the subjects. Although the speech signals can be detected at a distance of 20 m, to guarantee high quality speech signals, a distance of 5 m was selected as a representative distance. The volunteers were asked to speak one sentence of Mandarin Chinese "1-2-3-4-5-6". All of the experimental procedures were in accordance with the rules of the Declaration of Helsinki [31].

3.2. Evaluations

In order to test the performance of the proposed algorithm, both objective and subjective methods were applied to assess the results. Signal-noise ratio (SNR), speech spectrogram and mean opinion score (MOS) tests were conducted. In the experiments, three different kinds of background noise—white noise, pink noise and babble noise—were added to the original radar speech. All the types of noise were taken from the NOISEX-92 database, and the noisy radar speech with SNR_{in} of −5, 0, 5 and 10 dB. In addition, to further illustrate the effectiveness of the proposed algorithm, the results were compared to the spectral subtraction and wavelet shrinkage algorithms.

The SNR is used as an objective measure to evaluate the proposed method's performance, and the SNR_{in} of noisy speech is defined by:

$$SNR_{in} = 10\log_{10}\frac{\sum\limits_{n=1}^{N} s^2(n)}{\sum\limits_{n=1}^{N} [x(n) - s(n)]^2} \tag{8}$$

The SNR_{out} of the enhanced speech is given by:

$$SNR_{out} = 10\log_{10}\frac{\sum\limits_{n=1}^{N} s^2(n)}{\sum\limits_{n=1}^{N} [y(n) - s(n)]^2} \tag{9}$$

where $x(n)$ is the noisy speech, $s(n)$ is the clean speech, $y(n)$ is the enhanced speech, N indicates the number of samples in speech, and n represents the sample index.

The speech spectrogram and MOS test are used as the subjective measures to evaluate the proposed method's performance. From the speech spectrogram, it can be observed that the signal strength of different speech spectra over time, the abscissa of the speech spectrogram represents time, and the ordinate of the speech spectrogram represents frequency. The color depth shows the speech energy value; the deeper the color, the stronger the speech energy. For the MOS test, ten other volunteers were instructed to evaluate the intelligibility of the speech based on the criteria of the mean opinion score test, which is a five point scale (1: bad; 2: poor; 3: common; 4: good; 5: excellent). All listeners were healthy with no reported history of hearing disease.

4. Methods

4.1. Empirical Mode Decomposition

As the core component of the Hilbert Huang transforms (HHT), empirical mode decomposition (EMD) is an adaptive method for processing nonlinear and nonstationary signals [19]. Unlike previous signal processing methods [17,18], the EMD method is intuitive, direct and adaptive. In the whole process of decomposition, all the basis functions are derived from the signal itself. Therefore, the method is very well-suited to processing nonlinear and nonstationary signals [32], such as ECG and speech signal. Given a signal $x(t)$, EMD can adaptively decompose it into a series of oscillatory components called intrinsic mode functions (IMFs) through the "sifting" process, and each IMF is an oscillatory signal which consists of a subset of frequency components from the original signal. Figure 2 shows the flow chart of the EMD algorithm.

The sifting process can be described as follows:

1. Locate all the extrema (maxima/minima) of $x(t)$.
2. Interpolate the maxima and minima points by cubic splines to obtain an upper envelope $e_u(t)$ and a lower envelope $e_d(t)$, respectively.
3. Compute the average $m_1(t)$ of the upper and lower envelopes, subtracted from the original signal $x(t)$ to obtain $h_1(t) = x(t) - m_1(t)$.
4. Judging whether $h_1(t)$ is to satisfies the following two conditions of IMF:

    (a) In the whole data item, the number of extrema should be equal to the number of zero crossings, or one difference at the most.
    (b) At any point, the mean of the maxima envelope and the minima envelope should be zero. That is to say, signal is symmetric about the time axis.

    If $h_1(t)$ satisfies the conditions to be an IMF, it is regarded as the first $IMF_1(t)$, $IMF_1(t) = h_1(t)$.
5. If $h_1(t)$ does not satisfy the two conditions, the $h_1(t)$ is regarded as a new signal, steps 1–4 are repeated on $h_1(t)$ to generate the following $h_2(t)$. If $h_2(t)$ does not satisfy the two conditions, there is a standard deviation (SD) to terminate the sifting process. The stopping criterion is given by:

$$SD(i) = \sum_{t=0}^{N} \frac{|h_{i-1}(t) - h_i(t)|^2}{h_{i-1}^2(t)} \tag{10}$$

Usually, the value range of SD is between 0.2 and 0.3 [19]. If $h_2(t)$ satisfies the SD, then the $IMF_1(t) = h_2(t)$. If $h_2(t)$ does not meet the stopping criterion, and the $h_2(t)$ is regarded as a new signal, steps 1–5 are repeated on $h_2(t)$ to generate the following $h_i(t)$, until the $h_i(t)$ satisfies the two conditions of IMF or SD. Then, the $IMF_1(t) = h_i(t)$.
6. Once the $IMF_1(t)$ is generated and subtracted the original signal to get a residual $r_1(t)$: $r_1(t) = x(t) - IMF_1(t)$. The residual signal is treated as the original signal, and steps 1–5 are repeated to get

the next residual signal. Therefore, the residual signal can be expressed as $r_n(t) = r_{n-1}(t) - MF_n(t)$. At this point, the $r_n(t)$ is a monotonic sequence. After the sifting process, the original signal can be decomposed into several IMF components $IMF_1(t), IMF_2(t), \ldots IMF_n(t)$ and a residual sequence $r_n(t)$. Therefore, the original signal can be expressed as:

$$x(t) = \sum_{i=1}^{n} IMF_i(t) + r_n(t) \tag{11}$$

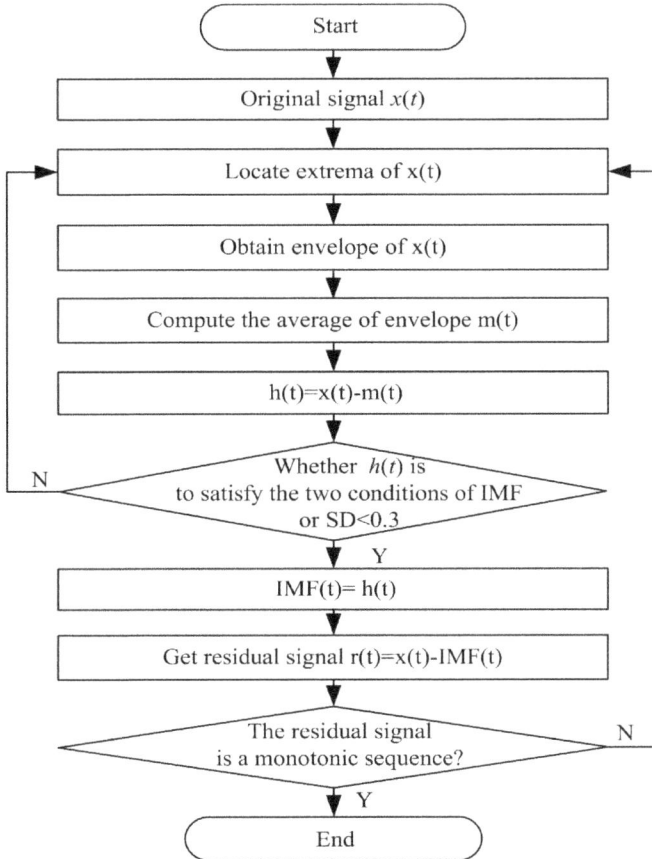

Figure 2. The flow chart of empirical mode decomposition algorithm.

4.2. Mutual Information Entropy

Mutual information entropy is an information theory measurement for quantifying how much information is shared between two or more random variables [33]. It can not only describe the linear correlation between these variables, but also can describe the nonlinear correlation between variables. The major advantage of MIE is that this method can indicate the correlation between two random events without any special requirements for the distribution of the types of variables.

In this paper, MIE is used as a cutoff point to determine the number of reconstructive components. MIE is always non-negative and can measure the relationship between two variables. The MIE $I(X;Y)$ between variables X and Y is defined as [34,35]:

$$I(X;Y) = \sum\sum p(x,y)\log_2(\frac{p(x,y)}{p(x)p(y)}) \qquad (12)$$

Entropy mainly measures the uncertainty of random variables, and the MIE can also be represented by the entropy as:

$$I(X;Y) = H(X) - H(X|Y) \qquad (13)$$

where:

$$H(X) = -\sum_{x\in\Omega_x} p(x)\log_2(p(x)) \qquad (14)$$

and:

$$H(X|Y) = -\sum\sum_{x\in\Omega_x} p(x,y)\log_2(p(x|y)) \qquad (15)$$

The more uncertain the event X is, the larger $H(X)$ is. Basically, the stronger the relationship between two variables is, the larger MIE they will have. Zero MIE means the two variables are independent or have no relationship [36].

4.3. Selecting the Reconstruction Components

Figure 3a shows original radar speech contaminated by white noise. Figure 3b shows the decomposition of the original radar speech signal by EMD. From top to bottom, the frequencies of IMFs decreased gradually. In general, the noise of the signal is spread across the IMFs. From Figure 3b, it is observed that the first three IMFs are mainly noise, and there are few useful original signals. From the fourth to the ninth IMFs, it is observed that there are many useful original signals and the IMFs are very similar to the original signal, but some noise components still remain. From the tenth to the last IMFs, the frequencies of the IMFs are lower and the amplitudes are smaller, and there is detailed information about the original signal. Thus, it is assumed that the original radar speech can be decomposed into high frequency modes, middle frequency modes and low frequency modes. The high frequency modes are mainly noise and interference signal, the middle frequency modes mainly include original useful signals and the low frequency modes mainly are the detailed information from the original signal. In short, the noise is mainly concentrated in the high frequency and middle frequency modes, and there is much less in the low frequency modes.

Some authors have used a wavelet soft-threshold method to remove the noise of IMFs. This method is often employed to process all the IMF components. However, with regard to radar speech, if all the frequency modes are denoised, we find that while the noise is suppressed, the intelligibility of the radar speech is poor. It is because the detailed information from the original signal is removed. Thus, in order to achieve a good tradeoff between radar speech distortion and noise reduction, the high and the middle frequency modes are denoised firstly, and then reconstruct speech signal with the processed IMFs and the remaining low frequency modes.

The mutual information values are sequentially calculated in the adjacent IMF components energy entropy. According to the information theory, the MIE of adjacent IMF components will be in order of large to small, and then back to large:

$$\begin{cases} \text{If } I(IMF_i, IMF_{i+1}) \downarrow \text{ and } I(IMF_{i+1}, IMF_{i+2}) \uparrow \\ \quad k = first(\arg\min_{1\leqslant i\leqslant n-1}[I(IMF_i, IMF_{i+1})]) \end{cases} \qquad (16)$$

The point which the minimum MIE appears is selected as the cutoff point to distinguish the high frequency and the middle frequency modes. In order to find the cutoff point of the middle frequency and the low frequency modes, the fixed threshold (FT) was defined as 10^{-1}. If the maximum amplitude of IMFs are lower than the FT, it can be assumed that these IMFs are low frequency modes.

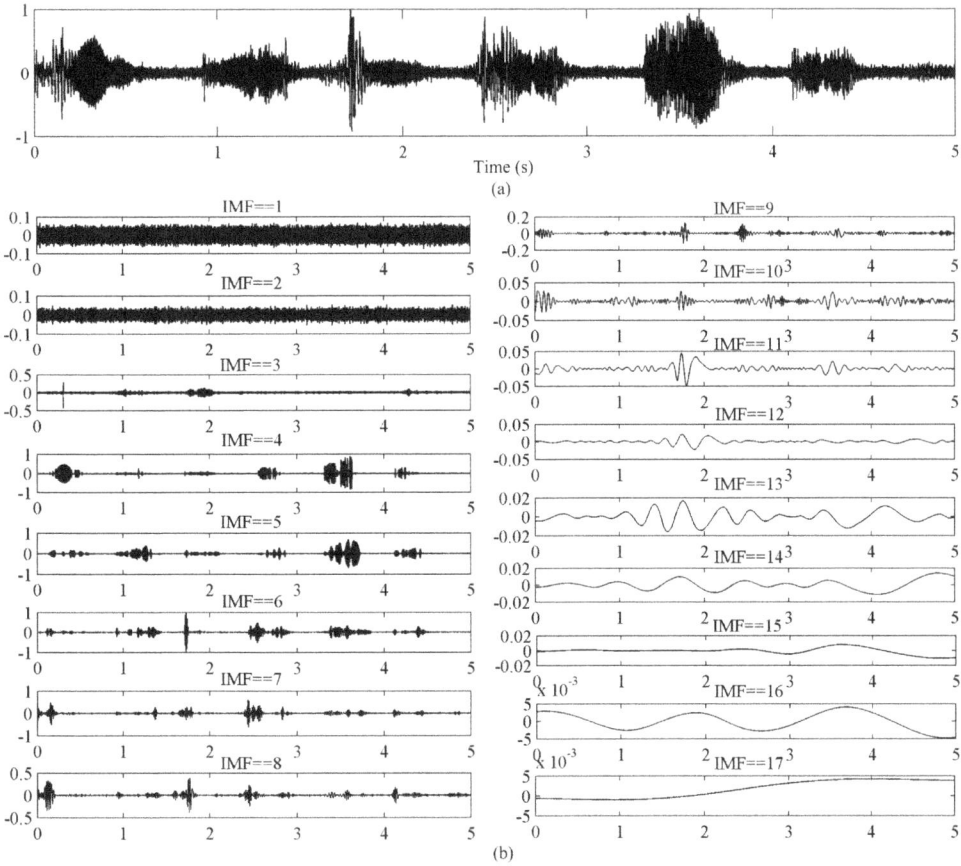

Figure 3. (a) The original radar speech signal contaminated by white noise; (b) the decomposition of the original radar speech corrupted by white noise using EMD.

4.4. The Proposed Algorithm for Radar Speech Enhancement

In the speech enhancement based on the proposed algorithm, the threshold plays an important role in removing noise from radar speech signal. The threshold was estimated by [17,23]:

$$Thr_i = \sigma_i \sqrt{2\log(N)} \tag{17}$$

where N is the signal length, σ is the estimated noise level and is defined by [22]:

$$\sigma = \frac{median\left\{|IMF_1(t) - median\left\{IMF_1(t)\right\}|\right\}}{0.675} \tag{18}$$

In this paper, the soft thresholding function is employed to denoise the high frequency and middle frequency modes for speech enhancement [18,23]:

$$IMF_i'(t) = \begin{cases} sign\left\{IMF_i(t)\right\}\left\{IMF_i(t) - Thr_i\right\} & |IMF_i(t)| \geqslant Thr_i \\ 0 & |IMF_i(t)| \leqslant Thr_i \end{cases} \tag{19}$$

Afterwards the high frequency and middle frequency modes are processed by the soft thresholding. Then, the enhanced speech $y(t)$ is reconstructed with the processed signal $IMF_i'(t)$ and the remaining low frequency modes. The $y(t)$ is given by:

$$y(t) = \sum_{i=1}^{k} IMF_i'(t) + \sum_{k+1}^{n} IMF_i(t) \tag{20}$$

where k is the number of the high frequency and middle frequency modes, and n is the number of IMFs. In conclusion, the proposed algorithm for radar speech enhancement includes the following steps:

1. Decompose the given signal $x(t)$ into IMFs using the sifting process.
2. Compute the energy entropy of each IMFs using Equations (14) and (15).
3. Compute the MIE of the adjacent IMF components using Equation (13).
4. Determine the cutoff point of high frequency and middle frequency modes using Equation (16).
5. Determine the cutoff point of the middle frequency and low frequency modes using the FT of IMF.
6. Denoise the high frequency and middle frequency modes using Equations (17)–(19).
7. Reconstruct the speech with the processed signal and remaining low frequency modes using Equation (20).

5. Results and Discussion

This section mainly presents the performance of the proposed algorithm. Speech time domain waveforms and spectrograms are appropriate tools for analyzing speech quality. They can evaluate the extent of noise reduction, residual noise and speech distortion by comparing the original radar speech and the enhanced speech. Figure 4 shows the time-domain waveforms and the spectrograms of the radar speech "1-2-3-4-5-6".

Figure 4a,e show the waveform and spectrogram of the original radar speech, respectively. It is observed that the original radar speech signals are contaminated by some noise. Figure 4b–d show the waveforms of the radar speech enhanced by the spectral subtraction algorithm, wavelet shrinkage algorithm and the proposed method, respectively. Figure 4f–h show the corresponding spectrograms of the radar speech enhanced using the three algorithms. Figure 4b,f show that the spectral subtraction algorithm is effective in reducing the combined noise of the radar speech, but the algorithm introduces some new musical noise to the enhanced speech, so the intelligibility of the radar speech was not improved. Figure 4c,g show that the wavelet shrinkage algorithm can also effectively reduce the noise of the radar speech, but in this case the change in the color depth illustrates that the essential information of the speech is removed. This results in severe radar speech distortion. Figure 4d,h show that the proposed EMD and MIE methods not only reduce the low frequency noise in which the combined noise are concentrated, but also eliminates the high frequency noise completely. In addition, to a large extent, the essential signal information of the radar speech is still preserved. These results suggest that the proposed algorithm outperforms the spectral subtraction and wavelet shrinkage algorithms, and that the proposed algorithm is an effective way to improve the quality of radar speech.

To test the proposed algorithm, a subjective MOS test was used to evaluate the quality of the enhanced radar speech. Ten listeners were selected to listen to the enhanced radar speech sentences using the three algorithms. The results of the averaged MOS under three types of noise at a SNR_{in} of 5 dB are presented in Table 1. It can be seen from the table that all the scores of the enhanced speech processed by using the three algorithms are improved, especially the proposed method obtained the highest score, between "3" and "4", followed by the wavelet shrinkage method, with a score of around "3", meanwhile the spectral subtraction algorithm achieved the lowest score. The results suggest that the proposed method presents the highest speech intelligibility and is more pleasant to the listeners.

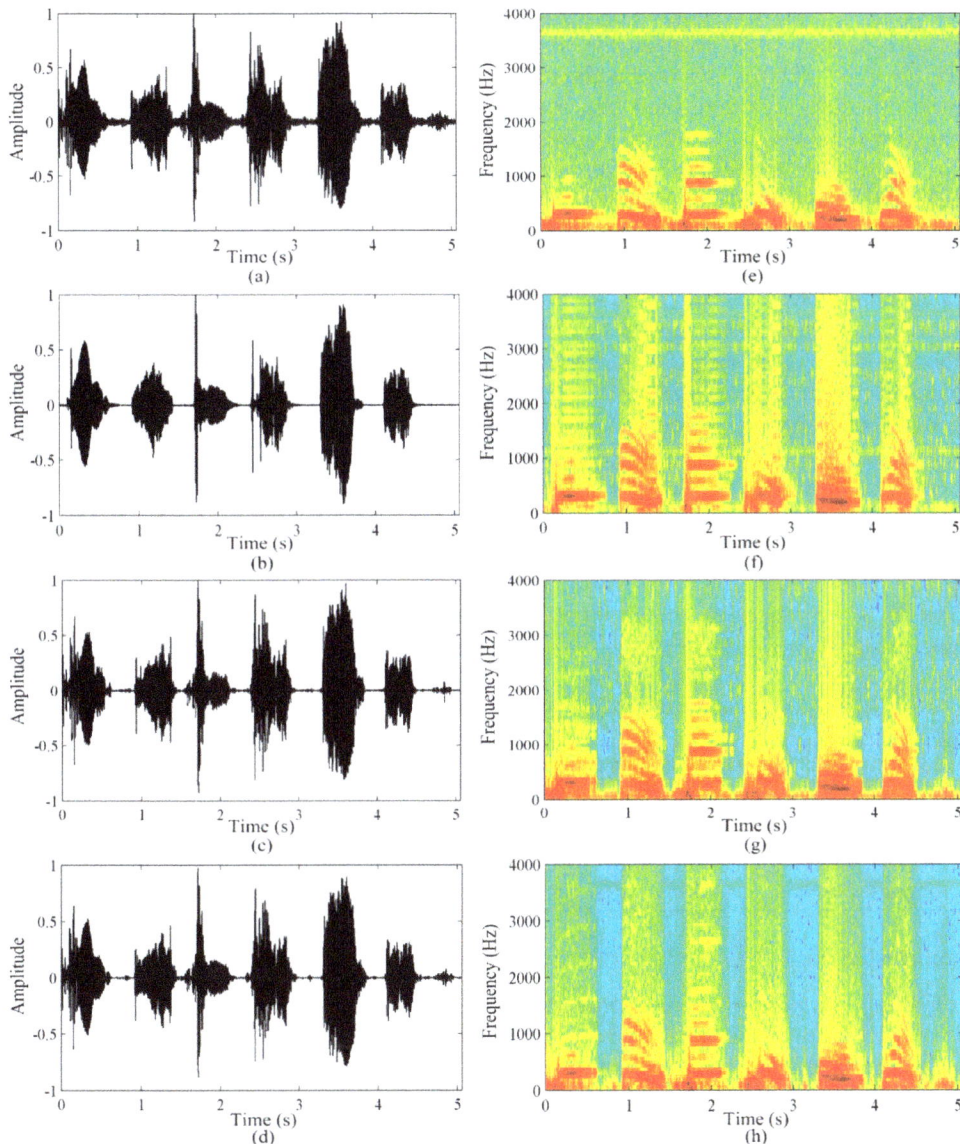

Figure 4. The time-domain waveforms and the spectrograms of the radar speech "1-2-3-4-5-6". (**a,e**) are the original radar speech; (**b,f**) are enhanced speech obtained by the spectral subtraction; (**c,g**) are enhanced speech obtained by the wavelet shrinkage; (**d,h**) are enhanced speech obtained by the proposed algorithm.

Table 1. Comparison of the results of averaged MOS with three types of noise at a SNR of 5 dB. The numbers in the brackets represent standard deviation for these mean opinion scores.

Enhancement Algorithms	White	Pink	Babble
Spectral subtraction	2.78 (0.30)	2.98 (0.38)	2.64 (0.35)
Wavelet shrinkage	3.25 (0.46)	3.37 (0.32)	3.21 (0.27)
Proposed method	3.59 (0.37)	3.71 (0.35)	3.56 (0.42)

The listening tests also indicated the EMD and MIE method is the most suitable for enhancing the radar speech. The method obtained a good tradeoff between the intelligibility and noise reduction. This is because EMD is an adaptive method for processing nonlinear and nonstationary signals, and it does not require presetting fixed basis functions, as all the basis functions are derived from the signal itself. The wavelet shrinkage algorithm will cause severe speech distortion when reducing noise. The spectral subtraction algorithm introduces some musical noise into the enhanced radar speech, so the perceptibility and intelligibility of the radar speech are not improved greatly, and the resulting speech sounds unpleasant to listeners. An objective measurement, the signal-noise ratio, was employed to evaluate the performance of the proposed method. We added babble noise, white noise and pink noise with SNR_{in} of −5, 0, 5 and 10 dB to the original radar speech. The results of the SNR_{out} obtained for different noise types and algorithms are seen in Table 2. It can be seen that the three methods lead to an increase of SNR_{out} values at different SNR_{in} levels, and the results demonstrate the effectiveness of the three methods. The SNR_{out} obtained by the proposed method is much higher than those obtained by the spectral subtraction and the wavelet shrinkage algorithms. Even for low SNR_{in} values, it can be observed the effectiveness of the proposed method in removing the noise components, and we can observe that the spectral subtraction algorithm achieved the worst speech enhancement. Especially at the SNR of 10 dB level, the spectral subtraction led to a decrease of SNR_{out}. This is due to musical noise being introduced to the speech. The wavelet shrinkage and the proposed algorithm performed better, and this is attributed to the time adaptive threshold strategy. However, the superiority of the proposed method over wavelet shrinkage is due to the adaptive decomposition of the speech signal provided by EMD, as it does not rely on the fixed basis functions.

Table 2. Comparison of the SNRs obtained by using three enhancement algorithms.

Enhancement Algorithms	White				Pink				Babble			
	−5	0	5	10	−5	0	5	10	−5	0	5	10
Spectral subtraction	4.1	7.1	8.9	9.7	3.7	6.8	7.4	9.2	2.3	3.7	7.1	8.7
Wavelet shrinkage	4.6	7.6	10.2	12.3	4.1	7.2	8.6	12.1	2.7	5.6	7.3	11.9
Proposed method	5.2	7.5	10.9	14.9	4.8	7.3	10.2	13.7	3.9	6.7	10.1	12.3

6. Conclusions

In this paper, a 94 GHz millimeter wave (MMW) radar sensor was employed to acquire speech. A superheterodyne quadrature receiver was designed to reduce the severe DC offsets and the associated $1/f$ noise at the baseband. An EMD and MIE algorithm was designed to enhance radar speech signals, and the performance of proposed algorithm was evaluated by both objective and subjective methods. The results show that human speech can be effectively acquired by a 94 GHz MMW radar sensor when the detection distance is 20 m. The results also show the advantages of the radar speech sensor in long distance detection, preventing acoustic disturbance and ensuring high directivity. Therefore, this novel radar sensor and signal processing method is expected to provide a promising alternative to current methods for various applications associated with speech.

Acknowledgments: This work was supported by the National Natural Science Foundation of China (NSFC, No. 61371163, No. 61327805). We also want to thank the participants from E.N.T. Department, the Xi Jing Hospital, the Fourth Military Medical University, for helping with data acquisition and analysis.

Author Contributions: Fuming Chen, Sheng Li did the data analysis and prepared the manuscript. Fuming Chen, Sheng Li and Chuantao Li developed the algorithms. Sheng Li, Fuming Chen, Chuantao Li and Miao Liu revised and improved the paper. Jianqi Wang, Fuming Chen, Zhao Li, Huijun Xue and Xijing Jing participated in the discussion about the method and contributed to the analysis of the results.

Conflicts of Interest: The authors declare no conflict of interest.

References

1. Avargel, Y.; Cohen, I. Speech measurements using a laser Doppler vibrometer sensor: Application to speech enhancement. In Proceedings of the Hands-Free Speech Communication and Microphone Arrays (HSCMA), Edinburgh, Scotland, 30 May–1 June 2011; pp. 109–114.

2. Holzrichter, J.F.; Burnett, G.C.; Ng, L.C.; Lea, W.A. Speech articulator measurements using low power EM-wave sensors. *J. Acoust. Soc. Am.* **1998**, *103*, 622–625. [CrossRef] [PubMed]

3. Eid, A.M.; Wallace, J.W. Ultrawideband Speech Sensing. *IEEE Antennas Wireless Propag. Lett.* **2009**, *8*, 1414–1417. [CrossRef]

4. Lin, C.S.; Chang, S.F.; Chang, C.C.; Lin, C.C. Microwave Human Vocal Vibration Signal Detection Based on Doppler Radar Technology. *IEEE Trans. Microw. Theory Tech.* **2010**, *58*, 2299–2306. [CrossRef]

5. Li, Z.W. Millimeter wave radar for detecting the speech signal applications. *Int. J. Infrared Mill. Wave.* **1996**, *17*, 2175–2183. [CrossRef]

6. Wang, J.; Zheng, C.; Lu, G.; Jing, X. A new method for identifying the life parameters via radar. *EURASIP J. Adv. Signal Process.* **2007**, *101*, 8–16.

7. Wang, J.Q.; Zheng, C.X.; Jin, X.J.; Lu, G.H.; Wang, H.B.; Ni, A.S. Study on a non-contact life parameter detection system using millimeter wave. *Hangtian Yixue yu Yixue Gongcheng/Space Med. Med. Eng.* **2004**, *17*, 157–161.

8. Li, S.; Wang, J.Q.; Niu, M.; Liu, T.; Jing, X.J. Millimeter wave conduct speech enhancement based on auditory masking properties. *Microw. Opt. Technol. Lett.* **2008**, *50*, 2109–2114. [CrossRef]

9. Li, S.; Wang, J.; Niu, M.; Liu, T.; Jing, X. The enhancement of millimeter wave conduct speech based on perceptual weighting. *Prog. Electromagn. Res. B* **2008**, *9*, 199–214. [CrossRef]

10. Tian, Y.; Li, S.; Lv, H.; Wang, J.; Jing, X. Smart radar sensor for speech detection and enhancement. *Sens. Actuator A Phys.* **2013**, *191*, 99–104. [CrossRef]

11. Jiao, M.; Lu, G.; Jing, X.; Li, S.; Li, Y.; Wang, J. A novel radar sensor for the non-contact detection of speech signals. *Sensors* **2010**, *10*, 4622–4633. [CrossRef] [PubMed]

12. Li, S.; Tian, Y.; Lu, G.; Zhang, Y.; Lv, H.; Yu, X.; Xue, H.; Zhang, H.; Wang, J.; Jing, X. A 94-GHz millimeter-wave sensor for speech signal acquisition. *Sensors* **2013**, *13*, 14248–14260. [CrossRef] [PubMed]

13. Mikhelson, I.V.; Bakhtiari, S.; Elmer, T.W., II; Sahakian, A.V. Remote sensing of heart rate and patterns of respiration on a stationary subject using 94-GHz millimeter-wave interferometry. *IEEE Trans. Biomed. Eng.* **2011**, *58*, 1671–1677. [CrossRef] [PubMed]

14. Bakhtiari, S.; Elmer, T.W., II; Cox, N.M.; Gopalsami, N.; Raptis, A.C.; Liao, S.; Mikhelson, I.; Sahakian, A.V. Compact millimeter-wave sensor for remote monitoring of vital signs. *IEEE Trans. Instrum. Meas.* **2012**, *61*, 830–841. [CrossRef]

15. Boll, S.F. Suppression of acoustic noise in speech using spectral subtraction. *IEEE Trans. Acous. Speech. Signal Process.* **1979**, *27*, 113–120. [CrossRef]

16. Proakis, J.G.; Manolakis, D.G. *Digital Signal Processing: Principles, Algorithms and Applications*; Prentice Hall: Upper Saddle River, NJ, USA, 1992.

17. Donoho, D.L.; Johnstone, I.M. Ideal spatial adaptation by wavelet shrinkage. *Biometrika* **1994**, *81*, 425–455. [CrossRef]

18. Donoho, D.L. De-noising by soft-thresholding. *IEEE Trans. Inform. Theory* **1995**, *41*, 613–627. [CrossRef]

19. Huang, N.E.; Shen, Z.; Long, S.R.; Wu, M.C.; Shih, H.H.; Zheng, Q.; Yen, N.C.; Tung, C.C.; Liu, H.H. The empirical mode decomposition and the Hilbert spectrum for nonlinear and non-stationary time series analysis. *Proc. A* **1998**. [CrossRef]

20. Flandrin, P.; Goncalves, P.; Rilling, G. Detrending and denoising with empirical mode decompositions. In Proceedings of the XII EUSIPCO, Vienna, Austria, 6–10 September 2004; pp. 1581–1584.

21. Kopsinis, Y.; McLaughlin, S. Development of EMD-Based Denoising Methods Inspired by Wavelet Thresholding. *IEEE Trans. Signal Process.* **2009**, *57*, 1351–1362. [CrossRef]

22. Khaldi, K.; Boudraa, A.O.; Bouchikhi, A.; Alouane, M.T.-H. Speech enhancement via EMD. *EURASIP J. Adv. Signal Process.* **2008**, *2008*, 1–8. [CrossRef]

23. Boudraa, A.O.; Cexus, J.C.; Saidi, Z. EMD-Based Signal Noise Reduction. *Int. J. Signal Process.* **2004**, *1*, 33–127.

24. Shannon, C.E. A mathematical theory of communication. *Bell Syst. Tech. J.* **1948**, *27*, 379–423. [CrossRef]

25. Kazemi, S.; Ghorbani, A.; Amindavar, H.; Li, C. Cyclostationary approach to Doppler radar heart and respiration rates monitoring with body motion cancelation using Radar Doppler System. *Biomed. Signal Process. Control* **2014**, *13*, 79–88. [CrossRef]

26. Li, C.; Chen, F.; Jin, J.; Lv, H.; Li, S.; Lu, G.; Wang, J. A Method for Remotely Sensing Vital Signs of Human Subjects Outdoors. *Sensors* **2015**, *15*, 14830–14844. [CrossRef] [PubMed]

27. Bakhtiari, S.; Liao, S.; Elmer, T.; Gopalsami, N.S.; Raptis, A.C. A Real-time Heart Rate Analysis for a Remote Millimeter Wave I-Q Sensor. *IEEE Trans. Biomed. Eng.* **2011**, *58*, 1839–1845. [CrossRef] [PubMed]

28. Sivannarayana, N.; Rao, K.V. I-Q imbalance correction in time and frequency domains with application to pulse doppler radar. *Sadhana* **1998**, *23*, 93–102. [CrossRef]

29. Chioukh, L.; Boutayeb, H.; Deslandes, D.; Wu, K. Noise and Sensitivity of Harmonic Radar Architecture for Remote Sensing and Detection of Vital Signs. *IEEE Trans. Microw. Theory Tech.* **2014**, *62*, 1847–1854. [CrossRef]

30. Lin, J.C. A new IEEE standard for safety levels with respect to human exposure to radio-frequency radiation. *IEEE Ant. Propag. Mag.* **2006**, *48*, 157–159. [CrossRef]

31. World Medical Association. World Medical Association Declaration of Helsinki: Ethical principles for medical research involving human subjects. *JAMA* **2013**, *310*, 2191–2194.

32. Boudraa, A.O.; Cexus, J.C. EMD-based signal filtering. *IEEE Trans. Instrum. Meas.* **2007**, *56*, 2196–2202. [CrossRef]

33. Omitaomu, O.A.; Protopopescu, V.A.; Ganguly, A.R. Empirical Mode Decomposition Technique with Conditional Mutual Information for Denoising Operational Sensor Data. *IEEE Sens. J.* **2011**, *11*, 2565–2575. [CrossRef]

34. Battiti, R. Using mutual information for selecting features in supervised neural net learning. *IEEE Trans. Neural. Netw.* **1994**, *5*, 537–550. [CrossRef] [PubMed]

35. Sugiyama, M. Machine learning with squared-loss mutual information. *Entropy* **2012**, *15*, 80–112. [CrossRef]

36. Fleuret, F. Fast binary feature selection with conditional mutual information. *J. Mach. Learn. Res.* **2004**, *5*, 1531–1555.

8

Integrating Sensors into a Marine Drone for Bathymetric 3D Surveys in Shallow Waters

Francesco Giordano, Gaia Mattei *, Claudio Parente, Francesco Peluso and Raffaele Santamaria

Academic Editors: Fabio Menna, Fabio Remondino and Hans-Gerd Maas

Dipartimento di Scienze e Tecnologie, Università degli Studi di Napoli "Parthenope", Centro Direzionale, Isola C4, 80143 Napoli, Italy; francesco.giordano@uniparthenope.it (F.G.); claudio.parente@uniparthenope.it (C.P.); francesco.peluso@uniparthenope.it (F.P.); raffaele.santamaria@uniparthenope.it (R.S.)
*   Correspondence: gaia.mattei@uniparthenope.it

Abstract: This paper demonstrates that accurate data concerning bathymetry as well as environmental conditions in shallow waters can be acquired using sensors that are integrated into the same marine vehicle. An open prototype of an unmanned surface vessel (USV) named MicroVeGA is described. The focus is on the main instruments installed on-board: a differential Global Position System (GPS) system and single beam echo sounder; inertial platform for attitude control; ultrasound obstacle-detection system with temperature control system; emerged and submerged video acquisition system. The results of two cases study are presented, both concerning areas (Sorrento Marina Grande and Marechiaro Harbour, both in the Gulf of Naples) characterized by a coastal physiography that impedes the execution of a bathymetric survey with traditional boats. In addition, those areas are critical because of the presence of submerged archaeological remains that produce rapid changes in depth values. The experiments confirm that the integration of the sensors improves the instruments' performance and survey accuracy.

Keywords: marine USV; open prototype; bathymetry; shallow waters; Gegraphic Iinformation System (GIS) application; 3D bathymetric data elaboration

1. Introduction

Bathymetric information is fundamental in all branches of oceanography, paleoclimate studies, and marine geology. It can be supplied by maps that indicate the water body depth as a function of the position (latitude and longitude), similar to topographic maps representing the altitude of the Earth's surface at different geographic coordinates [1].

Most techniques for obtaining these data are difficult to use in shallow waters where bathymetric surveys often entail expensive measurement costs . For most bathymetry acquisition techniques, it is not possible to obtain a better vertical accuracy than 0.5 m at the 95% confidence level. Airborne LiDAR and/or maritime vessels are the only options for surveys with an accuracy requirement of 0.5 m with a 95% confidence level. Other remote sensing techniques can also be used only if the accuracy requirements are relaxed to 2 m, 95% confidence [2].

Airborne laser (or lidar) bathymetry (ALB) is based on a scanning, pulsed laser beam to measure the depths of relatively shallow, coastal waters from the air. It is also named airborne lidar hydrography (ALH) when used principally for nautical charting [3].

The use of maritime vessels capable of carrying out bathymetric measurements is limited by the depth of the waters, so only small crafts are suitable in shallow waters. Because of their reduced dimensions, these vessels are not manned and are categorized as USVs (Unmanned Surface Vehicles) [4,5]. By analogy with avionics applications, they are also called marine drones [6]. Some such drones are also known as Autonomous Surface Crafts (ASCs) and Remotely Operated Vehicles (ROVs). According to [7], ASCs, also called autonomous surface vehicle (ASVs), are a kind of autonomous marine vehicle without the direct operation of humans, while ROVs are controlled by an operator who is not on-board. However, this distinction is not always observed and the terms are sometimes used with no difference in meaning.

In the last few years several specific crafts have been built for surveying in shallow waters, as reported in the literature.

In June 2006, the US Geological Survey Woods Hole Science Center (WHSC) integrated an ASV for hydrographic surveys in shallow waters (1–5 m), which was designed to map seafloor morphology and surficial sediment distribution and thickness. Named the Independently (or) Remotely Influenced Surveyor (IRIS) and designed as a catamaran-based platform (10 feet in length, 4 feet in width, and approximately 260 lbs in weight), this vehicle is equipped with a chirp dual-frequency side scan sonar (100/500 kHz) and seismic-reflection profiler (4–24 kHz), a wireless video camera and single-beam echosounder (235 kHz). IRIS is operated remotely through a wireless modem network enabling the real-time monitoring of data acquisition and navigated using RTK [8].

The ROAZ unmanned surface vehicle was proposed by the Autonomous Systems Laboratory (ASL) from Porto Polytechnic Institute (ISEP) for marine operations. It was designed to work in very shallow rivers and marine coastlines. Because of the possibility of transmitting the entire data collection on-board a base station, the operator receives online feedback on the vehicle's location and performance, as well as side-scan sonar imagery and bathymetry quality [9].

Another example of a craft used for bathymetric surveys in shallow water was developed by the Underwater Robotic Research Group's (URRG) who developed the URG—ASV, a battery-powered vessel [10].

CatOne is an example of a catamaran-robot that can operate in very shallow waters as well as in sensitive ecosystems because of its very low draft and an electric propulsion that guarantees zero pollution emission and low noise. It carries sonar and GPS on-board and can be equipped with other sensors to support different activities such as environment monitoring [11].

The purpose of this research was to create a marine drone, optimized for surveys in very shallow water, and benefitting from previous experiences in this field as noted above. The innovation of this project is twofold. First, the data and video are broadcast directly to several operators, enabling the visualization and the pre-processing of all data in real time, by means of several devices managed by experts from different disciplines (such as an archaeologist, a geophysicist, a topographer or a GIS expert). This feature was implemented in order to carry out interdisciplinary surveys in critical coastal areas In fact, in the two study cases (both in the Gulf of Naples) there are submerged archaeological remains in the survey area. Thus, each expert can verify that the data acquisition is correct from his/her point of view. In addition, in order to also obtain high precision bathymetric data in critical areas, a system of data quality control was implemented, using an inertial platform.

2. Experimental Section

The MicroVeGA drone is an Open Project of USV conceived, designed and built to operate in shallow water areas (0–20 m), where a traditional boat is poorly manoeuvrable. It was engineered by the DIST research group at the University of Naples and was designed to test the procedures and methods of morpho-bathymetric surveys in critical areas. In [7], the initial development phase of MicroVeGA is described.

The drone is a small and ultra-light catamaran that can be assembled in 30 min, with a few draught centimetres, therefore suitable to perform surveys up to the shoreline. It is driven by non-polluting electric motors, and is therefore suitable to perform surveys in marine protected areas. Table 1 lists the characteristics.

Table 1. Technical and physical characteristics of the drone.

Characteristics	Measures
Overall length	135 cm
Width	85 cm
Weight in navigation trim	20 kg
Motors	2 brushless 750 kV/140 W
Operating speed	0.5–2 m/s
Power autonomy	2–4 h

MicroVeGA is an evolving open project that has enabled surveys to be carried out already in its early stage of development (Figure 1). In this paper, two study cases are illustrated, the first created with the MicroVeGA prototype #1 (Figure 1b), the second with the MicroVeGA prototype #2 (Figure 1c).

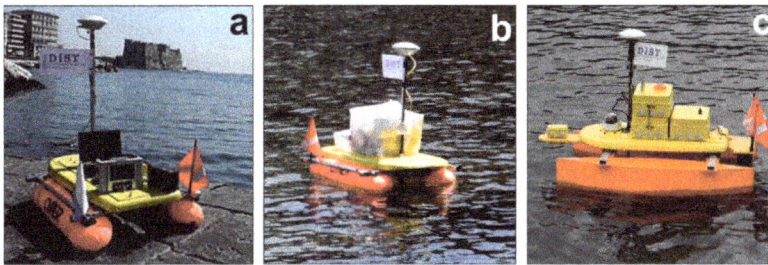

Figure 1. (**a**) Instruments on board of MicroVeGA; (**b**) Prototype #1 of MicroVeGA; (**c**) Prototype #2 of MicroVeGA.

This project is a low risk technology project. The spiral model of project management [12] is divided into smaller sections (Figure 2). Each prototype requires the following phases: requirements, design and refine, build; test, survey and analyse.

The current version of MicroVeGA (*i.e.*, Prototype #2) is remotely controlled by an operator and is equipped with a set of sensors for acquiring morpho-bathymetric high-precision data (see Section 2.2).

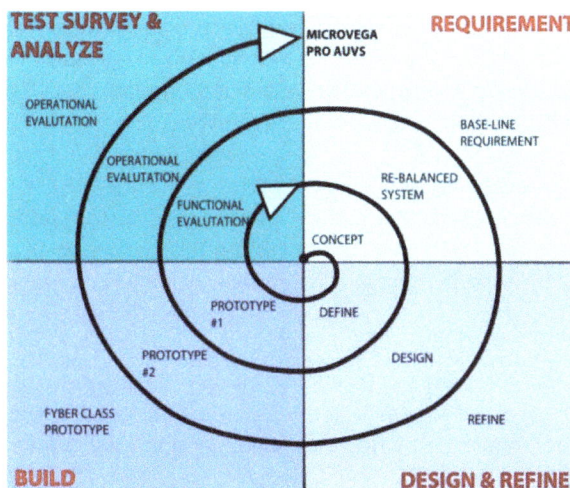

Figure 2. Spiral model of the project management.

2.1. System Architecture

The architecture of the data acquisition system (Figure 3) includes: (i) a base station, with a remote controlled PC and a video terminal; (ii) an on-board computerized system that manages the on-board instrumentation; (iii) a communication system via data link, to connect the UVS with the base station.

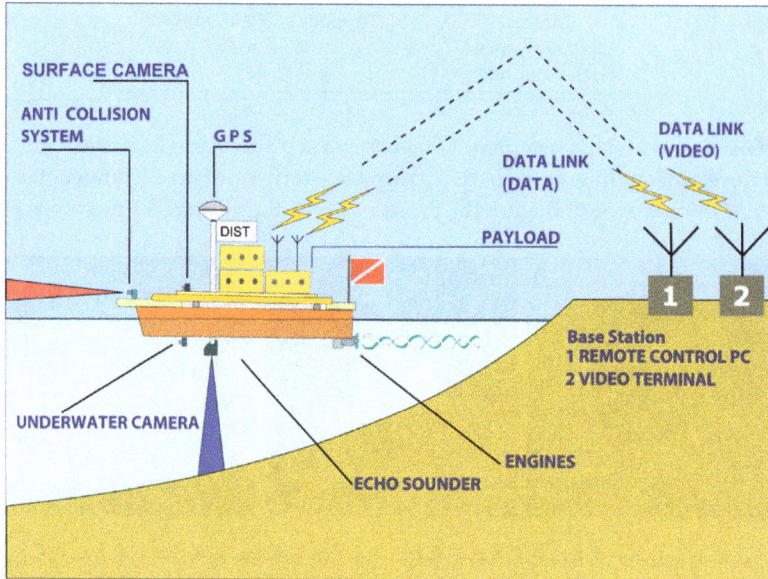

Figure 3. Data acquisition architecture: a base station, with a remote control PC and a video terminal; an on board computerized system that manages the on-board instrumentation; a communication system via data link, to connect the AUVS with the base station.

The operator responsible for the base station manages the mission data by means of TrackStar software by defining the navigation routes and monitoring the mission progress. The TrackStar software (described in Section 2.3 Data Acquisition and Software), implemented by our research group, manages the survey activities and automatically creates a measurement geodatabase.

The data is stored on board in RAW format by a computerized system that acquires and organizes the GPS, echo sounder, inertial platform, and obstacle-detection sensor data. This data is broadcast to the base station by a data link system, after which several operators can simultaneously receive the data in real time.

MicroVeGA data transmission is based on two wireless networks. The first transmits the telemetry data (*i.e.*, position, depth, atmospheric temperature and obstacle detection) from the vessel to Trackstar. The second network transmits the videos of the two on-board cameras to the base station. This information is managed by a specific app, and the images are viewed on a tablet in real time.

2.2. Sensors and Methods for Data Acquisition

The main instruments on-board are: (1) microcomputer; (2) differential GPS system and Single beam echo sounder; (3) integrated system for attitude control; (4) obstacle-detection system (SIROS1) with temperature control system; (5) video acquisition system (both above and below sea level) (Figure 4).

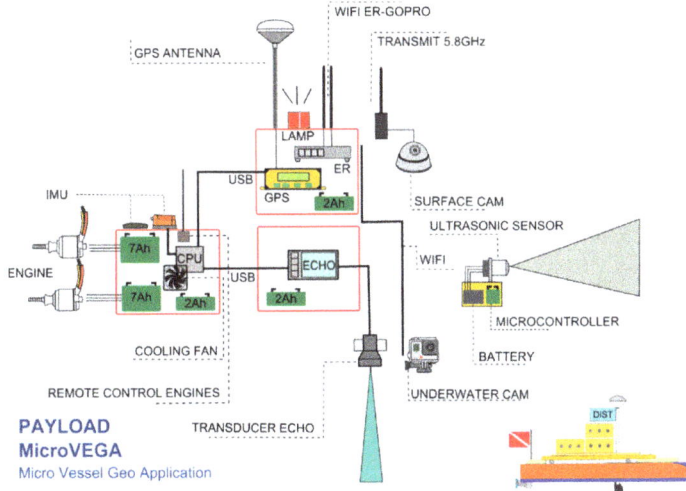

Figure 4. Payload of MicroVeGA drone.

2.2.1. Microcomputer

An OLinuXino microcomputer, with a Linux operating system, and three high-speed serials, manages all the survey phases, the data recording and its wi-fi transmission to the base station. An Arduino microcontroller controls the drone's engines, the temperature measurements, and the management of the obstacle-detection ultrasound systems (Figure 4).

2.2.2. GPS and Single Beam Echo Sounder (SBES)

The GPS receiver (Figure 5b), installed on board MicroVeGA, is the Trimble DSM™ 232 (24-channel L1/L2), which is a robust solution for dynamic positioning tasks in the marine environment. In fact, this device is easily installed and is able to withstand tough environmental conditions, and is thus suitable for surveys in very shallow waters. In addition, the GPS receiver and antenna are modular, and thus it was possible to install on board of MicroVeGA, the antenna vertically with respect to the SBES transducer.

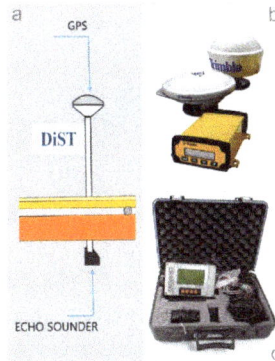

Figure 5. (a) Installation positions of the GPS and echo sounder on drone; **(b)** Trimble DSM232 GPS; **(c)** Omex Sonarlite echo sounder.

The Trimble DSM 232 GPS receiver enables the appropriate GPS correction method and accuracy to be selected. In this research, the DGPS option in post-processing was used, using Trimble software.

The SonarLite (Omex) is the SBES installed on-board (Figure 5c). This instrument is optimized for the bathymetric survey in shallow waters, and its transducer is positioned vertically above the GPS receiver in order to remove any offset (Figure 5a).

2.2.3. Inertial Platform Unit (IMU)

The inertial measurement unit used for measuring balance and direction on board of MicroVeGA is the Xsense MTi series G. This device is an integrated GPS and MEMS IMU with a Navigation and Attitude and Heading Reference System processor. It was used on the MicroVeGA drone because of it weighs very little.

The internal low-power signal processor runs a real-time Xsens Kalman Filter (XKF), providing inertial enhanced 3D position and velocity estimates [13,14].

The IMU data are stored in the survey geodatabase and increase the accuracy of the survey since measurements affected by attitude errors are removed [15]. In the case of errors due to pitch and roll, a quality control system that removes all measurements higher than a specific limit d was implemented (Figure 6):

$$d \leqslant spp \tag{1}$$

where:

$$d = Z' \sin \beta \tag{2}$$

and Z' = echo sounder measurement; spp = survey parameter precision related to survey scale, depth, survey target; β = angle between Z and $Z' = 90° - (90° - \alpha)$; α = pitch or roll.

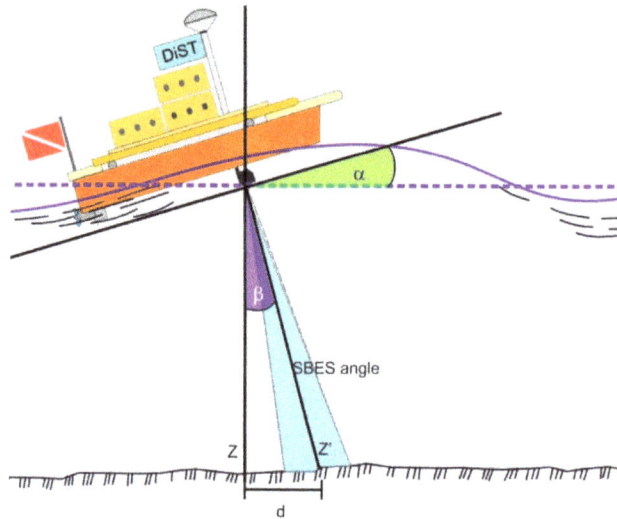

Figure 6. Horizontal error due to pitch or roll.

In the planning phase of the survey, the operator can establish the value of the spp survey parameter, thus defining the horizontal limit d that makes a measurement valid.

In both surveys described below (archaeological survey with rocky seabed and a cartographic scale of 1:1000), the threshold value spp was set equal to 1. As shown in Figure 7, if the measured depth increases, the roll angle becomes even more critical. In fact the same angle of roll (or pitch), equal to 10°, is associated with a valid measurement if the depth is −5 m, while it is associated with an invalid measurement if the depth is greater than −7 m (see also Table 2).

As the weather and sea conditions are essential for the proper execution of a bathymetric survey, surveys are not normally carried out when waves are beyond a certain strength. The validation system is primarily to prevent the storage of incorrect data due to occasional events, such as the passage of a vessel, and thus to improve the quality level of the whole survey.

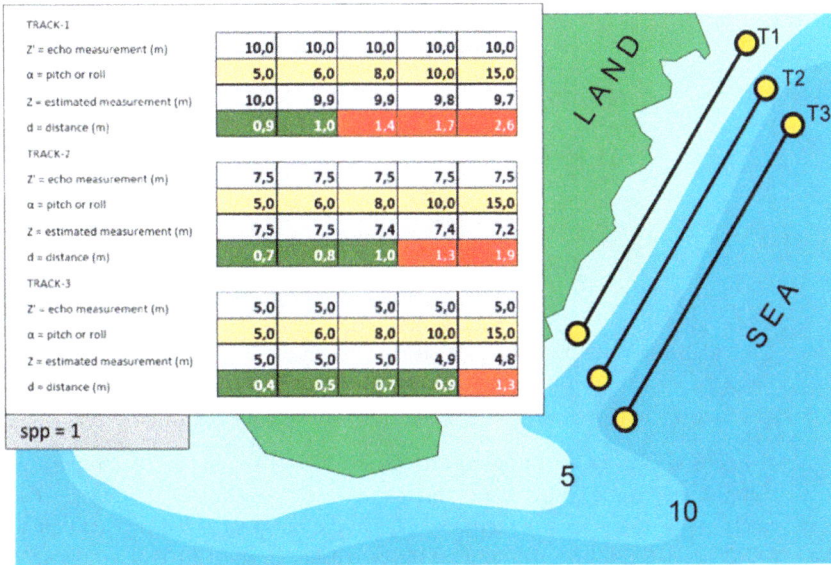

Figure 7. Example of three data acquisitions with spp = 1 constant and with α and Z′ variables.

Table 2. Variation of the distance d with the changing depth (see also Figure 7).

Measurement Parameters	T1	T2	T3
Z′ = Echo Measurement (m)	5.0	7.5	10.0
α = Pitch (or roll)	10.0	10.0	10.0
Z = Estimated Measurement (m)	4.9	7.4	9.8
d = distance (m)	0.9	1.3	1.7

The mission software—Trackstar—manages these calculations in real time highlighting the invalid measurements with a special color scale. This visualization allows the operator to evaluate the areal coverage of the survey, and to decide the possible repetition of a navigation line in real time.

The IMU data are also used to correct the depth with respect to the vertical error due to the wave effect (Figure 8):

$$CWL = Z \pm dZ \tag{3}$$

where: CWL = clam water level; Z = depth measured by SBES; dZ = vertical error measured by inertial platform.

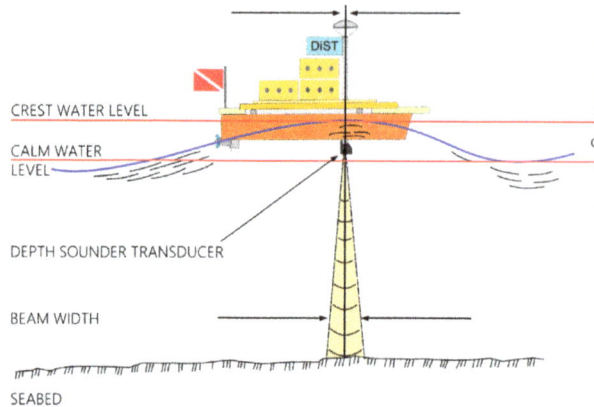

Figure 8. Vertical error in depth measuring due to the wave effect.

2.2.4. SIROS 1 (Obstacle-Detection System—In Italian: Sistema Rilevamento OStacoli)

The system is based on: an Arduino controller; an ultrasonic sensor; a temperature sensor; a servomechanism; an electronic component; and a software application. The main sensor used is the HY-SR05, which is able to detect emerged obstacles in the range of 2–450 cm, with an accuracy of 0.2 cm. The HY-SR05 uses a single output pin on the controller to send a trigger pulse to the sensor, and then another input pin to receive the pulse indicating the object's distance (Figure 9a).

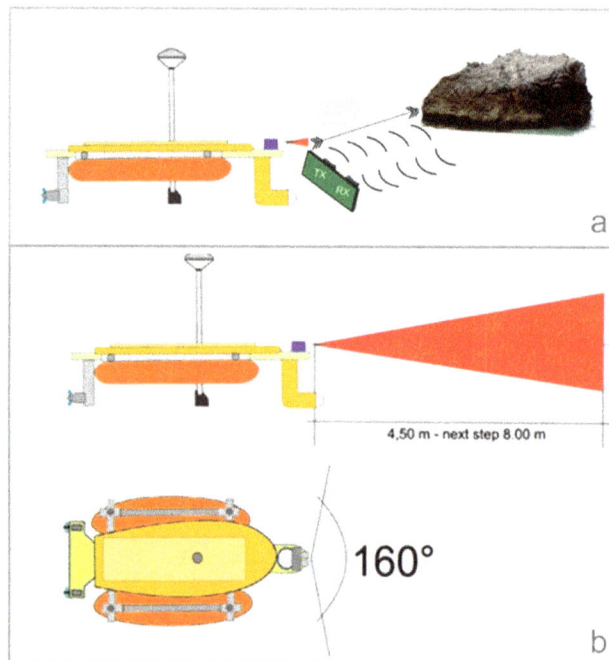

Figure 9. (**a**) Operation of the ultrasonic sensor; (**b**) Action range of obstacle-detection system.

Using a servomechanism, the obstacle-detection system can scan a prow sector of about 160° (Figure 9b). The software controls the ultrasonic sensor and using the servomechanism moves the azimuth of the same sensor in steps of 5°.

The distance detection is a function of the air temperature, and the obstacle-detection system is equipped with a temperature sensor (LM35) that compensates for temperature variations with an accuracy of ± 0.5 °C, making the obstacle-detection system more efficient.

According to the Laplace law, in the case of the air the speed of sound increases by 0.6 m/s for each increase of 1 °C air temperature:

$$v = 331.3 \frac{m}{s} + 0.606\ T_i \tag{4}$$

where v = sound velocity in the air; 331.5 m/s = the sound velocity at 0 °C; T_i = Air temperature value in a specified measure time.

Table 3 demonstrates the increasing accuracy of the measurements (dD column), by a comparison between the distance measured at the standard temperature of 20 °C (columns V1 and D1), and the distance measured at the current temperature (columns V2 (t) and D2), showing how the variation in the temperature influences the measured values.

Table 3. Comparison table between the distance measured at the standard temperature of 20 °C (columns V1 and D1), and the distance measured at the actual temperature (columns V2 (t) and D2).

V1 (20°) m/s	T (°C)	V2 (T) m/s	Time (s)	D1 (cm)	D2 (cm)	dD (cm)
343.4	5	334.3	0.010	171.7	167.2	4.5
343.4	10	337.4	0.010	171.7	168.7	3.0
343.4	20	343.4	0.010	171.7	171.7	0.0
343.4	30	349.5	0.010	171.7	174.7	−3.0

The obstacle-detection system, along with the camera's surface, is very useful when there are obstacles, such as scattered rocks, that are not marked on the cartography. This system enables the operator to navigate up to a few centimeters from the docks and piers, and thus is very useful in bathymetric surveys carried out in ports and harbours.

SIROS becomes active when the distance from an obstacle is <400 cm. As soon as this happens, TrackStar displays the progressive distances of the obstacle, thus alerting the operator. Normally the operator decreases the speed and, if necessary changes route. An operator controls the MicroVeGA drone remotely, and thus there are no automatic collision avoidance maneuvers. The only automatic actions of the system are:

- activate alarm visual and sound software management and control,
- activate flashing and sirens on board.

Especially in critical cases, the software automatically stops the engines and activates an alarm (go home command) to warn the operator about the need to stop the mission. Finally, SIROS 1 has a safety navigation system to support the operator in making browsing simpler, safer and fast.

2.2.5. Video Acquisition System

MicroVeGA has a complete system for video data acquisition, above and below sea level. Two GO PRO HERO 3 cameras are installed on-board, one above the water level and the other below. The cameras make a video recording during the whole survey, enabling the operator to check the environmental conditions and to manage the presence of obstacles in real-time. Video data is transmitted to the base station and is recorded on a hard disk.

For performance testing, two methods for transmitting video data from on board to the base station were used. One uses the wi-fi on board a GoPRO camera that (thanks to the Extended Range WiFi positioned on the MicroVeGA) transmits data to the shore. Here any wi-fi device (smart phone, tablet, or PC) can view content in realtime thanks to the app supplied with the GoPRO. The second method uses a 5.8 GHz 100 mW 8 channel video transmitter along with a RC805 5.8 GHz

AV Receiver. A small LCD, connected to the receiver, displays real-time video. In the next version of the drone, the second solution will be used, *i.e.*, without the Go-Pro wi-fi, as this will ensure low weight, the high flow rate, and the availability of more transmission channels.

2.3. Data Acquisition and Software

The TrackStar software, developed by our research group, manages the survey activities and automatically creates a measurement geodatabase.

The software displays in real time (Figure 10): the GPS navigation; the deviation of the vessel from the planned line; the SBES bathymetric measurements along the navigation line (bathymetric profiles); the distance from a detected obstacle; and the IMU measurements (pitch, roll, yaw and altitude).

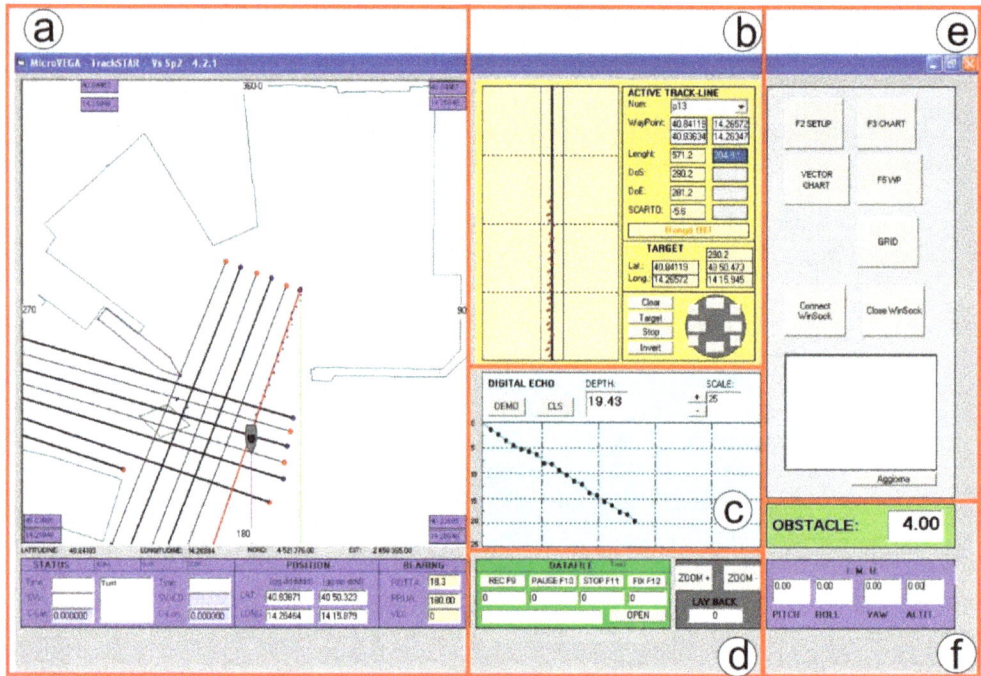

Figure 10. Trackstar desktop: (**a**) real time navigation; (**b**) deviation of the vessel from the planned line; (**c**) real time bathymetric profile; (**d**) real time data recording and datafile creation; (**e**) import of cartography and creation of navigation lines; (**f**) obstacle distance and attitude measurements.

The software also displays the data read from the IMU and, near to an emerged obstacle, shows the distance from the obstacle to the drone, thus facilitating the remote control of operations by the operator. All data from GPS, SBES and IMU are stored in a single datafile in ASCII format. The software was developed in Windows.

3. Results and Discussion

This section describes two cases of the MicroVeGA survey. The main characteristic of these areas is the coastal physiography that prevents any bathymetric surveys with traditional boats. There are also submerged archaeological remains that produce rapid changes in depth values.

The morpho-bathymetric survey carried out in each area was planned in order to obtain a GIS 3D model of the sea floor. The interpolation method used in the post-processing phase was the Inverse Distance Weighted (IDW) interpolation. This interpolator is one of the simplest and most readily available methods for interpolation. It is based on an assumption that the value at an unsampled point can be approximated as a weighted average of values at points within a certain cut-off distance, or from a given number of the closest points [16].

3.1. MicroVeGA Survey 1

The first bathymetric survey of MicroVeGA drone was carried out along the Sorrento Marina Grande coast in the nearshore area (0–3 m depths) between the tufa cliff and coastal protection works using Prototype #1. In Prototype #1 of the drone, the instruments were all contained in a plexiglas non waterproof case and the hulls of the catamaran consisted of two float tubes.

The navigation of the bathymetric survey (Figure 11) had a linear development of about 500 m, with a distance between the navigation lines of about 2 m. In the first instance, the positioning and the morphologic reconstruction were obtained of all the archaeological remains in the area [17], using the GPS, SBES and submerged camera.

Figure 11. (a) MicroVeGA drone prototype #1 used during the survey; (b) navigation lines of bathymetric survey in blue and position of archaeological targets located by submerged camera and SBES in red.

In addition, 3D data were processed in the ARCGIS environment, using 3D Analyst. The interpolation of the bathymetric data, through the IDW interpolator, transformed the point measurements into continuous measurements. The final product is a seafloor digital model of the area (Figure 12).

3.2. MicroVeGA Survey 2

The second bathymetric survey of MicroVeGA drone was carried out along the Posillipo Hill (Naples, Italy) coast in the nearshore area (0–10 m depths) of Marechiaro harbour, using Prototype #2. The instruments on board the second prototype were completely contained in a waterproof case and the obstacle-detection system was installed on a waterproofed wooden support on the drone bow, in addition, the catamaran's hulls were made of marine plywood (Figure 13), which widened the hull and lengthened the bearing surfaces side, thus increasing the stability of the drone in navigation by decreasing the pitch and roll movements (Table 4). In addition, the largest volume of the hulls, increasing the displacement, helped to improve the available payload (Table 4).

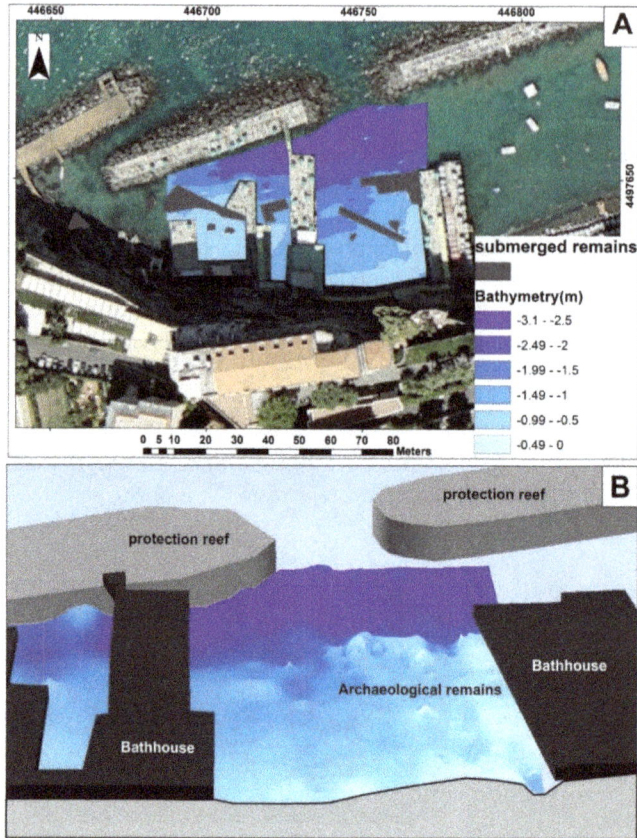

Figure 12. (**a**) 2D visualization of the sea floor digital model of the study area—Sorrento Marina Grande (Naples, Italy); (**b**) 3D visualization of the same sea floor.

Figure 13. MicroVeGA in action.

The transverse stability of the hull, in a catamaran like MicroVeGA, increases with the increase in the bearing surface on the water. In fact, while the longitudinal stability counteracts the pitching movements (the "fluctuations" of the vessel from bow to stern), the transverse stability counteracts the rolling motion (the lateral "oscillations" of the vessel). MicroVeGA can be approximated to a rectangular water plane, and the transversal (j_x) and longitudinal (j_y) moments of inertia, as shown in Figure 14, can be calculated as being equal to [18]:

$$j_x = \frac{a \cdot b^3}{12} \tag{5}$$

$$j_y = \frac{b \cdot a^3}{12} \tag{6}$$

Therefore in this version, the increase in the transverse and longitudinal stability increased the navigation safety (Table 4).

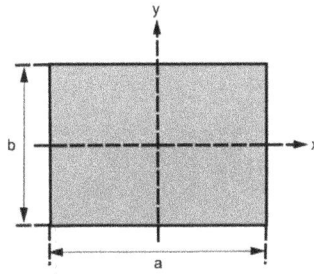

Figure 14. Schema of a rectangular vessel.

Table 4. Comparison between physical characteristics of Prototypes #1 and #2.

Prototype	Width (cm)	Length (cm)	J_x	J_y	Payload (kg)
MicroVeGA #1	72	92	0.029	0.047	12
MicroVeGA #2	86	120	0.064	0.124	22

The site of the second survey, was a port in the 1st century AD and several remains of a dock [19] are still present (red dashed line in Figure 15). MicroVeGA passed over these remains thanks to a few centimeters of draught.

The navigation of the bathymetric survey (Figure 14b) had a linear development of about 1500 m, with a distance between the navigation lines of about 5 m. In this survey, the tool that manages the inertial platform measurements eliminated 10% of depth measurement, due to the transition of some boats during the survey.

3D data were processed in ARCGIS, using the Geostatistical Analysis tool. The interpolation of the bathymetric data, through the IDW interpolator, transformed the point measurements into continuous measurements. The final product is a sea floor digital model of the area (Figure 16).

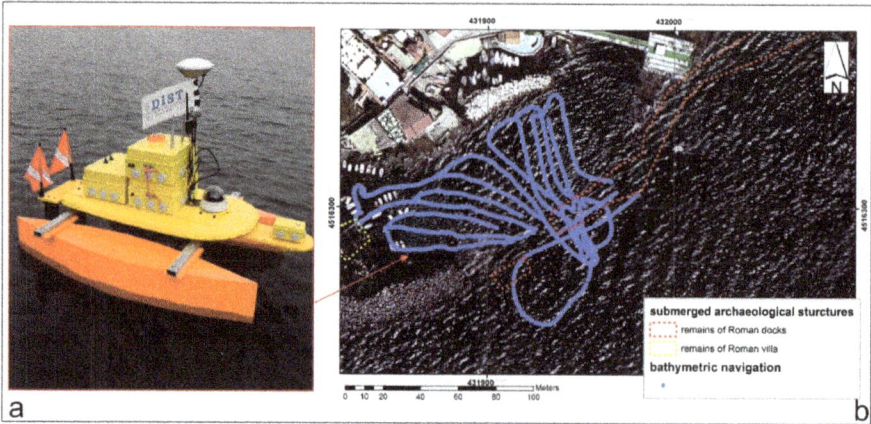

Figure 15. (a) MicroVeGA drone prototype #2 used during the survey; (b) navigation lines of bathymetric survey in blue and submerged archaeological structures in red.

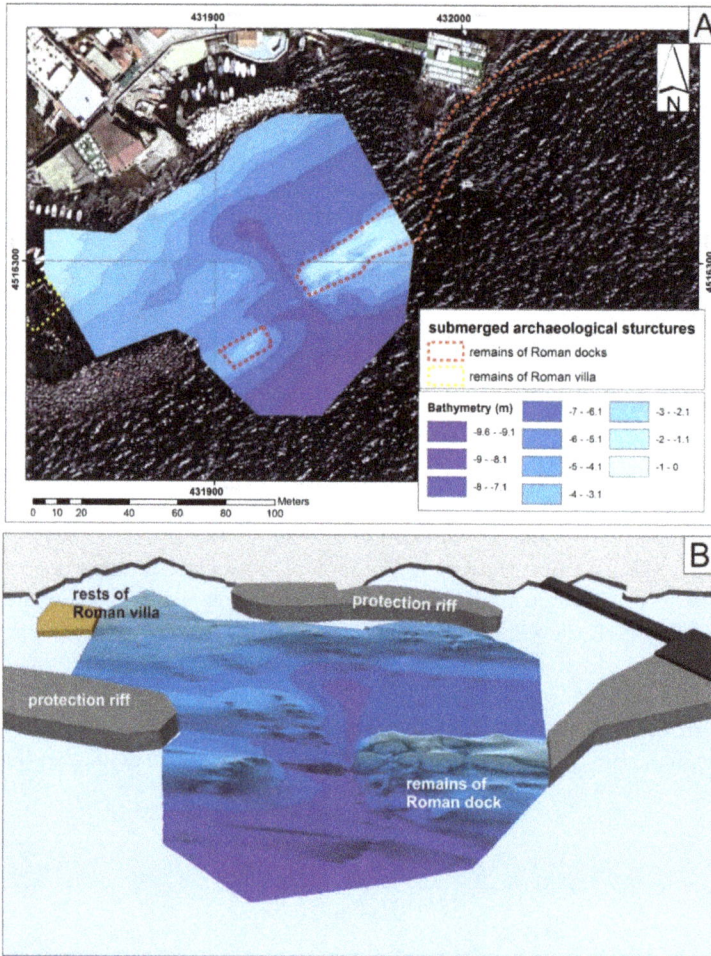

Figure 16. (a) 2D visualization of the sea floor digital model of the study area—Marechiaro harbour along Posillipo Hill (Naples, Italy); (b) 3D visualization of the same sea floor.

4. Conclusions

We have described a prototype of a marine drone optimized for very shallow water, which enables bathymetric surveys to be performed in areas that are not feasible for traditional boats. In the two study cases described in this paper, the various underwater structures would have created many navigation difficulties, if MicroVeGA had not had only a few centimeters of draught.

The experiments performed in the two coastal sites showed that integrating several existing technologies improved the final performance and the quality of the acquired data. The development of a specific software application (Trackstar) improved the accuracy of all the measured data, thus increasing the instruments' performance.

Trackstar improves survey accuracy using the inertial platform which extended the survey duration but always guaranteed a high quality control of measurements. In fact, during the planning phase of the survey, we established the survey precision parameter ssp as a function of survey scale, depth and survey target, reducing the attitude errors, as demonstrated in the Marechiaro survey where the effect of the sailing boats was deleted. The control of the speed and the possibility of navigating at a reduced speed also ensured a greater measurement accuracy.

In addition, the safety performance of the operation was improved by integrating the temperature sensor with the ultrasonic sensor, thus increasing the accuracy in the measurements of the distance from the obstacles, as demonstrated in Table 4.

Another important characteristic of this project is the low technology risk philosophy, guaranteed by the spiral model used to manage the drone construction phases. In fact, we had carried out a bathymetric survey in the Sorrento Marina Grande site, already using the first prototype.

Prototypes #1 and #2 provide the basic requirements of practicality and economy. Practicality is clear from the ease of performing the measurements (small footprint, highly portable, ultra lightweight and easy manoeuvrability). Low costs were achieved by assembling and integrating existing systems.

Finally, MicroVeGA is equipped not only with bathymetric sensor but also with an underwater camera which provides an overview of the investigated seabed and the surrounding underwater environment.

In the next (*i.e.*, third) phase of this project, the experience obtained in the current development phases will be used to design morpho—bathymetric surveys in critical areas. Future plans include new survey strategies and an industrial mock up in fiberglass (Figure 17).

Figure 17. An industrial mock up in fiberglass.

Acknowledgments: This paper brings together the results of works performed within the project PRIN 2010-11 financed by MIUR (Ministero dell'Istruzione, dell'Università e della Ricerca) and developed at the University of Naples "Parthenope" (Coordinator: Raffaele Santamaria). The authors would like to thank Alberto Greco, Ferdinando Sposito and PietroRossi for their active collaboration in the engineering of the drone and in all phases of the marine surveys.

Author Contributions: Francesco Giordano took part in the engineering of the drone and coordinated the marine surveys, took part in writing the paper; Gaia Mattei took part in the engineering of the drone and in the marine surveys; she carried out the planning of marine surveys and the data processing, and took part in writing the paper; Francesco Peluso engineered the marine drone and took part in the marine surveys and in writing the paper; Claudio Parente conducted the bibliographic research and supervised the planning of marine surveys and the data processing; he took part in writing the paper; Raffaele Santamaria supervised engineering of the drone, coordinated the marine surveys and supervised the data processing; took part in writing the paper.

Conflicts of Interest: The authors declare no conflict of interest.

References

1. Jawak, S.D.; Vadlamani, S.S.; Luis, A.J. A Synoptic Review on Deriving Bathymetry Information Using Remote Sensing Technologies: Models, Methods and Comparisons. *Adv. Remote Sens.* **2015**, *4*, 147–162. [CrossRef]

2. Quadros, N.D. CRSI UDEM2 Project4 Report Stage 2, Final Draft, 2013.   Available online: http://www.crcsi.com.au/assets/Resources/695d2af9-e397-4029-8cad-6dfe5510d581.pdf (accessed on 24 December 2015).

3. Guenther, G.C.; Cunningham, A.G.; LaRocque, P.E.; Reid, D.J. Meeting the Accuracy Challenge in Airborne Lidar Bathymetry. In Proceedings of the EARSeL-SIG-Workshop LIDAR, Dresden/FRG, Dresden, Germany, 16–17 June 2000; pp. 16–17.

4. Caccia, M.; Bibuli, M.; Bono, R.; Bruzzone, G.; Bruzzone, G.; Spirandelli, E. Unmanned surface vehicle for coastal and protected waters applications: The Charlie project. *Mar. Technol. Soc. J.* **2007**, *41*, 62–71. [CrossRef]

5. Bertram, V. Unmanned Surface Vehicles—A Survey. Available online: http://www.skibstekniskselskab.dk/public/dokumenter/Skibsteknisk/Download%20materiale/2008/10%20marts%2008/USVsurvey_DTU.pdf (accessed on 24 December 2015).

6. Giordano, F.; Mattei, G.; Parente, C.; Peluso, F.; Santamaria, R. MicroVeGA (micro vessel for geodetics application): A marine drone for the acquisition of bathymetric data for GIS applications. *ISPRS Int. Arch. Photogramm. Remote Sens. Spat. Inf. Sci.* **2015**, *1*, 123–130. [CrossRef]

7. Zhao, J.; Yan, W.; Jin, X. Brief review of autonomous surface crafts. *ICIC Expr. Lett.* **2011**, *5*, 4381–4386.

8. USGS Woods Hole Coastal and Marine Science Center (2014), IRIS—ASV (Autonomous Surface Vessel). Available online: http://woodshole.er.usgs.gov/operations/sfmapping/iris.htm (accessed on 24 December 2015).

9. Ferreira, H.; Almeida, C.; Martins, A.; Almeida, J.; Dias, N.; Dias, A.; Silva, E. Autonomous Bathymetry for Risk Assessment with ROAZ Robotic Surface Vehicle. In Proceedings of the IEEE Oceans 2009-Europe, Bremen, Germany, 11–14 May 2009; pp. 1–6.

10. Hassan, S.R.; Zakaria, M.; Arshad, M.R.; Aziz, Z.A. Evaluation of Propulsion System Used in URRG-Autonomous Surface Vessel (ASV). *Procedia Eng.* **2012**, *41*, 607–613. [CrossRef]

11. Romano, A.; Duranti, P. Autonomous Unmanned Surface Vessels for Hydrographic Measurement and Environmental Monitoring. In Proceedings of the FIG Working Week 2012, Knowing to Manage the Territory, Protect the Environment, Evaluate the Cultural Heritage, Rome, Italy, 6–10 May 2012; pp. 1–15.

12. Boehm, B.; Hansen, W.J. *Spiral Development: Experience, Principles, and Refinements*; (No. CMU/SEI-2000-SR-008); Software Engineering Institute: Pittsburgh, PA, USA, 2000; pp. 1–35.

13. Li, W.; Wang, J. Effective Adaptive Kalman Filter for MEMS-IMU/Magnetometers Integrated Attitude and Heading Reference Systems. *J. Navig.* **2013**, *66*, 99–113. [CrossRef]

14. Barshan, B.; Durrant-Whyte, H.F. Inertial navigation systems for mobile robots. *IEEE Trans. Robot. Autom.* **1995**, *11*, 328–342. [CrossRef]

15. Downing, G.C.; Fagerburg, T.L. *Evaluation of Vertical Motion Sensors for Potential Application to Heave Correction in Corps Hydrographic Surveys*; Army Engineer Waterways Experiment Station Vicksburg Ms Hydraulics Lab: Vicksburg, MS, USA, 1987; pp. 1–86.

16. Mitas, L.; Mitasova, H. Spatial interpolation. Geographical information systems: Principles, techniques. *Manag. Appl.* **1999**, *1*, 481–492.

17. Mingazzini, P. *Forma Italiae: Latium et Campania*; De Luca Ed: Surrentum, Italy, 1946.

18. Lodigiani, P. *Capire e Progettare Le Barche. Manuale Per Progettisti Nautici—Aero e Idrodinamica Della Barca a Vela*; Hoepli Editore: Milano, Italy, 2015.

19. Günther, R.T. *Posillipo Romana*; Electa: Napoli, Italy, 1908.

9

PIMR: Parallel and Integrated Matching for Raw Data

Zhenghao Li [1,2,*], Junying Yang [1], Jiaduo Zhao [1], Peng Han [2] and Zhi Chai [3]

Academic Editor: Vittorio M. N. Passaro

[1] Key Laboratory for Optoelectronic Technology and Systems of Ministry of Education,
    College of Optoelectronic Engineering, Chongqing University, Chongqing 400044, China;
    yangjunying_90@cqu.edu.cn (J.Y.); jdzhao@cqu.edu.cn (J.Z.)
[2] Chongqing Academy of Science and Technology, Chongqing 401123, China; hanpeng@cqu.edu.cn
[3] Beijing Institute of Environmental Features, Beijing 100854, China; ezhchai@163.com
* Correspondence: lizhenghao@cqu.edu.cn

Abstract: With the trend of high-resolution imaging, computational costs of image matching have substantially increased. In order to find the compromise between accuracy and computation in real-time applications, we bring forward a fast and robust matching algorithm, named parallel and integrated matching for raw data (PIMR). This algorithm not only effectively utilizes the color information of raw data, but also designs a parallel and integrated framework to shorten the time-cost in the demosaicing stage. Experiments show that compared to existing state-of-the-art methods, the proposed algorithm yields a comparable recognition rate, while the total time-cost of imaging and matching is significantly reduced.

Keywords: image sensor; raw data; image matching; image analysis

1. Introduction

Image matching is a crucial technique with many practical applications in computer vision, including panorama stitching [1], remote sensing [2], intelligent video surveillance [3] and pathological disease detection [4].

Hu's moment invariants as a shape feature has been widely used for image description due to its scaling and rotation invariance [5]. Wang *et al.* further proposed a two-step approach used for pathological brain detection by employing this feature combined with wavelet entropy [6]. Lowe's scale-invariant feature transform (SIFT) is a *de facto* standard for matching, on account of its excellent performance, which is invariant to a variety of common image transformations [7]. Bay's speeded up robust features (SURF) is another outstanding method performing approximately as well as SIFT with lower computational cost [8]. We proposed a lightweight approach with the name of region-restricted rapid keypoint registration (R^3KR), which makes use of a 12-dimensional orientation descriptor and a two-stage strategy to further reduce the computational cost [9]. However, it is still computationally expensive for real-time applications.

Recently, many efforts have been made to enhance the efficiency of matching by employing binary descriptors instead of floating-point ones. Binary robust independent elementary features (BRIEF) is a representative example which directly computes the descriptor bit-stream quite fast, based on simple intensity difference tests in a smoothed patch [10]. When combined with a fast keypoint detector, such as features from accelerated segment test (FAST) [11] or center surround extrema (CenSurE) [12], the method provides a better alternative for real-time applications. Despite the efficiency and robustness to image blur and illumination change, the approach is very sensitive to rotation and scale changes. Rublee *et al.* further proposed oriented FAST and rotated BRIEF (ORB) on the basis of BRIEF [13]. The approach acquires multi-scale FAST keypoints using a pyramid scheme, and computes the orientation of the keypoints utilizing intensity centroid [14]; thus, the descriptor is rotation- and scale-invariant. In addition, it also uses a learning method to obtain binary tests with lower correlation, so that the descriptor becomes more discriminative accordingly. Some researchers also try to increase the robustness of matching by improving the sampling pattern for descriptors. The binary robust invariant scalable keypoints (BRISK) method proposed by Leutenegger adopts a circular pattern with 60 sampling points, of which the long-distance pairs are used for computing the orientation and the short-distance ones for building descriptors [15]. Alahi's fast retina keypoint (FREAK) is another typical one leveraging a novel retina sampling pattern inspired by the human visual system [16]. Leutenegger uses a scale-space FAST-based detector in BRISK to cope with the scale invariance, which is employed by FREAK as well.

However, several inherent problems remain in existing methods. The digital image which consists of 24 bit blue/green/red (BGR) data is color-interpolated, which is known as demosaicing from raw data, including adjustment for saturation, sharpness and contrast, and sometimes compression for transmission [17,18]. This operation leads to irreversible information loss and quality degeneration. Furthermore, up until now, conventional image matching has been implemented after demosaicing, and such sequential operation restricts its application to general tasks. To address these limitations, in this paper we introduce an ultra-fast and robust algorithm, coined parallel and integrated matching for raw data (PIMR). The approach takes raw data instead of a digital image as the object for analysis, which is efficient for preventing information from being tampered with artificially. It is crucial to obtain high-quality features with the result that the approach can achieve comparable precision. Meanwhile, a parallel and integrated framework is employed to accelerate the entire image matching, in which two cores are used to respectively process the matching and demosaicing stages in parallel. Our experiments demonstrate that the proposed method can acquire more robust matches in most cases, even though it is much less time-consuming than traditional sequential image matching algorithms, such as BRIEF, ORB, BRISK and FREAK.

The rest of the paper is organized as follows. Section 2 gives the implementation details of the proposed method, which mainly includes the raw data reconstruction, and the parallel and integrated framework. In Section 3, we evaluate the performance of PIMR. Lastly, in Section 4, conclusions are presented.

2. Parallel and Integrated Matching for Raw Data

In Section 2.1, we give a brief account of raw data to make the reader understand our work more clearly. The key steps in PIMR are explained in Sections 2.2 and 2.3 namely the reconstruction step for raw data and the details of the parallel and integrated framework.

2.1. Raw Data

Raw data, which is the unprocessed digital output of an image sensor, represents an amount of electrical charges accumulated in each photographic unit of the sensor. The notable features of raw data are nondestructive white balance, lossless compression, and high bit depth (e.g., 16 bits), providing considerably wider dynamic range than the JPEG file [19–21]. Hence, raw data maximally retains the real information of the scene compared to the digital image. By directly analyzing raw data, not only does it prevent the original information from being tampered with artificially, which contributes to an increase in precision, but it also shortens the time-cost in the demosaicing part.

The most popular format of raw data is Bayer Tile Pattern [22,23], typically in GBRG mode. It is widely used in industrial digital cameras. As illustrated in Figure 1, it is basically a series of band pass filters that only allow certain colors of light through, which are red (R), green (G) and blue (B). Green detail, to which the human visual system is more responsive than blue and red, is sampled more frequently, and the number of G is twice that of B and R.

Figure 1. Bayer pattern of GBRG mode.

2.2. Reconstruction for Raw Data

At each pixel location, the sensor measures either the red, green, or blue value, and the values of other two are highly related to the neighbors around the pixel. On this basis, in order to make raw data more appropriate for further processing, we reconstruct it as follows. Initially, we define the 2×2 set of pixels as a cell. It starts with the target pixel $C(i, j)$, and the other three pixels belonging to the cell are $C(i + 1, j)$, $C(i, j + 1)$ and $C(i + 1, j + 1)$, respectively, where C represents an arbitrary color in G, B and R. Next, we utilize the color information in each cell to reconstruct the intensity I for the target pixel:

$$I_{MAX} = \max(G + G', B + R) \tag{1}$$

$$I_{MIN} = \min(G + G', B + R) \tag{2}$$

$$I = (w_1 I_{MAX} + w_2 I_{MIN})/2 \tag{3}$$

where G, G', B and R are the luminance values of each pixel in the defined cell, I_{MAX} and I_{MIN} are the maximum and minimum of the diagonal values in the cell, respectively, and w_1 and w_2 are the weight coefficients, which represent the contribution of diagonal color components to the cell. The greater the contribution, the larger the weight that will be assigned to the components. According to the results of multiple tests, $w_1 = 0.6$, $w_2 = 0.4$ in our method. All the cells in raw data are processed in this way.

In this procedure, the diagonal elements in each cell, one of which is two values of G and the others are B and R values, are combined to generate the target intensity which contains the essential information of the cell. Thus, the reconstruction operation making full use of abundant color information in raw data is conducive to enhancing the matching accuracy. Moreover, it reduces time complexity effectively as well, without the procedure of demosaicing for acquiring a full color image with traditional methods.

2.3. Parallel and Integrated Framework

With the focus on efficiency of computation, in our methodology, two cores in a multi-core processor are used to parallel the procedure of matching and imaging. As shown in Figure 2, thread 1 is responsible for handling the raw data matching, and thread 2 performs demosaicing to get a full color image with high quality. Eventually, the final results are obtained by merging the outcomes from the two different threads.

When handling the raw data matching, first of all, we preprocess it using the raw data reconstruction step whose detailed implementation is in Section 2.2. Since the intensity information of the pixels after reconstruction has a similar data form to gray images, we adopt the multi-scale FAST detector and the rotated BRIEF descriptor in ORB to complete feature detection and description for raw data, which achieves high robustness to general image transformations, including image blur, rotation, scale and illuminance change, and is of fast speed as well [13]. For the part of feature matching, we first find the k-nearest neighbors in the sensed image to the keypoints in the reference image via brute-force matching, and then employ the ratio test explained by Lowe [7] to select the best one from the k matches. The three parts mentioned above, *i.e.*, the feature detection, description and matching, are collectively referred to as the overall matching stage in this paper. Moreover, note that the similarity between descriptors is measured by Hamming distance [10], which can be computed rapidly via a bitwise exclusive or (XOR) operation followed by a bit count. However, as the development of the multimedia instruction set of the central processing unit (CPU) is fast, it is more effective to calculate the number of bits set to 1 using the newer instructions on modern CPUs compared with the previous bit count operation. Hence, the population count (POPCNT) instruction, which is part of SSE4.2, is applied to our method for speeding up feature matching [24].

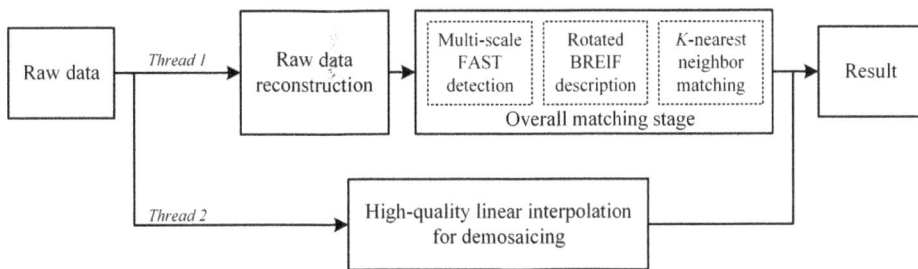

Figure 2. Flow chart of PIMR.

We use Malvar's algorithm for raw data demosacing [25]. The method computes an interpolation using a bilinear technique, computes a gradient correction term, and linearly combines the interpolation and the correction term to produce a corrected, high-quality interpolation of a missing color value at a pixel. A gradient-correction gain is also used to control how much correction is applied, and the gain parameters are computed by a Wiener approach and they contribute to the coefficients of the linear filter in the method. The approach outperforms most nonlinear and linear demosaicing algorithms with a reduced computational complexity. Due to the high image quality and low computational cost, the method is considered the major demosaicing approach in industrial digital cameras [26].

3. Performance Evaluation

3.1. Experimental Details

We evaluate PIMR using a well-known dataset, *i.e.*, the Affine Covariant Features dataset introduced by Mikolajczyk and Schimid [27].

It should be noted that our approach directly processes raw data captured by the digital camera without demosaicing. For this reason, we need to extract the corresponding color information at each pixel location from the original images in the chosen dataset, in accordance with the GBRG mode of the Bayer pattern. The new dataset, being made up of a series of raw data, mainly contains the following six sequences: wall (view point change), bikes and trees (image blur), leuven (illumination change), University of British Columbia (UBC) (JPEG compression), and graffiti (rotation). Each sequence consists of six sets of raw data just as six images, sorted in order of a gradually increasing degree of distortions with respect to the first image with the exception of the graffiti sequence. We take the raw data of image 1 in each sequence as the reference data, and match the reference one against the remaining ones, yielding five matching pairs per sequence (1|2, 1|3, 1|4, 1|5 and 1|6). Figure 3 shows the first image in the original six sequences from the standard dataset.

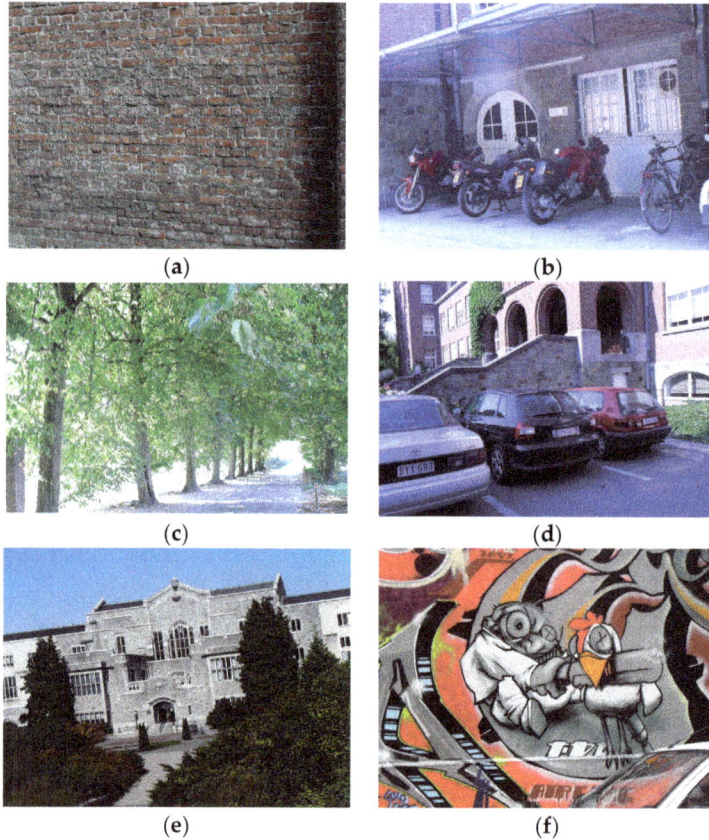

Figure 3. Image sequences from the Affine Covariant Features dataset. (**a**) wall; (**b**) bikes; (**c**) trees; (**d**) leuven; (**e**) UBC; (**f**) graffiti.

Since our work aims at realizing robust and fast feature detection, description and matching for raw data, we assess not only the accuracy but also the time-cost, comparing our PIMR with the state-of-the-art algorithms, including BRIEF, ORB, BRISK, and FREAK.

We compute 1000 keypoints on five scales per raw data with a scaling factor of 1.3 using PIMR. In order to ensure a valid and fair assessment, the full color image processed by the BRIEF, ORB, BRISK and FREAK algorithms is demosaiced from raw data using the same interpolation approach as PIMR. It must be emphasized that we combine the BRIEF descriptor with the CenSurE detector, and the multi-scale adaptive and generic corner detection based on the accelerated segment test (AGAST) detector proposed in BRISK with the FREAK descriptor, based on the settings in reference [10,16]. The implementation of these methods is built with OpenCV 2.4.8 which provides a common two-dimensional (2D) feature interface.

3.2. Accuracy

We use the recognition rate, namely the number of correct matches *versus* the number of total good matches, as the evaluation criterion. The results are shown in Figure 4. Each group of five bars with different colors represents the recognition rates of PIMR, ORB, BRIEF, BRISK and FREAK, respectively.

On the basis of these plots, we make the following observations. In general, our algorithm performs well for all test sequences, yielding comparable recognition accuracy with the state-of-the-art algorithms, and even outperforms them in most cases. There are only a few exceptions. For example, in the wall sequence, BRIEF achieves slightly higher precision than our methods, but in the other five sequences, PIMR outperforms the other algorithms.

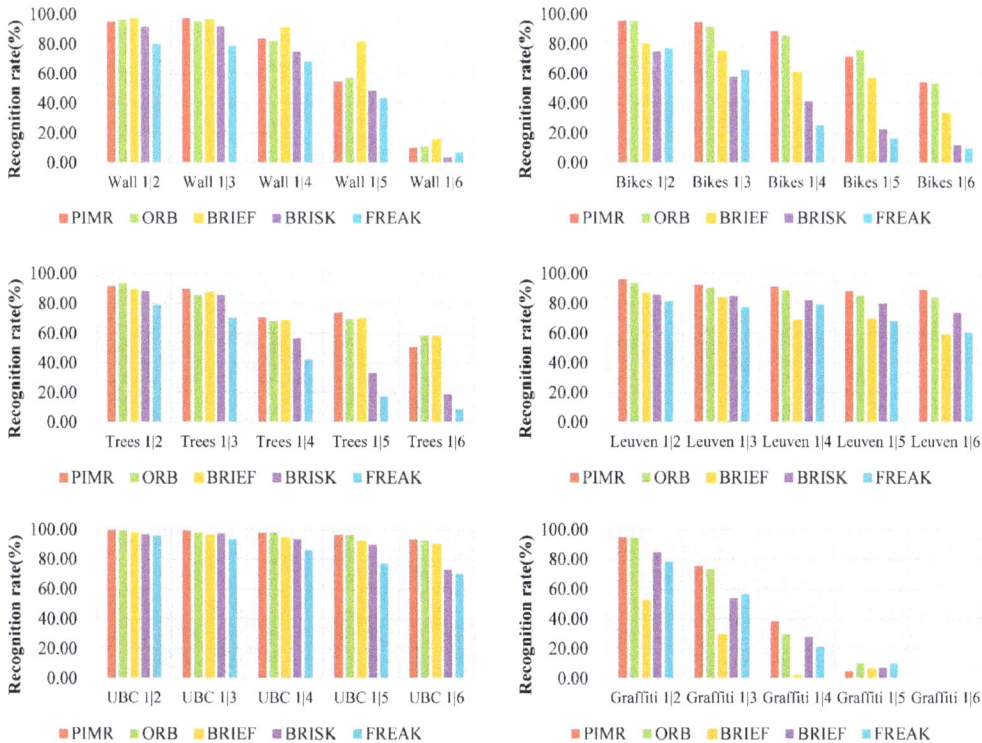

Figure 4. Recognition rates achieved by PIMR, ORB, BRIEF, BRISK and FREAK.

3.3. Time-Cost

Table 1 shows the time-cost of the bikes 1 | 3 employing different algorithms, measured on an Intel Core i7 processor at 3.4 GHz, in seconds. Since there is no raw data construction step in traditional methods, and the demosaicing is processed in parallel with the other stages within the proposed approach, the procedures marked with dashes in the table involve no extra time. Note that the time is averaged over 50 runs.

Considering the total time listed in the last column of Table 1, PIMR is about 1.4× faster than ORB and 7.1× faster than FREAK. In PIMR, the time-cost of demosaicing is excluded from the total time, as its demosaicing stage is faster than the sum of the other two stages. One potential reason for the high computational cost in BRISK and FREAK is the application of the scale-space FAST-based detector [15] which promises coping with the scale invariance.

Table 1. Times of matching for the bikes image 1 and 3.

Methods	Demosaicing (s)	Raw Data Reconstruction (s)	Overall Matching (s)	Total (s)
ORB	0.009	-	0.040	0.049
BRIEF	0.009	-	0.049	0.058
BRISK	0.009	-	0.225	0.234
FREAK	0.009	-	0.231	0.240
PIMR	-	0.004	0.030	0.034

3.4. Matching Samples with PIMR

Extensive evaluations have been made for PIMR above, and we also provide some matching samples with the chosen dataset in this part. It is considered that the invariance of rotation, scale and illuminance is the overriding concern in image matching algorithms. Thereby, we present the matching results of the graffiti (rotation and scale change) and leuven (illumination change) sequences to further demonstrate the robustness of the proposed approach. Figure 5a shows the matching result of the graffiti sequence (1 | 2), and Figure 5b shows the matching result of the leuven sequence (1 | 2). The keypoints connected by green lines indicate keypoint correspondences, namely valid matches. It can be seen that the method PIMR can acquire sufficient robust matches with few outliers.

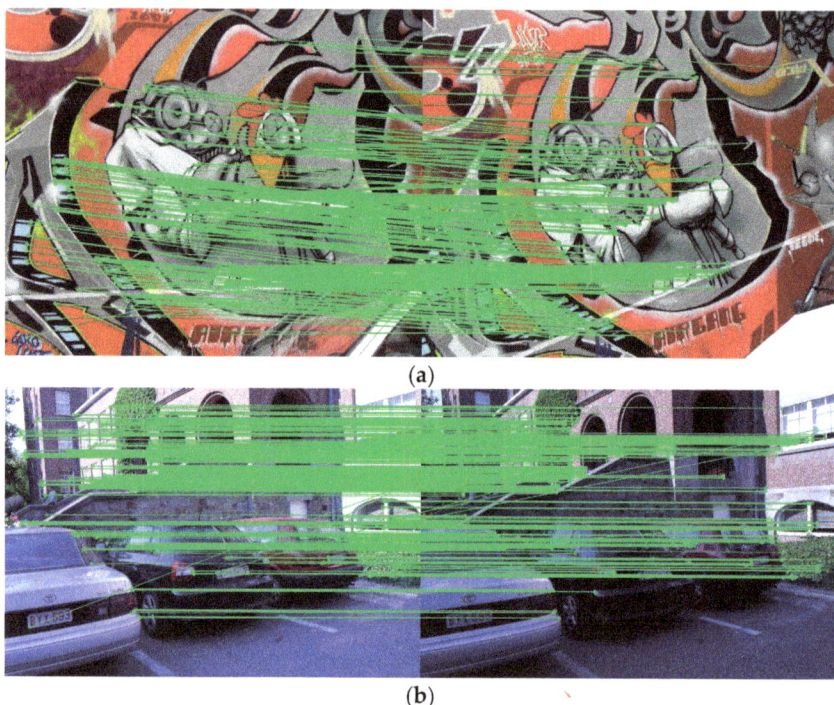

(a)

(b)

Figure 5. Matching samples using the PIMR. (a) The matching result of the graffiti sequence; (b) The matching result of the leuven sequence.

4. Conclusions

We have presented a parallel and integrated matching algorithm for raw data named PIMR, mainly to speed up whole image matching and to enhance the robustness as well. In most cases, it achieves better performance for most image transformations than current approaches, with a fairly low computational cost. Furthermore, our work is not only basic research in the field of rapid image matching, but also a preliminary study with respect to a method for processing raw data. In the future, an ultra-light-weight local binary descriptor will be studied further.

Acknowledgments: This research was supported by the National Natural Science Foundation of China (61105093), the Fundamental Research Funds for the Central Universities (106112013CDJZR120013), the Chongqing Postdoctoral Science Foundation (Xm2015014), the Zhejiang Provincial Natural Science Foundation of China (LY15F020042), the Opening Fund of Key Laboratory of Inland Waterway Regulation Engineering (Chongqing Jiaotong University), Ministry of Communications (NHHD-201503), and the Visiting Scholar Foundation of Key Laboratory of Optoelectronic Technology & Systems (Chongqing University), Ministry of Education.

Author Contributions: The work presented in this paper is a collaborative development by five authors. Li defined the research theme, designed the methods and experiments. Li and Yang performed the experiments. Yang developed the data collection modules and guided the data analysis. Zhao, Han and Chai performed the data collection and data analysis. Yang wrote this paper, Li, Zhao, Han and Chai reviewed and edited the manuscript.

Conflicts of Interest: The authors declare no conflict of interest.

References

1. Zaragoza, J.; Tat-Jun, C.; Tran, Q.H.; Brown, M.S.; Suter, D. As-projective-as-possible image stitching with moving dlt. *IEEE Trans. Pattern Anal. Mach. Intell.* **2014**, *36*, 1285–1298. [PubMed]
2. Mei, X.; Ma, Y.; Li, C.; Fan, F.; Huang, J.; Ma, J. A real-time infrared ultra-spectral signature classification method via spatial pyramid matching. *Sensors* **2015**, *15*, 15868–15887. [CrossRef] [PubMed]
3. Xu, X.; Tang, J.; Zhang, X.; Liu, X.; Zhang, H.; Qiu, Y. Exploring techniques for vision based human activity recognition: Methods, systems, and evaluation. *Sensors* **2013**, *13*, 1635–1650. [CrossRef] [PubMed]
4. Zhang, Y.D.; Dong, Z.C.; Wang, S.H.; Ji, G.L.; Yang, J.Q. Preclinical diagnosis of magnetic resonance (MR) brain images via discrete wavelet packet transform with tsallis entropy and generalized eigenvalue proximal support vector machine (GEPSVM). *Entropy* **2015**, *17*, 1795–1813. [CrossRef]
5. Hu, M. Visual-pattern recognition by moment invariants. *Ire Trans. Inf. Theory* **1962**, *8*, 179–187.
6. Zhang, Y.D.; Wang, S.H.; Sun, P.; Phillips, P. Pathological brain detection based on wavelet entropy and hu moment invariants. *Bio-Med. Mater. Eng.* **2015**, *26*, S1283–S1290. [CrossRef] [PubMed]
7. Lowe, D. Distinctive image features from scale-invariant keypoints. *Int. J. Comput. Vis.* **2004**, *60*, 91–110. [CrossRef]
8. Bay, H.; Ess, A.; Tuytelaars, T.; van Gool, L. Speeded-up robust features (SURF). *Comput. Vis. Image Underst.* **2008**, *110*, 346–359. [CrossRef]
9. Li, Z.; Gong, W.; Nee, A.Y.C.; Ong, S.K. Region-restricted rapid keypoint registration. *Opt. Express* **2009**, *17*, 22096–22101. [CrossRef] [PubMed]
10. Calonder, M.; Lepetit, V.; Ozuysal, M.; Trzcinski, T.; Strecha, C.; Fua, P. Brief: Computing a local binary descriptor very fast. *IEEE Trans. Pattern Anal. Mach. Intell.* **2012**, *34*, 1281–1298. [CrossRef] [PubMed]
11. Rosten, E.; Drummond, T. Machine learning for high-speed corner detection. In *Computer Vision—Eccv 2006*; Leonardis, A., Bischof, H., Pinz, A., Eds.; Springer: Berlin/Heidelberg, Germany, 2006; Volume 3951, pp. 430–443.
12. Agrawal, M.; Konolige, K.; Blas, M. Censure: Center surround extremas for realtime feature detection and matching. In *Computer Vision—Eccv 2008*; Forsyth, D., Torr, P., Zisserman, A., Eds.; Springer: Berlin/Heidelberg, Germany, 2008; Volume 5305, pp. 102–115.
13. Rublee, E.; Rabaud, V.; Konolige, K.; Bradski, G. ORB: An efficient alternative to SIFT or SURF. In Proceedings of the 2011 IEEE International Conference on Computer Vision (ICCV), Barcelona, Spain, 6–13 November 2011; pp. 2564–2571.
14. Rosin, P.L. Measuring corner properties. *Comput. Vis. Image Underst.* **1999**, *73*, 291–307. [CrossRef]
15. Leutenegger, S.; Chli, M.; Siegwart, R.Y. Brisk: Binary robust invariant scalable keypoints. In Proceedings of the 2011 IEEE International Conference on Computer Vision (ICCV), Barcelona, Spain, 6–13 November 2011; pp. 2548–2555.
16. Alahi, A.; Ortiz, R.; Vandergheynst, P. Freak: Fast retina keypoint. In Proceedings of the 2012 IEEE Conference on Computer Vision and Pattern Recognition (CVPR), Providence Rhode Island Convention Center Providence, RI, USA, 16–21 June 2012; pp. 510–517.
17. Schafer, R.W.; Mersereau, R.M. Demosaicking: Color filter array interpolation. *IEEE Signal Process. Mag.* **2005**, *22*, 44–54.
18. Li, X.; Gunturk, B.; Zhang, L. Image demosaicing: A systematic survey. *Proc. SPIE* **2008**, *6822*. [CrossRef]
19. Lu, Y.; Jian, S. High quality image reconstruction from raw and jpeg image pair. In Proceedings of the 2011 IEEE International Conference on Computer Vision (ICCV), Barcelona, Spain, 6–13 November 2011; pp. 2158–2165.
20. Macfarlane, C.; Ryu, Y.; Ogden, G.N.; Sonnentag, O. Digital canopy photography: Exposed and in the raw. *Agric. For. Meteorol.* **2014**, *197*, 244–253. [CrossRef]
21. Verhoeven, G.J.J. It's all about the format—Unleashing the power of raw aerial photography. *Int. J. Remote Sens.* **2010**, *31*, 2009–2042. [CrossRef]

22. Bayer, B.E. Color Imaging Array. U.S. Patent US3971065, 20 July 1976.
23. Maschal, R.A.; Young, S.S.; Reynolds, J.P.; Krapels, K.; Fanning, J.; Corbin, T. New image quality assessment algorithms for cfa demosaicing. *IEEE Sens. J.* **2013**, *13*, 371–378. [CrossRef]
24. *Intel's Software Network, Sofwareprojects. Intel. com/avx*; programming reference; SSE Intel.: Santa Clara, CA, USA, 2007; Volume 2.
25. Malvar, H.S.; He, L.-W.; Cutler, R. High-quality linear interpolation for demosaicing of bayer-patterned color images. In Proceedings of the IEEE International Conference on Acoustics, Speech, and Signal Processing (ICASSP'04), Montreal, QC, Canada, 17–21 May 2004; pp. 485–488.
26. Point Grey Research. Different Color Processing Algorithms. Available online: http://www.ptgrey.com/KB/10141 (accessed on 14 December 2015).
27. Mikolajczyk, K.; Schmid, C. A performance evaluation of local descriptors. *IEEE Trans. Pattern Anal. Mach. Intell.* **2005**, *27*, 1615–1630. [CrossRef] [PubMed]

10

Multi-Stage Feature Selection Based Intelligent Classifier for Classification of Incipient Stage Fire in Building

Allan Melvin Andrew *, Ammar Zakaria, Shaharil Mad Saad and Ali Yeon Md Shakaff

Academic Editor: Vittorio M.N. Passaro

Centre of Excellence for Advanced Sensor Technology (CEASTech), Universiti Malaysia Perlis, Jejawi, Arau, Perlis 02600, Malaysia; sag.unimap@gmail.com (A.Z.); shaharil85@gmail.com (S.M.S.); aliyeon@unimap.edu.my (A.Y.M.S.)
*   Correspondence: allanmelvin.andrew@gmail.com

Abstract: In this study, an early fire detection algorithm has been proposed based on low cost array sensing system, utilising off- the shelf gas sensors, dust particles and ambient sensors such as temperature and humidity sensor. The odour or "smellprint" emanated from various fire sources and building construction materials at early stage are measured. For this purpose, odour profile data from five common fire sources and three common building construction materials were used to develop the classification model. Normalised feature extractions of the smell print data were performed before subjected to prediction classifier. These features represent the odour signals in the time domain. The obtained features undergo the proposed multi-stage feature selection technique and lastly, further reduced by Principal Component Analysis (PCA), a dimension reduction technique. The hybrid PCA-PNN based approach has been applied on different datasets from in-house developed system and the portable electronic nose unit. Experimental classification results show that the dimension reduction process performed by PCA has improved the classification accuracy and provided high reliability, regardless of ambient temperature and humidity variation, baseline sensor drift, the different gas concentration level and exposure towards different heating temperature range.

Keywords: electronic nose; gas sensors; fire detection; feature selection; feature fusion; normalized data; Principal Component Analysis (PCA); Probabilistic Neural Network (PNN)

1. Introduction

Fires can be categorized into two main groups: direct burning and indirect burning. Residential fires may happen indoors or outdoors [1]. Most fires start from an incipient stage and develop further to smouldering, flaming and fire stages [2]. In incipient and smouldering cases, fires have less flames and smoke, while in the flaming and fire stages, fires have more flames and radiate extreme heat.

According to the work published in the recent decade, fire research can be categorized mainly into four types; namely, fire detection, fire prediction, fire data analysis and reduction of false fire alarms [2]. Predicting or perceiving fire at the early stage is very challenging and crucial for both personal and commercial applications. Over the years, several methods have been proposed which utilise various sensing technologies to provide early fire detection [2]. The research conducted by Rose-Pehrsson is able to provide early fire detection using a Probabilistic Neural Network and achieves higher classification accuracy [3]. However, they were only able to demonstrate it as early as the smouldering stage. As for data analysis alone, various methodologies have been utilised. The most common methods used are related to clustering techniques and classification algorithms.

Several fire data analysis algorithms have been proposed. According to the research, most of these algorithms are based on time-fractal approaches to characterize the temporal distribution of detected fire sequences [4]. Some of the research has focused on utilizing unsupervised ways to detect fire from the signals [5]. In their paper, Chakraborty and Paul proposed a hybrid clustering algorithm using a modified k-means clustering algorithm. Although it required very little processing time and managed to detect the fire flames at fast speed, the proposed algorithm can be only be used in video image processing based on RGB and HSI colour models. Bahrepour *et al.*, in their research, investigated the feasibility of spatial analysis of indoor and outdoor fires using data mining approaches for WSN-based fire detection purposes [6]. In their paper, they had investigated the most dominant feature in fire detection applications. Kohonen self-organizing map (kSOM) had been utilized as a feature reduction technique which can cluster similar data together. Experimentals result show that their method reduces the number of features representing the fire data features. They also performed analysis on residential fires and used artificial neural network, naive Bayes and decision tree classifiers to compute the best combination of sensor type in fire detectors. The outputs of various classifiers were fused using data fusion techniques to achieve higher fire detection accuracy. The reported results showed that 81% accuracy for residential fire detection and 92% accuracy for wildlife fire detection could be achieved.

Most of the proposed methods provide high classification rates in detecting fires, albeit they need to be in close vicinity to the source of the fire and only operate based on specific types of sensors [7–13]. Mimicking the human nose in early fire detection is still the biggest challenge for olfactory engineering. The present electronic nose systems have difficulties in detecting early fires, especially in large spaces, and cannot provide additional information regarding the burning stages and the scorching fire material. To overcome the mentioned weakness, bio-inspired approaches based on electronic nose technology is a promising method, which utilises artificial intelligence in detecting and predicting the possibility of fire occurrence. Although there are many proposed feature selection techniques and classifiers involved, the real question is whether it is possible to implement them in conventional fire detectors, yet to be determined, at a low cost. This paper focuses on investigating a multi-stage feature selection method using a bio-inspired artificial neural network and principal component analysis for data reduction, which can give the best detection accuracy, reduce misclassification and offer high reliability for indoor fire detection applications. This work is important to investigate the most suitable features and classification algorithm, which could be proved less computationally complex and having potential to be used in embedded applications.

The rest of this paper is organized as follows: Section 1 introduces the features of fires. Section 2 describes the proposed four-stage fire detection algorithm. Section 3 discusses the experimental results of the proposed method and compares the performance of the proposed method with those of other fire detection algorithms, and Section 4 presents the conclusions of our study.

2. Methods

In this section, the odour measurement technique, the feature extraction from sensor arrays using various data normalisation techniques, the artificial neural network-based feature selection, the feature reduction using PCA, and the classification stages are explained. Figure 1 shows the flowchart of the proposed multi-stage feature selection approach using PCA and PNN. The dashed line around PNN training on training dataset in Figure 1 indicates that the PNN training is conducted prior to the classification of fire sources. The training dataset is used by PNN in the fire sources classification process.

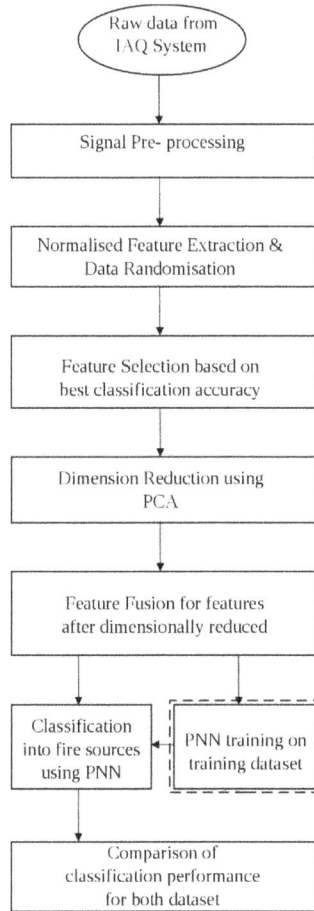

Figure 1. A flowchart of the proposed multi- stage feature selection approach using PCA and PNN.

2.1. Datasets

In this study, two datasets have been used. The first dataset consists of odour signals which have been obtained from an in-house metal oxide gas sensor-based low cost (IAQ) system, consisting of oxygen (O_2), volatile organic compound (VOC), carbon dioxide (CO_2), ozone (O_3), nitrogen dioxide (NO_2), particulate matter up to 10 micrometres in size (PM_{10}), temperature and humidity sensors. The prediction classifier for the early fire detection has been developed based on odours from various sample sources. The odour sources consist of five common fire sources and three common building construction materials. Information about the materials tested and their sample dimensions prepared according to the corresponding European Standard, is shown in Table 1. For each source, more than 100 odour measurement samples have been taken at seven different temperature points, starting from 50 °C up to 250 °C. About 200 ambient air measurement datapoints have been added to the dataset as a reference air sample. The ambient air samples are considered the 9th tested sample in this paper. The final IAQ system dataset is a matrix of 1000 rows and eight columns. The training set contains 600 samples (60% of the dataset), the validation set contains 100 samples (10% of the dataset), and the test set contains the remaining samples, which is 30% of the dataset. In order to estimate the true performance of the classifier, the test is based on the remaining samples which were not used during the training and validation process. The dataset has been referred as the IAQ dataset in this paper.

Table 1. The tested materials and its sample dimension prepared according to European Standard.

Sample	Materials	Material Type	Dimension
Sample 1	Paper	Common Fire Source	16 pieces 5 cm × 5 cm 90 gsm sheets stacked together
Sample 2	Plastic	Common Fire Source	4 cm × 2 cm × 40 cm (density 20 kg·m^{-3}) polyurethane
Sample 3	Styrofoam	Common Fire Source	4 cm × 2 cm × 40 cm styrofoam
Sample 4	Cotton	Common Fire Source	1 wick 18 cm long (approx. 0.17 g)
Sample 5	Cardboard	Common Fire Source	16 pieces 5 cm × 5 cm stacked together
Sample 6	Wood	Building Construction Material	1 cm × 1 cm × 2 cm beech wood
Sample 7	Brick	Building Construction Material	1 piece brick
Sample 8	Gypsum board	Building Construction Material	1 cm × 1 cm × 2 cm gypsum board

The second dataset obtained from a Portable Electronic Nose (PEN3) from Airsense Analytics GmbH (Schwerin, Germany) has been used as the control dataset. This set has 10 sensor inputs (10 columns). For each source, more than 100 samples of odour measurements have been taken at seven temperature points, starting from 50 °C up to 250 °C. Like IAQ, 200 ambient air measurement datapoints have been added to the dataset as a reference air sample. The final PEN3 dataset is a matrix of 1000 rows and 10 columns. The training set contains 600 samples (60% of the dataset), the validation set contains 100 samples (10% of the dataset), and the test set contains the remaining samples, which is 30% of the dataset, similar to the first dataset. A similar approach for performance analysis was followed for the above process as with IAQ. The dataset is referred to as PEN3 dataset in this paper.

2.2. Measurement of Odour Signals

In the IAQ dataset, the odour samples have been collected from the IAQ system placed at 2.1 m height in the testing room. The height of 2.1 m has been selected to deploy the in-house system in buildings based on few classification preliminary tests done at different heights in a standard sized room (33 m^3 in volume) in Malaysia. Heights of 0.7, 1.4 and 2.1 m have been tested in the preliminary tests. A height of 2.1 m was the most suitable and was been selected because the experimental results show that the gases generated at the incipient fire stage fill the top part of the room first since the density of the emitted gases are lesser than that of ambient air. For this experiment, the deployment of the sensor unit at this height gives the best chance in predicting an earlier fire event. Having the sensor units deployed at an inappropriate height in the building can cause it to miss useful data for fire data analysis and prediction, and thus, could trigger false fire alarms. That is also the main reason why conventional fire detectors are placed on the ceilings of buildings [14]. For realisation of a wireless sensing IAQ system, the data of the low cost system is sampled at the sampling rate of 10 sample/min [15]. The data has been recorded for 15 min each time. Each data measurement has been sent wirelessly to the server for processing and data storage using an available wireless sensor network. The data measurements have been recorded in websocket "*sqlite*" format and then converted to ".csv" format using a custom LabVIEW application. Afterwards, the odour signals have been translated into digital form by a custom MATLAB application.

In the PEN3 dataset, the data from PEN3 has been captured using a program supplied by AirSense Analytics GmbH. The PEN3 has been placed at 1.5 m distance from the smell source which has been heated in a vacuum oven. PEN3 has a sampling frequency of 1 sample/s. The data has been recorded for 15 min each. The data measurements have been recorded in ".nos" format and then converted to ".xls" format using a custom application. Then, the samples have been converted into digital format by a custom MATLAB application.

2.3. Normalised Feature Extraction

Baseline drift is a widespread phenomenon in signal analysis, which could also cause incorrect representation of data in subsequent feature extraction and feature selection processes of an odour signal, and baseline correction is the solution to the problem and the correct way of representing the signal when the analysis deals with sensor values from different conversion units. Baseline manipulation helps to pre-process the sensor output to free itself from the drift effect, the intensity dependence and, possibly, from non-linearity [7].

In this paper, for the feature extraction stage, five types of baseline correction algorithms have been executed on both datasets by converting the raw data value from Volts to unit ratio values. Unit ratio is a dimensionless unit. Each type of baseline correction has been considered as a feature. The ability to distinguish the fire event from the normalised data itself helps to reduce the computation complexity and classification time, thus it will be easier to implement it in the embedded system using C programming.

The first feature is Relative Logarithmic Sum Squared Voltage value (RLSSV). RLSSV is the division of logarithmic voltage by the logarithmic sum squared voltage value. The equation for calculating RLSSV is shown in Equation (1):

$$\text{RLSSV} = \frac{\log v_i}{\log(\sum v^2)} \tag{1}$$

where v_i is the voltage value at time i for each specific sensor.

The second feature is the Relative Logarithmic Voltage value (RLV). RLV is the ratio between the logarithmic voltage and the instantaneous voltage value. It can be calculated using Equation (2):

$$\text{RLV} = \frac{\log v_i}{v} \tag{2}$$

where v_i is the voltage value at time i for each specific sensor.

The next feature is Relative Sum Squared Voltage value, referred to as RSSV. RSSV is obtained by dividing the instantaneous voltage value by the square root value of sum of squared voltages. Equation (3) shows the formula used in computing the RSSV:

$$\text{RSSV} = \frac{v_i}{\sqrt{\sum v^2}} \tag{3}$$

where v_i is the voltage value at time i for each specific sensor.

The fourth feature is Relative Voltage value (RV). RV is calculated by finding the ratio of the voltage at time I and the average. It can be calculated using Equation (4):

$$\text{RV} = \frac{v_i}{v_0} \tag{4}$$

where v_i is the voltage value at time i and v_0 is the baseline voltage value for each specific sensor.

The final feature investigated is the Fractional Voltage Change value (FVC). FVC is directly proportional to the difference between the averaged baseline value and current value and indirectly proportional to the averaged baseline value, as shown in Equation (5):

$$\text{FVC} = \frac{\overline{v}_0 - v_i}{\overline{v}_0} \tag{5}$$

where v_i is the actual sensor value at time i and $\overline{v}_0$ is the baseline value of each specific sensor.

A raw data example of the scorching smell generated by paper at 250 °C and its waveform after the RLSSV feature has been extracted are presented in Figure 2a,b, respectively.

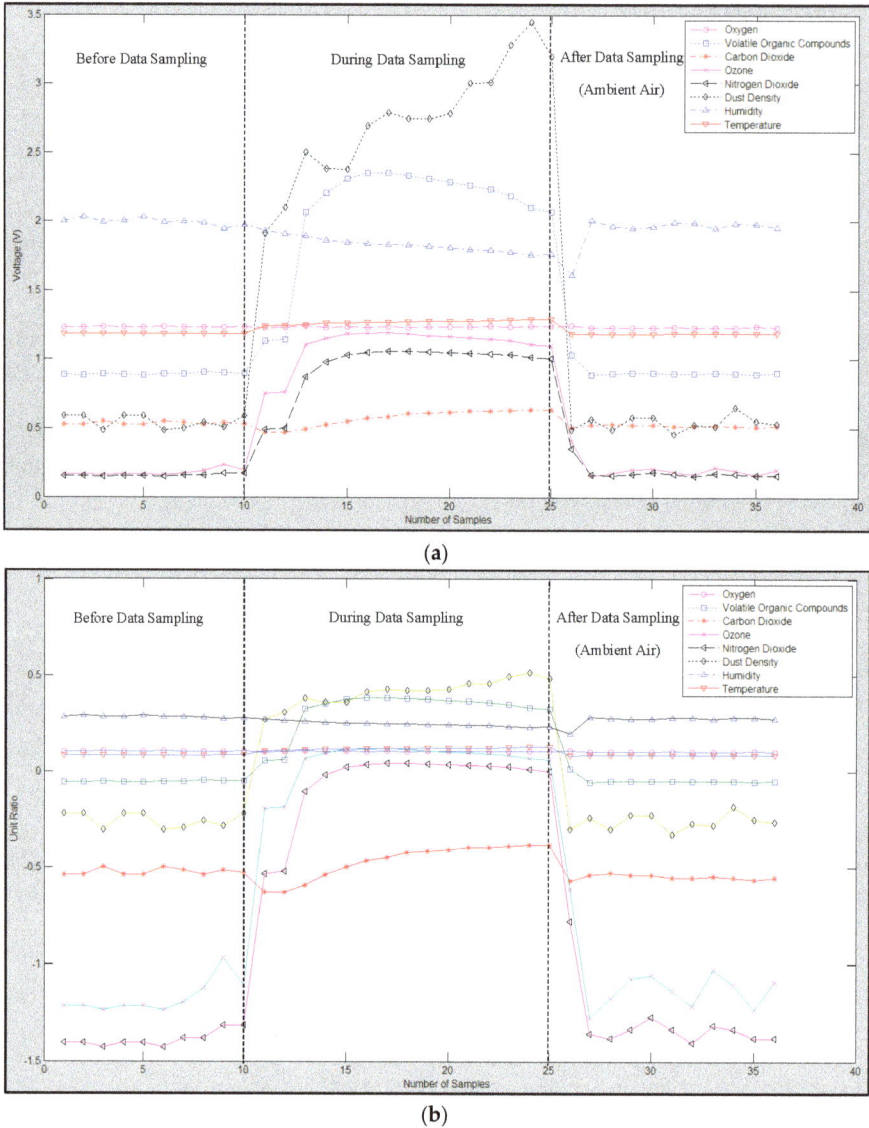

Figure 2. (a) Example of raw data for a scorching smell generated by paper at 250 °C; **(b)** The RLSSV feature extracted from the scorching smell of paper at 250 °C in (a).

2.4. Feature Selection

In this feature selection stage, the relative logarithmic sum squared voltage, the relative logarithmic voltage value, the relative sum squared voltage value, the relative voltage value, and the fractional voltage value, of the signal have been obtained. The selected features are chosen to investigate their performance on early fire data. The features have been tested for their reliability by examining the classification accuracy with a Probabilistic Neural Network (PNN). PNN and its function in this paper is explained further in Section 2.7. Out of the five features, the three best features with the highest classification accuracy are selected for dimensional reduction using PCA.

2.5. Dimension Reduction Using PCA

PCA is a linear technique which transforms a dataset from its original m-dimensional form into a new and compressed n-dimensional form where n < m. Dimension reduction has been implemented to investigate its effects on classification. Since the number of observations is reduced after the dataset is dimensionally reduced, the training period of PNN classifier will be minimized [16]. Thus, PCA is helpful not only in reducing the input variables of a dataset, but it also indirectly increases the classification ability of a classifier.

PCA gives the same number of principal components as the number of input variables. For example, if the data matrix has a dimension of 100 rows and 10 columns, the data matrix could be reduced to a 100 rows and three column matrix of principal components, without removing any important information from the original dataset. The data is arranged according to the variances between the classes, starting from highest variances descending from first column up to n numbered columns. However, out of the n reduced principal components, not all the principal components are needed to represent the data. Thus, the principal components need to be tested to find the appropriate number of principal components required for feature fusion. As explained in previous studies the optimal number of principal components can be obtained using a few criteria, such as the Broken stick model, Velicer's partial correlation procedure, cross-validation, Bartlett's test for equality of eigenvalues, Kaiser's criterion, Cattell's scree test and cumulative percentage of variance [17], which basically explais how much variances we are about to retain in the data. Based on this, in this study, eight principal components have been selected to observe the effect on the classification accuracy of PNN. For each selected feature in IAQ dataset, eight principal components have been obtained from eight input variables while for PEN3 dataset, 10 principal components have been obtained from 10 input variables. The latent, proportion and cumulative percentage corresponding to the principal component value from the principal components for the relative voltage value feature in the IAQ dataset and PEN3 dataset are given in Tables 2 and 3 respectively.

Table 2. Latent, proportion, and cumulative values of selected principal components for relative voltage value feature in the IAQ dataset.

Principal Component	Latent	Proportion	Cumulative
1	0.1064	0.4813	0.4813
2	0.0474	0.2141	0.6954
3	0.0335	0.1517	0.8471
4	0.0144	0.0650	0.9121
5	0.0096	0.0435	0.9556
6	0.0073	0.0329	0.9886
7	0.0019	0.0085	0.9970
8	0.0007	0.0030	1.0000

Table 3. Latent, proportion, and cumulative values of selected principal components for relative voltage value feature in the PEN3 dataset.

Principal Component	Latent	Proportion	Cumulative
1	7.8692	0.5338	0.5338
2	3.5164	0.2385	0.7723
3	1.8546	0.1258	0.8981
4	0.7612	0.0516	0.9497
5	0.4236	0.0287	0.9784
6	0.2476	0.0170	0.9954
7	0.0461	0.0030	0.9984
8	0.0176	0.0012	0.9996
9	0.0041	0.0003	0.9999
10	0.0015	0.0001	1.0000

2.6. Feature Fusion

In the feature fusion stage, the dimensionally reduced features have been fused to form the proposed IAQ-PCA hybrid feature for the IAQ dataset and the proposed PEN3- PCA hybrid feature for the PEN3 database. A similar approach was also reported by Luo who proposed an adaptive sensory fusion method for fire detection and isolation for intelligent building systems [18]. The proposed features have been tested and compared with the other normalised features mentioned in Section 2.3. The result of classification trials will be shown in Section 3. The feature fusion process for the IAQ-PCA hybrid features is shown in Figure 3. A similar process was also repeated for the PEN3- PCA hybrid features.

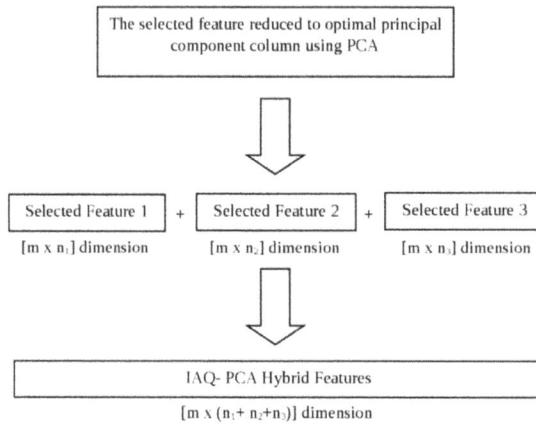

Figure 3. Feature Fusion Process for IAQ- PCA Hybrid Features.

2.7. Probabilistic Neural Network

Probabilistic Neural Network is highly regarded as a biologically inspired approach in classification as it functions similar to the human cognitive system. It requires less computational time and processing power compared to other classifiers. The human brain receives the input pattern from the nerves, compares it to the pattern in memory, and sums it together with other input patterns to find the probability that an the event will occur [3]. Thus, in this work, PNN has been selected and used as a core classifier.

PNN can be used for classifying different input patterns. It was proposed by Specht based on Bayesian classification and the probability density function using classical estimators. Compared to the conventional multi-layer perceptron (MLP) classifier which uses a sigmoidal activation function, PNN uses an exponential activation function in its algorithm. The computational time for PNN is also much less than for the MLP classifier [3]. For example, let us consider a simple two class problem:

Classifying two classes problem, class A and class B.

The estimator for the probability density function as given in Equation (6) has been used in PNN:

$$f_A(X) = \frac{1}{(2\pi)^{n/2}} \frac{1}{m_A} \sum_{i=1}^{m_A} \exp\left[-\frac{(X - X_{Ai})^T (X - X_{Ai})}{2\sigma^2} \right] \qquad (6)$$

where, X_{Ai} is the ith training pattern from class A, n is the dimension of the input vectors, m_A is the number of training patterns in class A, T is the transpose of the value and σ is a smoothing parameter corresponding to the standard deviation of the Gaussian distribution. This is the standard probability density function estimator used commonly in PNN and other neural networks. There are also some works highlighting on the modification in the exponential power of Equation (6), for example, normal, log- normal, Rayleigh and Weibull probability density functions which intend to provide better estimations of unknown stochastic processes, which do not require either an *a priori* choice of a mathematical model or the elaboration of the data histogram, but only the computation of the variability range of each components of available data samples [19].

Similar to our biological brain, the probabilistic neural network has four operational units known as input units, pattern units, summation units and output units. When PNN is given an input, the pattern unit will calculate the distance between the input vector and the trained input vectors. A vector with the information regarding the distance between the input and the training input is produced and passed to the summation unit. The contributions for each class of input are summed by the summation unit and a net output is generated. The net output has the information of the maximum of the probabilities to indicate a 1 for the specific class or a 0 for the other class.

The steps involved in the PNN algorithm are described below:

Step 0: Initialize the weights
Step 1: For each training input to be classified, do Step 2 to 4
Step 2: Pattern units:
  Compute the net input to the pattern units:

$$Z_{inj} = x(w_j) = x^T w_j \tag{7}$$

  Compute output Equation (8) using Equation (7):

$$Z_{outj} = \exp\left[\frac{z_{inj} - 1}{\sigma^2}\right] \tag{8}$$

Step 3: Summation unit:
  Sum the inputs from the pattern units to which they are connected. The summation unit for class B multiplies its total input by Equation (9):

$$V_B = -\frac{P_B C_B m_A}{P_A C_A m_B} \tag{9}$$

Where:

  P_A & P_B are the priori probalility of occurrence of patterns in Class A and Class B,
  C_A & C_B are the cost associated with classifying vectors in Class A and B, and
  m_A & m_B are the number of training patterns in Class A and Class B.

Step 4: Output (decision) unit:

The output unit sums the signals from f_A and f_B. The input vector is classified as Class A if the total input to the decision unit is positive. Based on the above example, the PNN network can classify two different classes when the input patterns of both classes are given to it. However, training the network with more sample inputs improves the ability of PNN. The degree of nonlinearity of the decision boundaries of PNN can be controlled by varying the spread factor, σ. Large values of σ make the decision boundary approach a hyperplane, while having a relatively small value approaching zero for σ gives a good approximation for highly nonlinear decision surfaces of PNN [3].

Consequently, in this paper, PNN is used to select the dominant features and to test the classification accuracy of the proposed and dominant features in distinguishing various materials involved in incipient fire cases. The PNN architecture is shown in Figure 4.

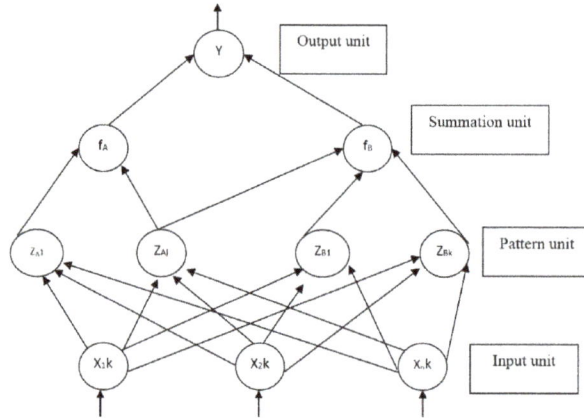

Figure 4. PNN Architecture.

The overall process flow of proposed multi-stage feature selection and fusion for both datasets is shown in Figure 5.

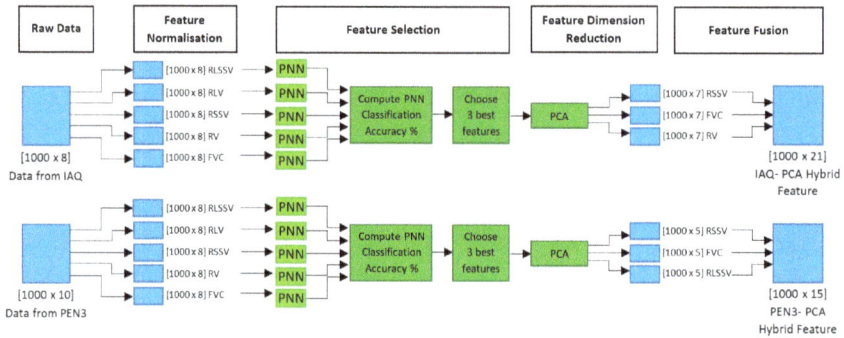

Figure 5. Multi-stage Feature Selection and Fusion Process Flow.

3. Results and Discussion

A Probabilistic Neural Network has been applied for classification of scorching smells generated from the different materials. In this application, both raw datasets have been subjected to the PNN classifier to select the most dominant features, prior to dimension reduction.

Table 4. PNN architectures.

Parameters	Value for the IAQ Dataset	Value for PEN3 Dataset
Number of input neurons	8	10
Number of output neurons	9	9
Spread factor	0.08	0.08
Testing Tolerance	0.001	0.001
Number of training samples	600	600
Number of validation samples	100	100
Number of testing samples	300	300
Total number of samples	1000	1000

The parameters used in PNN are shown in Table 4. As mentioned earlier in Section 2.7, the spread factor can be varied to control the degree of nonlinearity of the decision boundaries. It is the most important factor which influences the classification performance of the classifier. Therefore, the spread factor has been varied in these experiments to obtain the best classification performance [15]. The best value for spread factor for both datasets is recorded to be 0.08.

Classification performances have been computed for the nine classes for the IAQ dataset and PEN3 dataset as shown in Table 5. The classification accuracy of the each feature is clearly shown in the table. The classification result has been obtained by averaging the classification accuracy for 50 repetitions.

Table 5. Average PNN classification accuracies of features for IAQ and PEN3 datasets.

Features	IAQ			PEN3		
	Minimum Classification Accuracy (%)	Maximum Classification Accuracy (%)	Average Classification Accuracy (%)	Minimum Classification Accuracy (%)	Maximum Classification Accuracy (%)	Average Classification Accuracy (%)
RLSSV	97.11	99.41	98.75	97.15	99.54	99.29
RLV	97.64	98.65	98.31	97.43	99.02	98.84
RSSV	97.31	99.16	98.90	98.16	100.00	99.75
RV	97.36	99.43	98.81	98.19	99.45	99.12
FVC	97.42	99.14	98.84	98.41	99.55	99.51

For each dataset, the three best features with the highest classification accuracy have been selected for dimensional reduction with PCA. For the IAQ dataset, it is observed that RSSV, FVC and RV give the best accuracies, 98.90%, 98.84% and 98.81%, respectively. The PEN3 dataset, on the other hand, has RSSV, FVC and RLSSV with 99.75%, 99.51% and 99.29%, respectively, as its best features.

The three selected features have eight columns each (inputs from eight gas and electrochemical sensors). At this stage, the dimension of each feature has been reduced to remove the redundant data and to select only the optimal number of features with high variance between classes, which is sufficient to represent the fire signature. Reducing the dimensions of the original data indirectly increases the classification accuracy and reduces the processing time of the classifier. The selection of principal component values in PCA will determine how much the dimensions of the m-dimension dataset will be reduced. The performance of the classifier has been investigated by varying the principal component values and the results have been recorded in Table 6.

Table 6. Average PNN classification results in % for selecting principal component values in PCA for the IAQ and PEN3 datasets.

Principal Component Value	IAQ			PEN3		
	RSSV	FVC	RV	RSSV	FVC	RLSSV
1	74.07	75.30	74.47	83.26	82.58	82.12
2	82.43	83.11	83.56	87.51	87.39	87.03
3	87.74	87.27	88.28	91.97	91.67	90.97
4	90.17	90.21	90.21	98.28	97.95	97.49
5	95.62	95.66	95.45	100.00	99.91	99.76
6	98.30	98.13	97.70	98.75	98.66	98.12
7	99.02	99.02	98.96	97.35	97.12	96.81
8	98.88	98.80	98.86	96.74	96.55	96.26

As seen in Table 6, 6–8 principal components give the most successful classification results for the IAQ dataset, while 4–6 principal components give the most successful classification results for the PEN3 dataset. The range of classification accuracies range from a minimum of 98.13% to a maximum 99.02% for the IAQ dataset, and from a minimum of 97.49% to maximum of 100.00% for the PEN3 dataset. Out of this range, the best classification accuracies for the IAQ dataset have been observed to occur when the principal component value is seven, while, for the PEN3 dataset, the optimal principal component value has been observed to be five. Thus, the dimensions of the IAQ and PEN3 datasets have been reduced to seven and five principal components scores, respectively. The dimensionally reduced features have been fused to form the proposed IAQ-PCA hybrid feature for the IAQ dataset and the proposed PEN3-PCA hybrid feature for the PEN3 database. The fused feature for the IAQ dataset is a matrix of 1000 rows and 21 columns, while the fused feature for the PEN3 dataset is a matrix of 1000 rows and 15 columns.

The confusion matrixes of PNN of both the IAQ-PCA hybrid feature and the PEN3-PCA hybrid feature for classification trials and its respective mean classification accuracy for 50 repetitions have been tabulated in Tables 7 and 8. Both tables consist of the true positive, true negative, false positive and false negative counts, which are useful in computing performance evaluation of the PNN classifier. M1 denotes material 1, and NA denotes normal air.

Table 7. Confusion Matrix of PNN of proposed IAQ-PCA hybrid feature for 50 repetitions.

| | | Actual | | | | | | | | | |
		M1	M2	M3	M4	M5	M6	M7	M8	NA	Mean Classification Accuracy (%)
Predicted	M1	40	0	0	0	0	0	0	0	0	100.00
	M2	0	39	0	0	0	1	0	0	0	99.52
	M3	0	0	40	0	0	0	0	0	0	100.00
	M4	0	0	1	39	0	0	0	0	0	99.12
	M5	0	0	0	0	39	0	1	0	0	99.01
	M6	1	0	0	0	0	39	0	0	0	99.51
	M7	0	0	0	0	0	0	40	0	0	100.00
	M8	0	0	0	0	0	0	1	39	0	99.15
	NA	0	0	0	2	0	0	0	0	78	99.24

Table 8. Confusion Matrix of PNN of proposed PEN3-PCA hybrid feature for 50 repetition.

| | | Actual | | | | | | | | | |
		M1	M2	M3	M4	M5	M6	M7	M8	NA	Mean Classification Accuracy (%)
Predicted	M1	40	0	0	0	0	0	0	0	0	100.00
	M2	0	40	0	0	0	0	0	0	0	100.00
	M3	0	0	40	0	0	0	0	0	0	100.00
	M4	0	0	0	40	0	0	0	0	0	100.00
	M5	0	0	0	0	40	0	0	0	0	100.00
	M6	0	0	0	0	0	40	0	0	0	100.00
	M7	0	0	0	0	0	0	40	0	0	100.00
	M8	0	0	0	0	0	0	0	40	0	100.00
	NA	0	0	0	0	0	0	0	0	80	100.00

The performance evaluation of a classifier can be performed by examining a few statistical measures obtained by calculating the sensitivity, specificity and accuracy scores for the classifier [20]. The sensitivity is the division of the correctly selected decisions over the total decisions which are actually the deserved selections, as shown in Equation (10). The specificity (Equation (11)) indicates the division of correctly rejected decisions by the total decisions which actually deserve rejection. The accuracy is the score of correctly decided decisions over the total decisions made. The accuracy formula is shown in Equation (12):

$$Sensitivity = \frac{TP}{TP + FN} \times 100\% \tag{10}$$

$$\text{Specificity} = \frac{TN}{TN + FP} \times 100\,\%, \text{and} \qquad (11)$$

$$\text{Accuracy} = \frac{TP + TN}{TP + TN + FP + FN} \times 100\% \qquad (12)$$

where, the TP indicates the true positive decisions, FP is the false positive decisions, TN is the true negative decisions and FN is the false negative decisions. Based on Table 7, TP is 315, FP is 5, TN is 78 and FN is 2.

Both hybrid features have been compared with the other best features selected as discussed earlier through Table 5 for both the IAQ and PEN3 datasets. Tables 9 and 10 show that the proposed IAQ-PCA and PEN3-PCA hybrid features have better performances compared to the standard normalised features. The IAQ-PCA hybrid feature recorded a highest accuracy value of 98.25%, while the PEN3-PCA hybrid feature recorded a highest accuracy of 100%.

Table 9. Average PNN classification results comparison between the best features for the IAQ dataset.

Feature	Sensitivity (%)	Specificity (%)	Accuracy (%)
IAQ-PCA Hybrid Feature	99.37	93.98	98.25
RSSV	99.05	91.67	97.50
FVC	98.74	91.57	97.25
RV	99.04	89.53	97.00

Table 10. Average PNN classification results comparison between the best features for the PEN3 dataset.

Feature	Sensitivity (%)	Specificity (%)	Accuracy (%)
PEN3-PCA Hybrid Feature	100.00	100.00	100.00
RSSV	99.85	96.17	99.75
FVC	99.63	96.85	99.51
RLSSV	99.51	96.09	99.29

The proposed features have been compared with other common available classifiers. Feed-forward Neural Network (FFNN), Elman Neural Network (ENN) and k-Nearest Neighbour (kNN) classifiers have been selected for this purpose. The comparison results between the classifiers for the proposed PCA-based hybrid features are presented in Table 11.

Table 11. Average classification results comparison between different classifiers for proposed PCA based hybrid features.

Classifier	IAQ			PEN3		
	Sensitivity (%)	Specificity (%)	Accuracy (%)	Sensitivity (%)	Specificity (%)	Accuracy (%)
PNN	99.75	92.63	98.25	100.00	100.00	100.00
FFNN	98.71	91.53	97.16	99.88	95.47	99.75
ENN	98.53	91.64	97.65	99.78	94.57	99.74
kNN	99.41	91.42	97.89	99.89	95.91	99.85

For FFNN and ENN, the number of hidden layers, the learning rate, the momentum factor, and the type of activation functions have been modified to obtain the best classification performance. The architectures of the classifiers have been modelled to have 21 input neurons, 45 hidden neurons and nine output neurons for the IAQ-PCA hybrid feature, and 15 input neurons, 32 hidden neurons and nine output neurons for the PEN3-PCA hybrid feature, respectively. The learning rate has been set at 0.001 and the momentum factor is 0.85 for both classifiers. In addition, the activation function, the testing tolerance and the maximum iteration have been tuned to log-sigmoid, 0.00001 and 1000, respectively. The backpropagation algorithm has been utilised for the weights training.

For the kNN classifier, the k value has been set to 3 for the IAQ-PCA feature. For the PEN3-PCA feature, the k value is set at 1. The k value in the kNN classifier is extremely training data dependent. Having cross-validation methods such as K- fold and leave-one-out are useful to find the k value which leads to the highest classification generalizability. In these paper, all the parameters involved in these classifiers have been selected based on trial and error to get the best classification accuracy. As seen on Table 11, the sensitivity, specificity and accuracy of each classifier have been tabulated for both features. From the table, it can be clearly seen that the dimensional reduction and fusion of the features to form hybrid features has deliberately increased the classification accuracy of the classifiers. The success rate of PCA-based hybrid features in the PNN classifier surpasses the performance of other common classifiers.

4. Conclusions

Feature selection and feature reduction have been demonstrated in detail. Both combined features from IAQ and PEN3 gives better classification accuracy. In this paper, a PCA-PNN-based feature selection technique has been proposed and investigated. The data has gone through various stages of processing such as normalised feature extraction, feature verification, binary data normalisation, PCA and data randomisation, before it is fed to the classifier. For investigation purposes, PNN has been selected as the classifier and the results have been further tested using other classifiers on the two datasets, The IAQ dataset from the in-house system and the PEN3 dataset from a commercial electronic nose system. As a result, the PEN3 dataset has better classification performance compared to the IAQ dataset for all the comparisons. This could be due to the sensitivity of the PEN3 electronic nose's gas sensors and the data capturing ability of the Winmuster software, which is used commercially. It is also observed from the analysis that the performance of the IAQ electronic nose is almost comparable to that of the PEN3 electronic nose. Thus, it is proven to be useful for early fire detection and prediction of various incipient stage scorching materials.

Acknowledgments: This research work is supported by Malaysian Technical University Network (MTUN) COE research grant (grant number: 9016-00010), Ministry of Education Malaysia under the *Skim Latihan Tenaga Pengajar Akademik IPTA* (SLAI) scholarship and Centre of Excellence for Advanced Sensor Technology (CEASTech), Universiti Malaysia Perlis, Malaysia.

Author Contributions: All authors have agreed with the design of experiment for the research which is prepared according to the European Standard. Ammar Zakaria and Ali Yeon Md Shakaff have contributed with the correction and critical comments on the manuscript. Shaharil Mad Saad has designed the low cost electronic nose (IAQ) unit incorporating gas and electrochemical sensors. Ali Yeon Md Shakaff has given permission to use the PEN3 unit for the experiments.

Conflicts of Interest: The authors declare no conflict of interest.

References

1. Chen, T.; Yuan, H.; Su, G.; Fan, W. An automatic fire searching and suppression system for large spaces. *Fire Saf. J.* **2004**, *39*, 297–307. [CrossRef]
2. Mahdipour, E.; Dadkhah, C. Automatic fire detection based on soft computing techniques: Review from 2000 to 2010. *Artif. Intell. Rev.* **2014**, *42*, 895–934. [CrossRef]
3. Rose-Pehrsson, S.L.; Hart, S.J.; Street, T.T.; Williams, F.W.; Hammond, M.H.; Gottuk, D.T.; Wright, M.T.; Wong, J.T. Early warning fire detection system using a probabilistic neural network. *Fire Technol.* **2003**, *39*, 147–171. [CrossRef]
4. Lasaponara, R.; Santulli, A.; Telesca, L. Time-clustering analysis of forest-fire sequences in southern italy. *Chaos Solitons Fractals* **2005**, *24*, 139–149. [CrossRef]
5. Chakraborty, I.; Paul, T.K. A hybrid clustering algorithm for fire detection in video and analysis with color based thresholding method. In Proceedings of the International Conference on Advances in Computer Engineering, Bangalore, Karnataka, India, 20–21 June 2010; pp. 277–280.

6. Bahrepour, M.; Van der Zwaag, B.J.; Meratnia, N.; Havinga, P. Fire data analysis and feature reduction using computational intelligence methods. In Proceedings of the Intelligent Systems Design and Applications, Cairo, Egypt, 29 November–1 December 2010; pp. 1–10.

7. Romain, A.C.; Nicolas, J.; Wiertz, V.; Maternova, J.; Andre, P.H. Use of a simple tin oxide sensor array to identify five malodours collected in the field. *Sens. Actuators B Chem.* **2000**, *62*, 73–79. [CrossRef]

8. Yu, C.Y.; Zhang, Y.M.; Fang, J.; Wang, J.J. Texture analysis of smoke for real-time fire detection. In Proceedings of the 2nd International Workshop on Computer Science and Engineering, Qingdao, China, 28–30 October 2009; pp. 511–515.

9. Bahrepour, M.; Meratnia, N.; Havinga, P. Use of AI techniques for residential fire detection in wireless sensor networks. In Proceedings of the Artificial Inteligence Applications and Innovations (AIAI 2009) workshops proceedings, Thessaloniki, Greece, 23–25 April 2009; pp. 311–321.

10. Choudhury, J.R.; Banerjee, T.P.; Das, S.; Abraham, A.; Snášel, V. Fuzzy rule based intelligent security and fire detector system. *Comput. Intell. Secur. Inf. Syst. AISC* **2009**, *63*, 45–51.

11. Huseynov, J.J.; Baliga, S.; Widmer, A.; Boger, Z. Infrared flame detection system using multiple neural networks. In Proceedings of the International Joint Conference on Neural Networks (IJCNN 2007), Orlando, Florida, USA, 12–17 August 2007; pp. 608–612.

12. Xu, L.M.; He, W. Application of fuzzy neural network to fire alarm system of high-rise building. *J. Commun. Comput.* **2005**, *2*, 18–21.

13. Arrue, B.C.; Ollero, A.; Martinez de Dios, J.R. An intelligent system for false alarm reduction in infrared forest-fire detection. *IEEE Intell. Syst.* **2000**, *15*, 64–73. [CrossRef]

14. Yuen, R.K.K.; Lee, E.W.M.; Lo, S.M.; Yeoh, G.H. Prediction of temperature and velocity profiles in a single compartment fire by an improved neural network analysis. *Fire Saf. J.* **2006**, *41*, 478–485. [CrossRef]

15. Lim, Y.S.; Lim, S.; Choi, J.; Cho, S.; Kim, C.K.; Lee, Y.W. A fire detection and rescue support framework with wireless sensor networks. In Proceedings of the International Conference on Convergence Information Technology, Gyeongju, South Korea, 21–23 November 2007; pp. 135–138.

16. Saracoglu, R. Hidden markov model-based classification of heart valve disease with pca for dimension reduction. *Eng. Appl. Artif. Intell.* **2012**, *25*, 1523–1528. [CrossRef]

17. Ferre, L. Selection of components in principal component analysis: A comparison of methods. *Comput. Stat. Data Anal.* **1995**, *19*, 669–682. [CrossRef]

18. Luo, R.C.; Su, K.L.; Tsai, K.H. Fire detection and isolation for intelligent building using adaptive sensory fusion method. In Proceedings of the 2002 IEEE International Conference on Robotics & Automation, Washington, DC, USA; 2002; pp. 1777–1781.

19. Rayneri, L.; Colla, V.; Vannucci, M. Estimate of a Probability Density Function through Neural Networks. *Adv. Comput. Intell.* **2011**, *6691*, 57–64.

20. Truong, T.X.; Kim, J.M. Fire flame detection in video sequences using multi-stage pattern recognition techniques. *Eng. Appl. Artif. Intell.* **2012**, *25*, 1365–1372. [CrossRef]

11

Novel H⁺-Ion Sensor Based on a Gated Lateral BJT Pair

Heng Yuan [1,*], **Jixing Zhang** [1], **Chuangui Cao** [1], **Gangyuan Zhang** [1] and **Shaoda Zhang** [2,*]

Academic Editor: W. Rudolf Seitz

[1]   Science and Technology on Inertial Laboratory, Beihang University, No. 37 Xueyuan Road, Beijing 100191, China; zhangjixing@buaa.edu.cn (J.Z.); caochuangui@buaa.edu.cn (C.C.); zhanggangyuan@buaa.edu.cn (G.Z.)
[2]   Pen-Tung Sah Institute of Micro-Nano Science and Technology, Xiamen University, No. 422 South Siming Road, Xiamen 361005, China
*   Correspondences: hengyuan@buaa.edu.cn (H.Y.); shaodazhang@hotmail.com (S.Z.)

Abstract: An H⁺-ion sensor based on a gated lateral bipolar junction transistor (BJT) pair that can operate without the classical reference electrode is proposed. The device is a special type of ion-sensitive field-effect transistor (ISFET). Classical ISFETs have the advantage of miniaturization, but they are difficult to fabricate by a single fabrication process because of the bulky and brittle reference electrode materials. Moreover, the reference electrodes need to be separated from the sensor device in some cases. The proposed device is composed of two gated lateral BJT components, one of which had a silicide layer while the other was without the layer. The two components were operated under the metal-oxide semiconductor field-effect transistor (MOSFET)-BJT hybrid mode, which can be controlled by emitter voltage and base current. Buffer solutions with different pH values were used as the sensing targets to verify the characteristics of the proposed device. Owing to their different sensitivities, both components could simultaneously detect the H⁺-ion concentration and function as a reference to each other. Per the experimental results, the sensitivity of the proposed device was found to be approximately 0.175 μA/pH. This experiment demonstrates enormous potential to lower the cost of the ISFET-based sensor technology.

Keywords: gated lateral BJT; MOSFET–BJT hybrid; ion sensor; ISFET

1. Introduction

The reference electrode that supplies the reference value for sensor data analysis is one of the most important components in a classical ion-sensitive transistor (IST) system. The IST has much been used in not only environmental detection, but also in biosensors [1–4]. With the development of sensor technology, semiconductor-based sensors have been extensively studied because of their perceived advantages such as low cost, small size, and ease of mass production. However, although the ion-sensitive field-effect transistor (ISFET) is based on the semiconductor-based sensor technology, the above advantages are not imputable to the current reference electrode technology. The primary reason is that the classical ISFET, which based on the metal-oxide semiconductor field-effect transistor (MOSFET) structure is difficult to operate under room temperature without gate bias, and the ISFET requires the reference electrode to supply a stable bias to the gate with a long lifetime, which requires Pt and Ag/AgCl reference electrodes [5]. These classical reference electrodes are additional components that increase the size of the device and make it inconvenient to use. Many researchers have improved the reference electrode by focusing on fabrication of the surface of the device [6,7], which is a well-known method for ISFET-based sensor fabrication, but this type of reference electrode is difficult to fabricate by the standard complementary metal-oxide semiconductor (CMOS) mass production process due to the bulky and brittle nature of the reference electrode materials. Moreover, the other fabrication processes cause a sharp increase in the production cost. In recent years, several new devices for ion detection without the reference electrode have been suggested owing to developments in organic materials and nano-technology [8,9]. They have successfully realized ion sensors without reference electrodes, but their cost and mass production fabrication issues have not been resolved yet.

The lateral BJT structure was first proposed in 1964 by Lin using a BJT process [10]. This special device combines a MOSFET and a BJT, and was developed for various power device applications [11]. In previous works, we developed a gated lateral BJT using the standard CMOS process for several sensor applications to achieve specific characteristics such as large dynamic range, high transconductance, and low gate bias [12–17].

In the CMOS process, the silicide treatment process has been widely used to form electrical contacts between the semiconductor material and the supporting interconnect structure [18–20]. In recent works, it was found that the silicide layer affects the characteristics of the gated lateral BJT. In this paper, the impact of the silicide layer on the gated lateral BJT was discussed. Based on this impact, we propose a H^+-ion sensor that can be operated without the classical reference electrode. We fabricated a gated lateral BJT pair, which contains one gated lateral BJT with a silicide layer and another without the silicide layer. After comparing these two components, pH value detection experiments were performed and are discussed in this paper. According to the results, the proposed sensor can be operated without the reference electrode and the sensitivity was found to be approximately 0.175 µA/pH. The success of the proposed sensor has important implications for the development of advanced IST-based pH sensors and biosensors.

2. Experimental Setup

2.1. Fabrication of the Gated Lateral BJT Pair

The proposed sensor consists of one p-type gated lateral BJT with a silicide layer and one p-type gated lateral BJT without the silicide layer, both of which are of the same size and fabricated using a standard 0.35 µm CMOS process, as shown in Figure 1a. In this figure, GLBJT refers to the gated lateral BJT. A floating gate was fabricated on the top of the two components and is shared by the two components. A Si_3N_4 layer was fabricated on the top of the device for H^+-ion concentration detection. The equivalent circuit of the proposed device is shown in Figure 1b. In this figure, the terms V_{E1}, I_{B1}, V_{E2}, and I_{B2} denote the emitter (source) bias voltage and the base current of the two gated lateral BJT components, respectively. The symbols E, B, and C stand for the emitter (source), base, and lateral collector (drain) of the gated lateral BJT structures, respectively. The gated lateral BJT components with/without the silicide layer also share the n-well and a p-substrate. This type of structure can reduce the terminal volume and the errors, induced by the different base supply. In each component, there is one lateral p-n-p structure and one p-type MOSFET structure that can be operated. Besides, a vertical p-n-p BJT can be formed, but cannot be operate because the lateral collectors (drains) and the substrate were grounded. Each gated lateral BJT component could be operated under the MOSFET operation mode, BJT operation mode, and the MOSFET-BJT hybrid operation mode, which have been described in detail in previous works [12–17]. In other words, the MOSFET functional part is switched on by the gate bias supply, and the BJT functional part is switched on by the base current input. If the MOSFET and the BJT functional parts are switched on simultaneously, the gated lateral BJT is operated under the MOSFET-BJT hybrid operation mode. Joarder has proved that, in this mode, the emitter (source) current consists of a proportional MOSFET channel current and a bipolar current [21].

The fabricated device was embedded into a printed circuit board (PCB). All terminals were connected to the pins of the PCB.

(a)

(b)

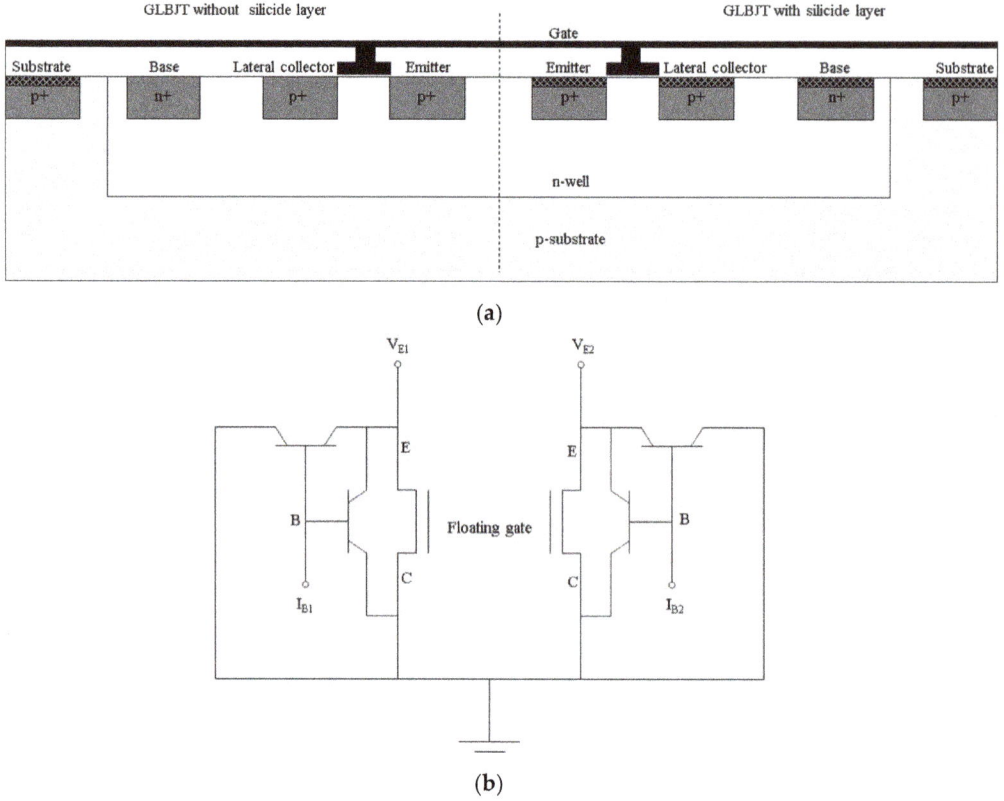

Figure 1. (**a**) Schematic and (**b**) equivalent circuits of the gated lateral BJT components with/without the silicide layer. The dotted line in (**a**) differentiates between the gated lateral BJT components with/without silicide layer.

2.2. pH Value Detection System Setup

The schematic diagram of the pH value detection system setup is shown in Figure 2. It consists of the proposed sensor with PCB connections, a test fixture, a semiconductor test analyzer, and a control system implemented in a computer. The semiconductor test analyzer was used for supplying and detecting the input/output signals. The test fixture was not only used for connecting the semiconductor test analyzer to the sensor, but also to decrease the detection noise.

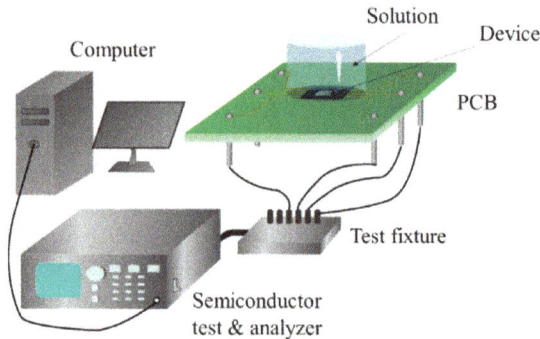

Figure 2. Schematic diagram of the experimental setup.

2.3. Comparison of the Electrical Characteristics of the Two Components with/without the Silicide Layer

According to the transconductance value, the gated lateral BJT has sensing capability [15], so the transconductance of the two components was verified. The gate bias and the base current were varied from –3 V to 5 V and from –50 μA to 20 μA, respectively. The emitter (source) bias supplied was 1 V. The lateral collectors (drains) and the vertical collectors (substrates) were grounded. The two gated lateral BJT components were then operated under the MOSFET–BJT hybrid mode, MOSFET mode and the BJT mode.

2.4. pH Value Detection

pH buffer solutions (with pH values of 4.00, 5.00, 7.00, 9.18, and 10.01) were used as the sensing target. For the setup to work, the two gated lateral BJT components functioned as references for each other, the bases and the floating gate were connected because of the specific structure, and both the emitters (sources) and the collectors (drains) were connected. The base currents were kept constant (–50 μA), and an emitter (source) bias was supplied with 1 V.

The sensing process can be divided into two steps: the first step follows the mechanism of the site-binding model theory that is based on the Si_3N_4 membrane [22–26]. Therefore, when the concentration of H^+-ions varies, the potential of the floating gate is altered. This change affects the sensing process in the MOSFET channel. The second step is the sensing of the H^+-ion concentration. The distinct advantage of this step is that the gated lateral BJT can operate below room temperature without the gate bias [16]. Therefore, based on the structural difference between the gated lateral BJT with the silicide layer and the gated lateral BJT without the silicide layer, the sensing properties are different. Finally, the sensing results are obtained by analyzing the data.

3. Results and Discussion

3.1. Comparison of the V_G-I_E Curve of the Gated Lateral BJTs with/without the Silicide Layer

The transconductance curves of the gated lateral BJTs with/without the silicide layer were obtained from the V_G-I_E curves of the two components, as shown in Figure 3. The classical operation modes of the two components are illustrated in this figure. The lower left curves refer to the MOSFET operation mode. The gate bias and the base current can affect the MOSFET functional parts and the BJT function parts were shut down or switched on, respectively. The portion on the right of the curves, in which the emitter currents have almost 0 μA output, indicate that the gated lateral BJTs were shut down.

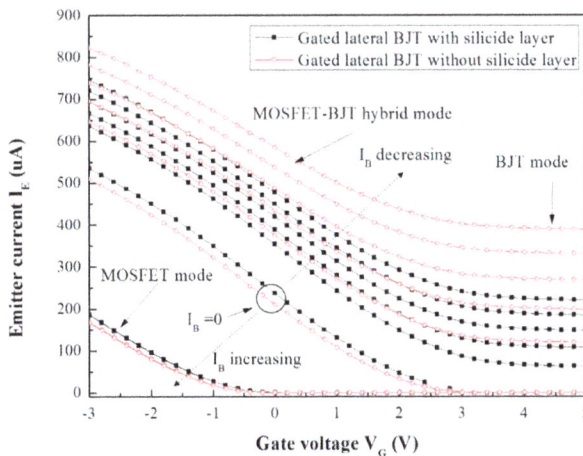

Figure 3. V_G-I_E curve of the gated lateral BJT with/without silicide layer.

In the case of the MOSFET functional parts, owing to the fact the gate bias primarily affects the channel current of the p-type MOSFET, the gate bias increases and the MOSFET channel current is gradually switched off, as shown in Figure 3. Moreover, in the case of the BJT functional parts, after the gate bias is increased, the percentage of the BJT current in the emitter current is increased too. In addition, according to the fact the transconductance of the BJT is higher than that of the MOSFET, the BJT effect and the emitter (source) current increase to produce higher V_G-I_E curves as the base currents decrease. That is because the BJT has higher transconductance characteristics than the MOSFET. Furthermore, in the MOSFET operation mode, the curves did not change according to the base current because the BJT functional parts were switched off.

Following the results, the gated lateral BJT component without the silicide layer exhibited lower threshold voltage in the MOSFET operation mode and higher emitter (source) current during the operation of the BJT functional parts. This is because, first, the silicide layer reduces the doping depth, which reduces the BJT effect. Besides, the MOSFET functional part in the gated lateral BJT component without the silicide layer is nearer to the gate than that in the gated lateral BJT component with the silicide layer. Second, the percentage of the BJT current in the gated lateral BJT component without the silicide layer shows more dominance than the gated lateral BJT component with the silicide layer.

The V_G-I_E curve difference between the gated lateral BJTs with/without the silicide layer can be illustrated as shown in Figure 4.

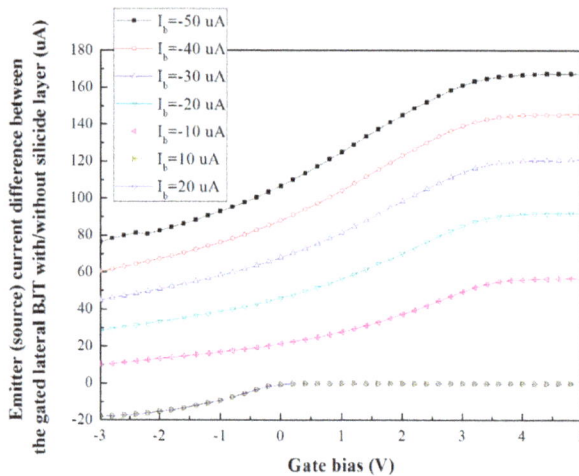

Figure 4. V_G-I_E curve difference between the gated lateral BJT components with/without silicide layer.

In accordance with these results, the difference between the two components increased as the base current decreased. A base current of -50 μA was used in the following ion detection experiment.

3.2. The V_G-g_m Curve Properties of the Two Sensor Components

According to the V_G-I_E curve and Equation (1), the transconductance curve of the proposed device can be obtained, as shown in Figure 5:

$$g_m = \partial I_e / \partial V_{GC} \tag{1}$$

where g_m is the transconductance, ∂_{IE} and ∂V_{GC} are the changes of the emitter (source) current and the gate voltage, respectively.

According to Figure 5, the gated lateral BJT with the silicide layer had a higher transconductance value than the one without the silicide layer at constant base current. However, its transconductance difference was smaller than the one without the silicide layer under varying base current conditions. This is because the BJT effect was dominant when the gated lateral BJT lacked the silicide layer, as described in Section 3.1. Since the sensitivity of the two components was primarily decided by the MOSFET functional parts, the gated lateral BJT component without the silicide layer has lower sensitivity. Moreover, in this case, the percentage of carriers in the BJT and MOSFET functional parts could be controlled easily. It was due to this phenomenon that the transconductance difference of the gated lateral BJT component without the silicide layer was smaller than that with the silicide layer.

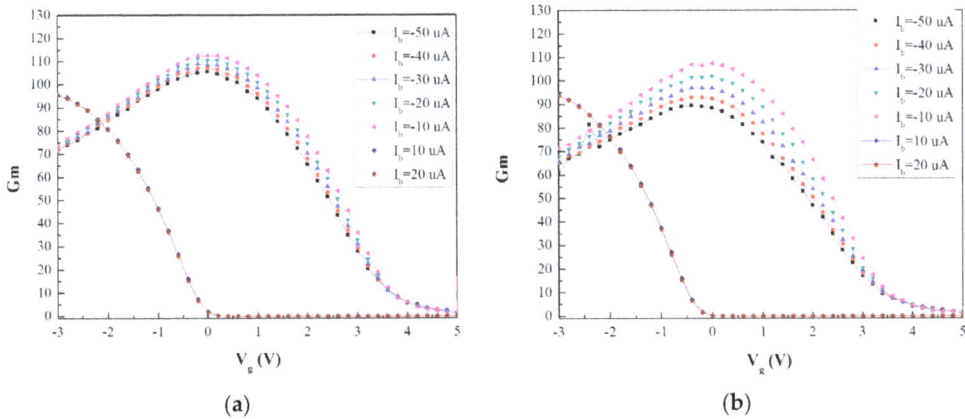

Figure 5. V_G-g_m curve of the gated lateral BJT components with (**a**) and without (**b**) silicide layer.

The sensitivity of a sensor can be calculated as Equation (2):

$$S = \partial I_{output}/\partial V_{input} \tag{2}$$

According to the experimental setup, the input signal was the gate bias, which changed with the H+-ion concentration of the target solution; the output signal was the emitter (source) current. Therefore, the transconductance behavior can reflect the sensitivity of the sensor. Subsequently, the sensitivity of the gated lateral BJT components with/without silicide layer is difference on the basis of Figure 5.

3.3. pH Value Detection Based on Proposed Device

pH value detection experiments were carried out using the variation of the pH value of the target solution with time. The difference between the two gated lateral BJT components was defined as the sensitivity of the sensor. The experiments were performed from a pH value of 7.00 down to the pH value of 4.00, and then, the pH was increased up to a pH value of 10.01, and then back to the pH value of 7.00, as shown in Figure 6a. The results between two pH values were obtained on the device as and when there was a change in the concentration of H+-ions. The reversibility of the proposed device was also supported.

In order to analyze the results, Figure 6b was extracted from Figure 6a. Figure 6b shows the resultant emitter (source) current difference curve against the pH value without the reference electrode. According to the results, the proposed device can be used for H+-ion detection without the reference electrode. The sensitivity was calculated as approximately 0.175 µA/pH. As the pH value increased, the concentration of the H+-ion in the buffer solutions decreased. This induced a negative potential in the floating gate. According to Figure 4, the emitter (source) current difference between the two gated lateral BJT components decreased.

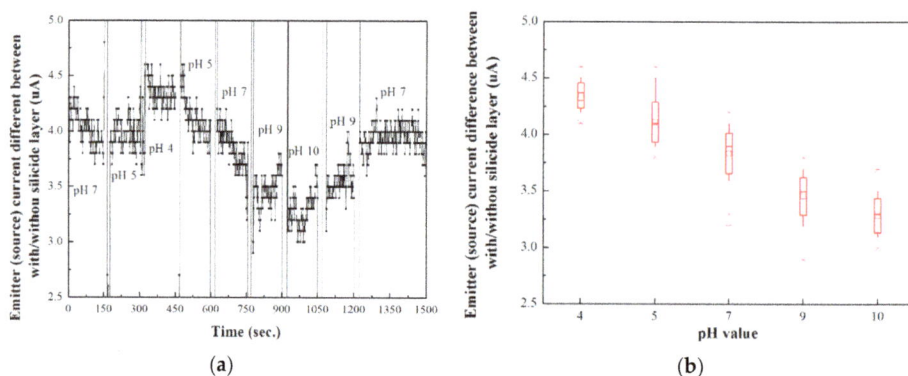

Figure 6. Plot of the H^+-ion detection using (**a**) the proposed gated lateral BJT pair device against time and (**b**) against pH value.

4. Conclusions

A gated lateral BJT pair that consists of two gated lateral BJT components, one with a silicide layer and the other without the silicide layer, was demonstrated for H^+-ion detection. The proposed sensor can be operated without a reference electrode. In order to prove the possibilities of this approach, the V_G-I_E and the transconductance of the gated lateral BJT components with/without the silicide layer were compared. The results show that the BJT effect of the gated lateral BJT without the silicide layer was more distinct than that of the one with the silicide layer. In addition, the gated lateral BJT with the silicide layer had a higher transconductance value and a smaller transconductance difference with varying base currents. After the phenomenon was discussed, an ion detection experiment was performed using a pH buffer solution and a Si_3N_4 sensing membrane. The two gated lateral BJT components function as references for each other. The experiments proved that the proposed device could successfully detect H^+-ions. This device this achieved the detection of H^+-ions without the reference electrode and was fabricated by the standard CMOS process. It demonstrated a significant cost reduction for the ISFET-based sensor technology. Furthermore, this device can be used as a prototype in similar research fields. In the future, the structure of the device will be optimized and apply to the biosensors.

Acknowledgments: This work is supported by the Projects of National Science Foundation of China under grant (No. 61403014).

Author Contributions: Heng Yuan and Shaoda Zhang conceived and designed the device and the experiments; Heng Yuan, Chuangui Cao, and Gangyuan Zhang fabricated the device; Heng Yuan and Jixing Zhang analyzed the data and performed the experiments. Heng Yuan, Jixing Zhang and Shaoda Zhang wrote the paper.

Conflicts of Interest: The authors declare no conflict of interest.

References

1.   Ariga, K.; Yamauchi, Y.; Ji, Q.; Yonamine, Y.; Hill, J.P. Research Update: Mesoporous sensor nanoarchitectionics. *APL Mater.* **2014**, *2*, 1–11. [CrossRef]
2.   Ariga, K.; Ji, Q.; Mori, T.; Naito, M.; Yamauchi, Y.; Abe, H.; Hill, J.P. Enzyme nanoarchitectonics: Organization and device application. *Chem. Soc. Rev.* **2013**, *42*, 6203–6568. [CrossRef] [PubMed]
3.   Ishihara, S.; Labuta, J.; Rossom, W.V.; Ishikawa, D.; Minarni, K.; Hill, J.P.; Ariga, K. Porphyrin-based sensor nanoarchitectonics in diverse physical detection modes. *Phys. Chem.* **2014**, *16*, 9713–9746. [CrossRef] [PubMed]
4.   Ariga, K.; Yamauchi, Y.; Rydzek, G.; Ji, Q.; Yonamine, Y.; Wu, K.C.W.; Hill, J.P. Layer-by-layer nanoarchitectonics: Invention, innovation, and evolution. *Chem. Lett.* **2014**, *43*, 36–68. [CrossRef]
5.   Chen, C.C.; Chen, H.I.; Liu, H.Y.; Chou, P.C.; Liou, J.K.; Liu, W.C. On a GaN-based ion sensitive field-effect transistor (ISFET) with a hydrogen peroxide surface treatment. *Sens. Actuators B* **2015**, *209*, 658–663. [CrossRef]

6.   Deng, X.; Wang, F.; Chen, Z. A novel electrochemical sensor based on nano-structured film electrode for monitoring nitric oxide in living tissues. *Talanta* **2010**, *82*, 1218–1224. [CrossRef] [PubMed]

7.   Goda, T.; Yamada, E.; Katayama, Y.; Tabata, M.; Matsumoto, A.; Miyahara, Y. Potentiometric responses of ion-selective microelectrode with bovine serum albumin adsorption. *Biosens. Bioelectron.* **2016**, *77*, 208–214. [CrossRef] [PubMed]

8.   Kokot, M. Measurement of sub-nanometer molecular layers with ISFET without a reference electrode dependency. *Sens. Actuators B* **2011**, *157*, 424–429. [CrossRef]

9.   Dong, Z.; Wejinya, U.C.; Chalamalasetty, S.N.S. Development of CNT-ISFET based pH sensing system using atomic force microscopy. *Sens. Actuators A* **2012**, *173*, 293–301. [CrossRef]

10. Lin, H.C.; Tan, T.B.; Chang, G.Y.; Leest, V.V.D.; Formigoni, N. Lateral complementary transistor structure for the simultaneous fabrication of functional blocks. *Proc. IEEE* **1964**, *10*, 1491–1495. [CrossRef]

11. Hefner, A.R.; Diebolt, D.M. An experimentally verified IGBT model implemented in the saber circuit simulator. *IEEE Trans. Power Electron.* **1994**, *9*, 532–542. [CrossRef]

12. Kwon, H.C.; Kwon, D.H.; Sawada, K.; Kang, S.W. The characteristics of H+ ion-sensitive transistor driving with MOS hybrid mode operation. *IEEE Electron Device Lett.* **2008**, *29*, 1138–1141. [CrossRef]

13. Yuan, H.; Kwon, H.C.; Yeom, S.H.; Wang, B.; Kim, K.J.; Kwon, D.H.; Kang, S.W. Volatile organic compound gas sensor using a gated lateral bipolar junction transistor. *J. Korean Phys. Soc.* **2011**, *59*, 478–481.

14. Yuan, H.; Kwon, K.C.; Yeom, S.H.; Kwon, D.H.; Kang, S.W. MOSFET-BJT hybrid mode of the gated lateral bipolar junction transistor for C-reactive protein detection. *Biosens. Bioelectron.* **2011**, *28*, 434–437. [CrossRef] [PubMed]

15. Yuan, H.; Kwon, H.C.; Kang, B.H.; Kang, I.M.; Kwon, D.H.; Kang, S.W. Highly sensitive ion sensor based on the MOSFET-BJT hybrid mode of a gated lateral BJT. *Sens. Actuators B* **2013**, *181*, 44–49. [CrossRef]

16. Yuan, H.; Kang, B.H.; Jeong, H.M.; Kwon, H.C.; Yeom, S.H.; Lee, J.S.; Kwon, D.H.; Kang, S.W. Room temperature VOC gas detection using a gated lateral BJT with an assembled solvatochromic dye. *Sens. Actuators B* **2013**, *187*, 288–294. [CrossRef]

17. Yuan, H.; Zhang, J.X.; Zhang, C.; Zhang, N.; Xu, L.X.; Ding, M.; Patrick, J.C. Low gate voltage operated multi-emitter-dot H+ ion-sensitive gated lateral bipolar junction transistor. *Chin. Phys. Lett.* **2015**, *32*, 020701:1–020701:4. [CrossRef]

18. Zhang, S.L.; Ostling, M. Metal silicide in CMOS technology: Past, present, and future trends. *Crit. Rev. Solid State Mater. Sci.* **2003**, *28*, 1–129. [CrossRef]

19. Dalapati, G.K.; Tan, C.C.; Panah, S.M.; Tan, H.R.; Chi, D. Low temperature grown highly texture aluminum alloyed iron silicide on silicon substrate for opto-electronic applications. *Mater. Lett.* **2015**, *159*, 455–458. [CrossRef]

20. Sadia, Y.; Aminov, Z.; Mogilyansky, D.; Gelbstein, Y. Texture anisotropy of higher manganese silicide following arc-melting and hot-pressing. *Intermetallics* **2016**, *68*, 71–77. [CrossRef]

21. Joardar, K. An improved analytical model for collector currents in lateral bipolar transistors. *IEEE Trans. Electron Devices* **1994**, *41*, 373–382. [CrossRef]

22. Malgras, V.; Ji, Q.; Kamachi, Y.; Mori, T.; Shieh, F.K.; Wu, K.C.W.; Ariga, K.; Yamauchi, Y. Templated synthesis for nanoarchitectured porous materials. *Bull. Chem. Soc. Jpn.* **2015**, *88*, 1171–1200. [CrossRef]

23. Fung, C.D.; Cheung, P.W.; Ko, W.H. A generalized theory of an electrolyte-insulator-semiconductor field-effect transistor. *IEEE Electron Device Lett.* **1986**, *33*, 8–18. [CrossRef]

24. Koch, S.; Woias, P.; Meixner, L.K.; Drost, S.; Wolf, H. Protein detection with a novel ISFET-based zeta potential analyzer. *Biosens. Bioelectron.* **1999**, *14*, 413–421. [CrossRef]

25. Fernandes, P.G.; Stiegler, H.J.; Zhao, M.; Cantley, K.D.; Obradovic, B.; Chapman, R.A.; Wen, H.C.; Mahmud, G.; Vogel, E.M. SPICE macromodel of silicon-on-insulator-field-effect-transistor-based biological sensors. *Sens. Actuators B* **2012**, *161*, 163–170. [CrossRef]

26. Chermiti, J.; Ali, M.B.; Dridi, C.; Gonchar, M.; Jaffrezic-Renault, N.; Korpan, Y. Site-binding model as a basis for numerical evaluation of analytical parameters of capacitance-biosensors for formaldehyde and methylamine detection. *Sens. Actuators B* **2013**, *188*, 824–830. [CrossRef]

12

Bio-Inspired Stretchable Absolute Pressure Sensor Network

Yue Guo [1,*], Yu-Hung Li [2], Zhiqiang Guo [3], Kyunglok Kim [1], Fu-Kuo Chang [4] and Shan X. Wang [1,2]

Academic Editor: Vittorio M.N. Passaro

[1] Department of Electrical Engineering, Stanford University, 350 Serra Mall, Stanford, CA 94305, USA; kyunglok.kim@gmail.com (K.K.); sxwang@stanford.edu (S.X.W.)
[2] Department of Materials Science and Engineering, Stanford University, 476 Lomita Mall, Stanford, CA 94305, USA; liyuhung@stanford.edu
[3] Department of Mechanical Engineering, Stanford University, 440 Escondido Mall, Stanford, CA 94305, USA; zguo@stanford.edu
[4] Department of Aeronautics and Astronautics, Stanford University, 496 Lomita Mall, Stanford, CA 94305, USA; fkchang@stanford.edu
* Correspondence: yueguo@stanford.edu

Abstract: A bio-inspired absolute pressure sensor network has been developed. Absolute pressure sensors, distributed on multiple silicon islands, are connected as a network by stretchable polyimide wires. This sensor network, made on a 4″ wafer, has 77 nodes and can be mounted on various curved surfaces to cover an area up to 0.64 m × 0.64 m, which is 100 times larger than its original size. Due to Micro Electro-Mechanical system (MEMS) surface micromachining technology, ultrathin sensing nodes can be realized with thicknesses of less than 100 μm. Additionally, good linearity and high sensitivity (~14 mV/V/bar) have been achieved. Since the MEMS sensor process has also been well integrated with a flexible polymer substrate process, the entire sensor network can be fabricated in a time-efficient and cost-effective manner. Moreover, an accurate pressure contour can be obtained from the sensor network. Therefore, this absolute pressure sensor network holds significant promise for smart vehicle applications, especially for unmanned aerial vehicles.

Keywords: stretchable network; absolute pressure sensor; MEMS; smart skin

1. Introduction

Recently, bio-inspired multifunctional sensor networks, the fundamental component of smart artificial skin, have attracted increasing attention due to their emerging broad application perspectives [1,2]. A variety of sensor networks with different sensors to detect physical parameters, such as temperature, strain, and vibration, have been studied [3–5]. Absolute pressure is also one useful parameter to be tracked, especially in the fields of automotive, nautical, and aerospace applications, where next-generation self-aware smart vehicles rely heavily on accurate absolute pressure sensing [6–9]. In the case of Unmanned Aerial Vehicles (UAVs), monitoring the absolute pressure distribution surrounding the entire aircraft in real-time is highly desirable for assessing its health condition and adjusting itself to avoid hazardous stalling [10]. Thus, an absolute pressure sensor network consisting of a large amount of sensing elements with high sensitivity is required for the purpose of achieving adequate sensing resolution. The sensor network also needs flexibility to be mounted on curvature surfaces, such as an airfoil in smart autonomous vehicle applications. Additionally, small size, light weight, and low cost are highly preferred due to practical application needs.

Micro Electro-Mechanical System (MEMS) sensors with piezoresistive and capacitive pressure sensing mechanisms have been widely explored in previous studies, not only on rigid silicon wafers but different flexible substrates as well [11–17]. Although high sensitivity can be achieved in traditional MEMS pressure sensors, those sensors, generally built on rigid silicon substrates, do not feature useful stretchability and flexibility. In this work, a stretchable absolute pressure sensor network has been developed, aimed at application in smart autonomous vehicles, with decision-making capability and self-adaptive controllability. Absolute pressure sensors are located on multiple silicon islands and connected as a network by routing aluminum wires supported by flexible stretchable straps made of polyimide. This sensor network, originally fabricated on a 4″ silicon wafer, can be expanded up to 0.64 m × 0.64 m, which is 100 times larger than its original size. In terms of transducer mechanisms, piezoresistance is used to guarantee performance of sensitivity and linearity, as well as low complexity and cost. Thanks to MEMS surface micromachining technology, very thin pressure sensors can be fabricated on Silicon-On-Insulator (SOI) wafers. Since the MEMS sensor fabrication process has been well integrated with a flexible polymer substrate process, the entire sensor network can be made efficiently in one process on standard silicon wafers, then subsequently stretched, and finally mounted on various surfaces, where no further sensor alignment and interconnection are needed. Furthermore, an accurate pressure contour can be achieved from a sufficient number of sensing nodes, and a self-awareness intellectual system can be improved by taking advantages of absolute pressure distribution data captured from entire vehicle surfaces. Therefore, this new pressure sensor network holds significant promise for smart vehicle applications.

Figure 1 illustrates a schematic overview of smart skin materials in smart vehicle applications. The stretchable pressure sensor network, along with other types of networks such as temperature and vibration, can be integrated into a smart skin. In order to sense the absolute pressure, this pressure sensor network needs to be placed near the top surface of the smart skin structure. Voltage signals measured from the pressure network can be collected locally and sent to a central controller by a wireless transmitter.

Figure 1. Schematic overview of stretchable pressure sensor network integrated into smart skin materials.

2. Experimental Section

A fabrication process, combining surface micromachining and a flexible polymer substrate process together, has been developed to realize the stretchable pressure sensor network. Figure 2 illustrates the main fabrication steps. The piezoresistive absolute pressure sensing elements include a sealed vacuum cavity, a single-crystal silicon diaphragm in a square shape, and top electrodes. An SOI wafer is used as a starting substrate with a 2 μm-thick buried oxide layer. Multiple tiny holes with a diameter of 0.6 μm, which serve as venting holes, are first etched through the top silicon device layer using deep reactive-ion etching (DRIE). The buried oxide layer is then etched by a subsequent vapor-phase hydrofluoric acid (HF) process for 5 h, to release the top active layer. Next, these venting holes are sealed by the growth of single-crystal silicon at 1150 °C in an Applied Materials Centura epitaxial system. A high-quality single-crystal silicon membrane is preferred for obtaining large piezoresistivity, in order to achieve a sensitive and linear response to pressure variations. The silicon diaphragm has a square shape with a dimension of 550 μm by 550 μm. After desired mask patterns are defined lithographically, ion implantation of boron and drive-in annealing are performed to change active silicon areas to p-type silicon with a boron doping concentration of ~5×10^{18} cm^{-3}. By optimizing process parameters, a good electrical insulation is achieved between the active piezoresistors and the n-type substrate.

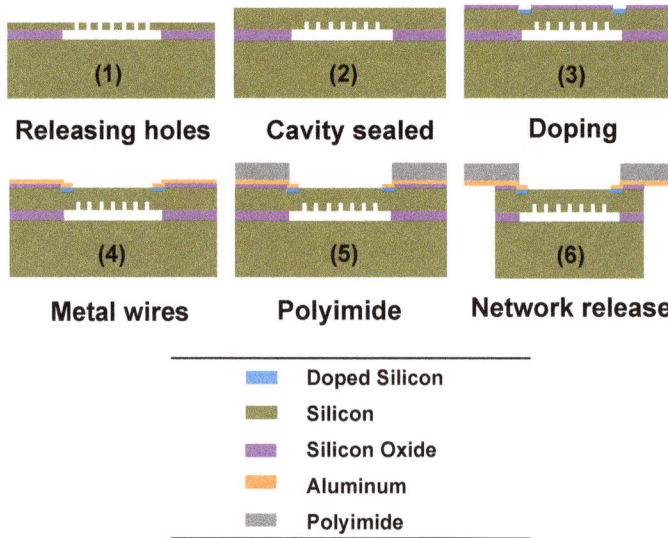

Figure 2. Main fabrication processes of the stretchable pressure sensor network.

The steps discussed above are suitable for building high-performance sensing elements, whereas the following steps serve to connect elements as a stretchable sensor network. An aluminum layer, which serves as electrodes and wires, is first sputtered on top of the diaphragm. Next, a polyimide layer of 15 μm thickness is spin coated onto the aluminum layer, followed by another aluminum deposition as a mask layer. The stretchable wires are then lithographically patterned and dry etched by oxygen plasma. Finally, the device wafer is mounted upside down onto another carrier wafer, and the process concludes with a through silicon etch by DRIE to release the sensor network. Therefore, the MEMS absolute pressure sensors can be fully integrated with the stretchable polyimide substrate.

3. Results and Discussion

In Figure 3, the stretchability and flexibility of the absolute pressure sensor network is demonstrated. Figure 3a shows a patterned pressure sensor network on a 4″ silicon wafer before release. There are seven pressure-sensing elements on this network, located on silicon islands aligned in the diagonal direction, whereas the remaining islands are used only for routing signals out in this demonstration with a single layer of aluminum wire. Further, more pressure sensing nodes can be available on a network with multilayer aluminum wires. Figure 3b depicts a released network. Small anchors are adopted to hold together the silicon islands at the edges. Figure 3c is a zoomed-in view of stretchable wires and silicon islands from the backside, which illustrates the good condition of wires after releasing. The polyimide wires are robust enough to hold the silicon islands that have a dimension of 4 mm by 4 mm, and each individual wire connects two islands. An expanded pressure sensor network is shown in Figure 3d. It has 77 nodes and covers an area up to about 0.64 m × 0.64 m, which is 100 times larger than its original size. Sensing elements are located at the center of rigid silicon islands, with a sealed vacuum cavity for the purpose of absolute pressure sensing. Due to the rigidity of individual silicon nodes, performances of individual absolute pressure sensing elements on the stretchable network are similar to separate micromachined transducers. A closer view of the island is given in Figure 3e. By leveraging the benefits from MEMS surface micromachining process, 4 mm × 4 mm silicon islands can be thinned down to less than 100 μm. Compared with typical commercial MEMS pressure sensors that are usually made by bulk micromachining technology, the ultrathin sensing elements make them well suited for installation on vehicle surfaces without affecting the airflow. In Figure 3f, a stretched pressure network is mounted onto a PVC soft film surface, and is easily held by hand. As shown, the proposed stretchable absolute sensor network holds great potential to be mounted onto various surfaces in integration.

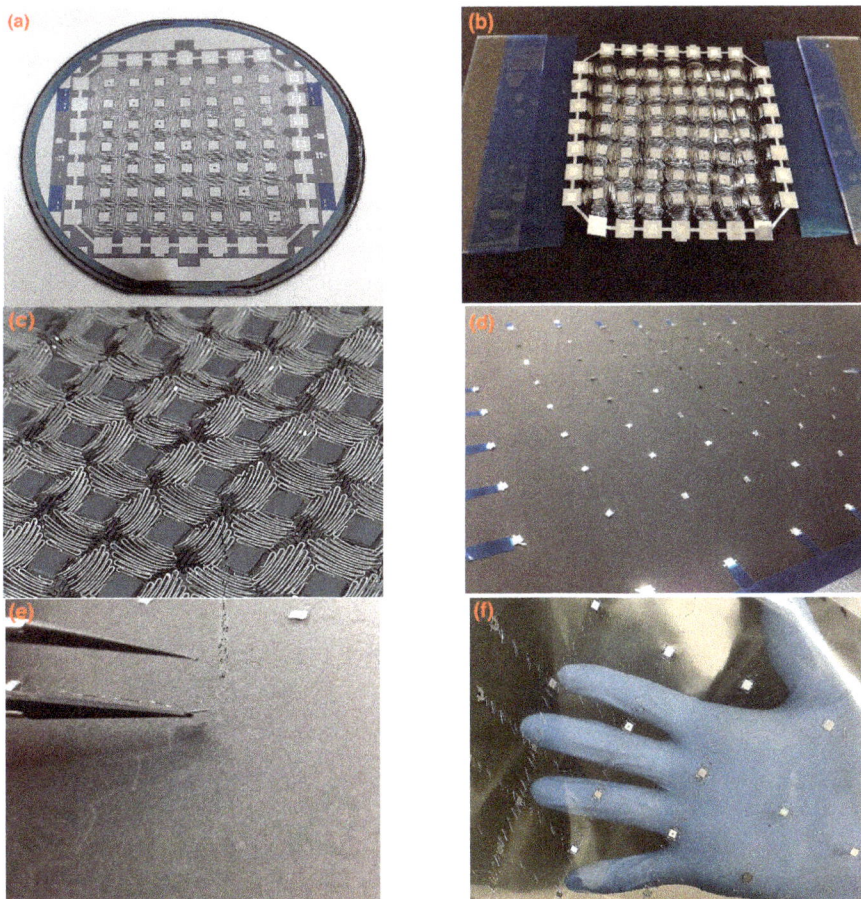

Figure 3. (a) Patterned pressure sensor network before releasing; (b) Released network before stretching; (c) Zoomed-in view of stretchable wires; (d) Stretched pressure sensor network; (e) Zoomed-in view of a silicon island and stretchable wires; (f) Stretched pressure network mounted onto a PVC soft film.

In terms of transducer mechanisms, both piezoresistive and capacitive sensing approaches can be used on the stretchable absolute pressure sensor network. Table 1 lists characteristic performance metrics for comparison between piezoresistive and capacitive sensing mechanisms [13,18–20]. Compared with capacitive sensing, piezoresistive mechanism has three advantages when applied to stretchable absolute pressure sensor networks: excellent linearity, simple interface electronics, and low cost. Firstly, an excellent linearity can typically be achieved on the network sensing nodes with small diaphragms. By contrast, capacitance changes nonlinearly with diaphragm displacement and its corresponding applied pressure in a capacitive absolute pressure sensor with parallel plates. Secondly, a piezoresistive absolute pressure sensor has relatively simple interface electronics, whereas a capacitive pressure sensor typically requires additional interface electronics to convert the sensor capacitance value to a voltage output. Thirdly, piezoresistive pressure sensors have fewer lithography steps in their fabrication process, thereby greatly reducing their complexity and cost. In the case of capacitive sensors, both top and bottom electrodes are needed, complicating the micromachining process. Because of these benefits discussed above, the piezoresistive transducer mechanism is employed in this work to construct the stretchable absolute pressure sensor network.

Table 1. Comparison between piezoresistive and capacitive sensing mechanisms.

Characteristic	Piezoresistive Sensing	Capacitive Sensing
Linearity	Good	Fair
Accuracy	$\pm 1\%$	$\pm 0.2\%$
Resolution	1 part in 10^5	1 part in 10^4 to 10^5
Temperature error	$\sim 1600 \times 10^{-6}/°C$	$\sim 4 \times 10^{-6}/°C$
Cost	Low	Medium
Electronics	Simple	Complex

In order to understand the mechanical characteristics of the sensing elements, both analytical calculations and multi-physics finite element analysis were conducted [21–23]. When a pressure difference exists between the inside and outside of the cavity, the top silicon diaphragm deforms correspondingly. Figure 4 plots finite element simulation results from the Ansys software.

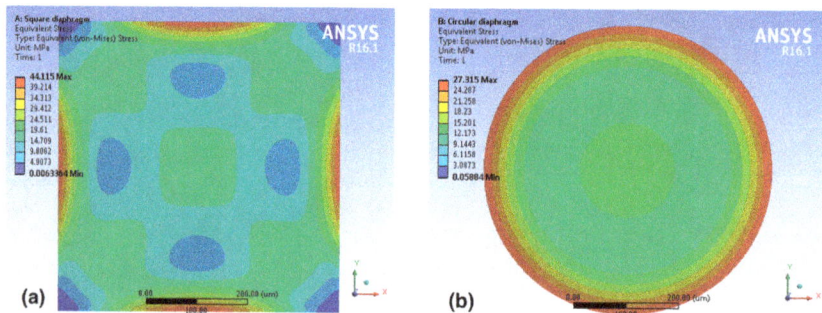

Figure 4. Contour plot of the Von Mises stress distribution on (**a**) a square and (**b**) a circular diaphragm.

The color contours illustrate Von Mises stress distribution in MPa on a square and a circular diaphragm, respectively. Both square and circular diaphragms have the same thickness and edge/diameter length. In comparison with circular diaphgrams, square diaphragms show better performance. When a uniform pressure of 100 kPa is applied, the maximum stress of the square diaphragm is about 1.62 times as large as that of the circular diaphragm, which is also consistent with prior results in literature [24]. Therefore, square diaphragms are employed in sensing elements of the stretchable sensor network. Based on our simulation results, maximum stresses occur near the edges of the square diaphragm. Hence, four piezoresistors are patterned at the most sensitive regions, and two different pairs of piezoresistor geometries are adopted at the opposite edges to increase the sensitivity. Also, sensors are connected in a full Wheatstone bridge configuration, which enables differential sensing and helps reduce errors from temperature changes. Assuming the full Wheatstone bridge is well balanced, and all four piezoresistors have the same resistance value, the relative output voltage change of the sensing elements can be expressed as follows [25]:

$$\frac{\Delta V}{V} = \frac{\Delta R}{R} = \pi_l \sigma_l + \pi_t \sigma_t \tag{1}$$

where π_l and σ_l are the piezoresistive coefficient and stress in the longitudinal direction of piezoresistors, and π_t and σ_t are the values in the transverse direction. In order to achieve optimum sensitivity, the piezoresistors are aligned along the <110> direction on a p type (100) silicon wafer. Since piezoresistive coefficient is highly dependent on doping concentration [12,19], it drops to about 70% of its maximum value at a doping concentration of ~5 × 10^{18} cm^{-3}. Figure 5 shows simulated results of a pressure sensing element at 1V input supply, and illustrates that a linear voltage output with a sensitivity of 18.7 mV/V/bar can be expected from sensing elements on the stretchable pressure sensor network.

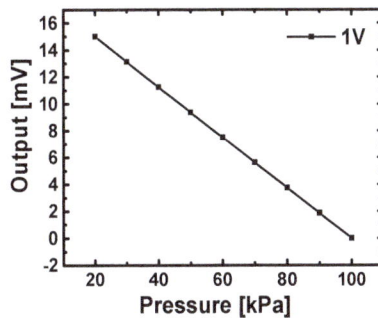

Figure 5. Simulated results of a pressure sensing element at 1 V input supply.

Figure 6a is a scanning electron microscopy (SEM) top-view of the MEMS absolute pressure transducer for the stretchable sensor network. The dark black regions near the edges of the top silicon square diaphragm in the plot are four piezoresistors with a concentration of boron doping (~5 × 10^{18} cm^{-3}), which gives a resistance value about 2 kΩ. The doping concentration was optimized not only to achieve high piezoresistivity but also for good isolation. Furthermore, an aluminum layer, which connects those piezoresistors to each other via ohmic contact, is deposited at the same time as routing wires for the sensor network. Figure 6b provides an SEM cross-section view showing a cavity sealed by a single-crystal silicon membrane. This high-quality top membrane, epitaxially grown at a high temperature of 1150 °C, offers high piezoresistivity, which then confers high sensitivity.

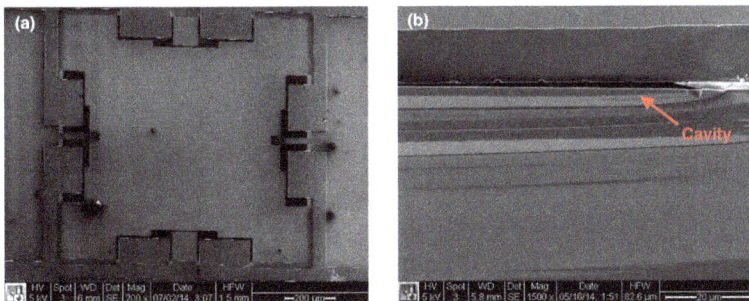

Figure 6. (a) Top view of a piezoresistive pressure sensing element by SEM; (b) SEM cross-section view of a sensing element.

The pressure sensors are characterized in a vacuum chamber, whose pressure can be controlled by a regulating valve. A high precision Druck PACE1000 pressure indicator (General Electric, Fairfield, CT, USA) is utilized to monitor reference chamber pressures. The pressure indicator has a resolution of 6.5 Pa in the measurement range of 3.5 kPa to 130 kPa [26]. Figure 7a provides measurement results of an absolute pressure sensing element at different input voltage supplies. The sensing elements on the stretchable network have a sensitivity of 14 mV/V/bar. When pressure decreases from 100 kPa down to 30 kPa, the output voltages (in mV) increase linearly. Also, the output signals are proportional to the input voltages, as a 5 V voltage input gives five times larger output signals than does 1 V. The pressure sensor outputs illustrate a good linearity, as shown in Figure 7b, which gives zoomed-in data from 95 kPa to 100 kPa at input voltages of 3 V and 5 V. As shown, the trends between output voltages and pressures are essentially straight lines. All plotted data are raw data measured directly from multimeter outputs, without any further signal amplification, conditioning, or averaging, other than removing the output offsets. Furthermore, measured output voltage *versus* pressure at the input voltage of 1 V is fitted to a straight line with an adjusted R-square value of 0.999, and the corresponding linear fitting deviations are shown in Figure 7c. Fitting deviations are the differences between observed data and predicted values from linear regression fitting, and then are normalized by the maximum output voltage in the measured range of 30 kPa to 100 kPa. The absolute pressure sensing element has normalized fitting deviations within ±0.15%, illustrating a good linearity. In addition, the performance of three sensing elements from different locations on the wafer measured at 1 V input voltage supply are shown in Figure 7d, where sensitivity is calculated from derivative of the raw output voltage with respect to the input pressure at each individual measured point. All three sensors give a very similar pressure response and a sensitivity of ~14 mV/V/bar. Therefore, the fabricated absolute pressure sensing elements show good uniformity.

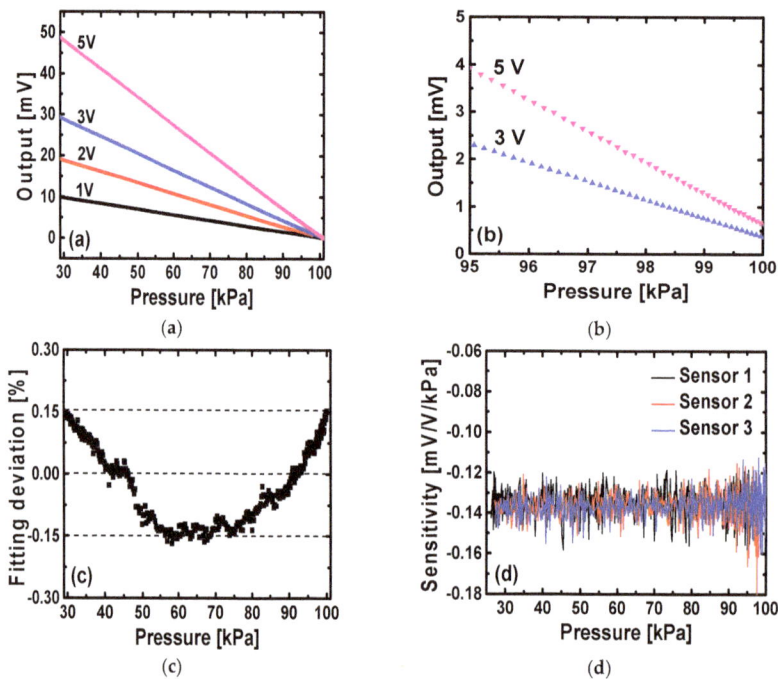

Figure 7. (a) Measured output voltages of a pressure sensing element at different input supplies; (b) Zoomed-in plot from 95 kPa to 100 kPa at the input voltages of 3 V and 5 V; (c) Linear fitting deviation of measured data at the input voltage of 1 V; (d) Sensitivity of three sensing elements from different locations on the wafer.

Temperature characteristics are also studied. In the experiments, temperature is controlled by a model 5310 temperature controller from Arroyo Instrument (San Luis Obispo, CA, USA), which monitors platinum resistance temperature detectors and drives Peltier thermoelectric modules, simultaneously. In order to minimize measurement error, resistance temperature detectors are attached next to pressure sensors, and thermoelectric modules are placed underneath the pressure sensors. As depicted in Figure 8, the ranges of output voltages shrink linearly with the increase in temperature from 25 °C to 60 °C at the input voltage of 1 V, and the corresponding temperature coefficient of sensitivity is −0.176%/°C. This temperature coefficient can be improved on our sensing elements by further reducing the mismatch between piezoresistors and adding temperature compensation circuitry. In smart skin applications as illustrated in Figure 1, the stretchable pressure sensor network can be integrated with a temperature sensor network. Since pressure outputs can be corrected by outputs from such a reference temperature network, minuscule pressure differences can be distinguished from temperature drifts.

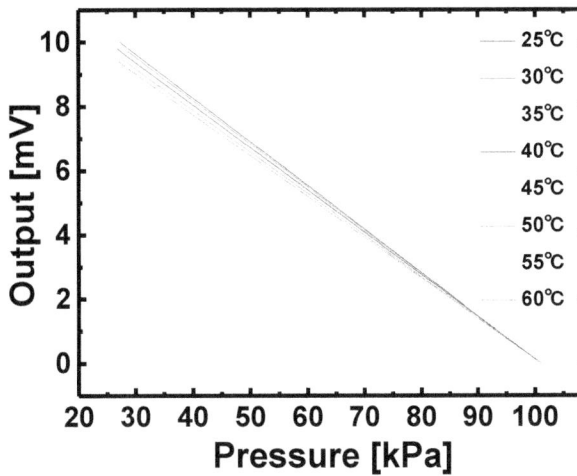

Figure 8. Temperature characteristics of network sensors from 25 °C to 60 °C.

Figure 9a illustrates a measurement setup for the stretchable pressure network prototype. The sensor network is mounted on the surface of a flat foam board. A compressed air gun was employed in the measurement in order to apply positive pressure to the sensing nodes. As the air gun moves along the diagonal direction of the network, five sensing nodes (depicted in red) were influenced sequentially. The measured pressure data were collected by a National Instrument data acquisition board and then sent to a desktop computer for signal processing. Figure 9b depicts a network response of five nodes to the moving compressed air gun. In this measurement, the air gun, with a tiny outlet size of less than 2 mm in diameter, has a volumetric flow rate around 100 standard cubic feet per hour (SCFH). Five downward peaks represent the pressure increase on top of sensing elements caused by the compressed air gun flow in the vertical direction. Thus, the absolute pressure network is capable of detecting pressure variations at different locations. With multiple layers of interconnecting wires, more sensing nodes can be built into one network, resulting in a more accurate absolute pressure contour on smart skins. The output data can have very broad applications. For example, lift force may be directly calculated by the absolute pressure distribution data over airfoil surfaces, which holds great significance in flight monitoring and stall detection for self-awareness UAVs. Further study on data processing and algorithms for various applications is ongoing.

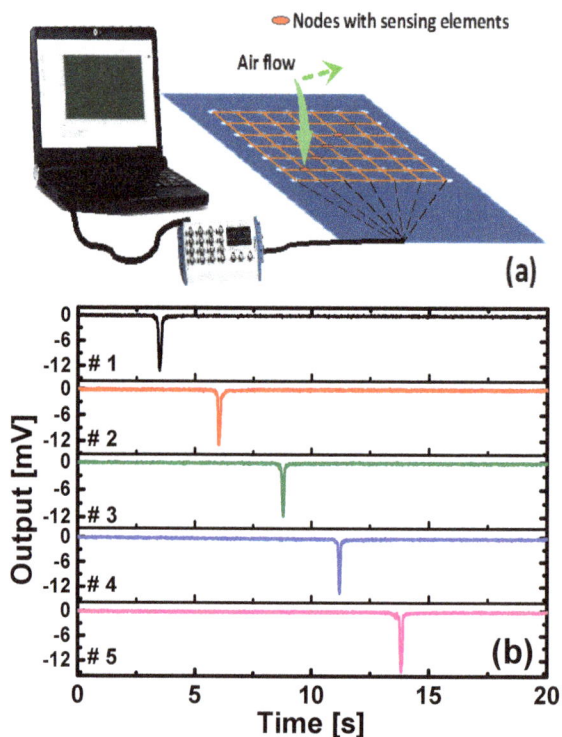

Figure 9. (**a**) Measurement setup of the stretchable pressure network prototype; (**b**) Network response from five nodes to a moving compressed air gun.

4. Conclusions

In summary, a bio-inspired stretchable absolute pressure sensor network with benefits of good performance and cost effectiveness has been investigated. This sensor network is flexible and thin, and can be attached to diverse surfaces to measure absolute pressure distribution without disturbing the airflows. Preliminary measurements have been performed on the prototype pressure network to demonstrate its excellent performance. Therefore, the stretchable absolute pressure sensor network is a promising technology toward building smart functional skins for autonomous vehicles.

Acknowledgments: The authors appreciate the support of this work through the MURI program by the U.S. Air Force Office of Scientific Research (AFOSR). The authors thank Thomas Kenny and Yushi Yang for their assistance with sensor measurements. The authors also thank Dawson Wong and Chin Chun Ooi for critical reading of the manuscript.

Author Contributions: Yue Guo, Shan X. Wang and Fu-Kuo Chang conceived and designed the experiments. Yue Guo fabricated the devices and performed the measurements. Yu-Hung Li participated in the network design and fabrication. Zhiqiang Guo participated in the method development. Kyunglok Kim contributed to scientific discussions. Fu-Kuo Chang and Shan X. Wang supervised and coordinated the research activity.

Conflicts of Interest: The authors declare no conflict of interest.

References

1.   Lanzara, G.; Salowitz, N.; Guo, Z.; Chang, F.K. A spider-web-like highly expandable sensor network for multifunctional materials. *Adv. Mater.* **2010**, *22*, 4643–4648. [CrossRef] [PubMed]
2.   Hammock, M.L.; Chortos, A.; Tee, B.C.; Tok, J.B.; Bao, Z. 25th anniversary article: The evolution of electronic skin (e-skin): A brief history, design considerations, and recent progress. *Adv. Mater.* **2013**, *25*, 5997–6038. [CrossRef] [PubMed]

3. Shih, W.P.; Tsao, L.C.; Lee, C.W.; Cheng, M.Y.; Chang, C.; Yang, Y.J.; Fan, K.C. Flexible temperature sensor array based on a graphite-polydimethylsiloxane composite. *Sensors* **2010**, *10*, 3597–3610. [CrossRef] [PubMed]
4. Katragadda, R.B.; Xu, Y. A novel intelligent textile technology based on silicon flexible skins. *Sens. Actuators A Phys.* **2008**, *143*, 169–174. [CrossRef]
5. Salowitz, N.; Guo, Z.; Li, Y.H.; Kim, K.; Lanzara, G.; Chang, F.K. Bio-inspired stretchable network-based intelligent composites. *J. Compos. Mater.* **2012**, *47*, 97–105. [CrossRef]
6. Callegari, S.; Zagnoni, M.; Golfarelli, A.; Tartagni, M.; Talamelli, A.; Proli, P.; Rossetti, A. Experiments on aircraft flight parameter detection by on-skin sensors. *Sens. Actuators A Phys.* **2006**, *130–131*, 155–165. [CrossRef]
7. Que, R.; Zhu, R. Aircraft Aerodynamic Parameter Detection Using Micro Hot-Film Flow Sensor Array and BP Neural Network Identification. *Sensors* **2012**, *12*, 10920–10929. [CrossRef] [PubMed]
8. Zagnoni, M.; Golfarelli, A.; Callegari, S.; Talamelli, A.; Bonora, V.; Sangiorgi, E.; Tartagni, M. A non-invasive capacitive sensor strip for aerodynamic pressure measurement. *Sens. Actuators A Phys.* **2005**, *123–124*, 240–248. [CrossRef]
9. Kottapalli, A.G.P.; Asadnia, M.; Miao, J.M.; Barbastathis, G.; Triantafyllou, M.S. A flexible liquid crystal polymer mems pressure sensor array for fish-like underwater sensing. *Smart Mater. Struct.* **2012**, *21*, 115030. [CrossRef]
10. Mohamed, A.; Watkins, S.; Clothier, R.; Abdulrahim, M.; Massey, K.; Sabatini, R. Fixed-wing mav attitude stability in atmospheric turbulence—Part 2: Investigating biologically-inspired sensors. *Prog. Aerosp. Sci.* **2014**, *71*, 1–13. [CrossRef]
11. Narducci, M.; Yu-Chia, L.; Fang, W.; Tsai, J. CMOS MEMS capacitive absolute pressure sensor. *J. Micromech. Microeng.* **2013**, *23*, 055007. [CrossRef]
12. Mohammed, A.A.; Moussa, W.A.; Lou, E. High-performance piezoresistive mems strain sensor with low thermal sensitivity. *Sensors* **2011**, *11*, 1819–1846. [CrossRef] [PubMed]
13. Fragiacomo, G.; Reck, K.; Lorenzen, L.; Thomsen, E.V. Novel designs for application specific mems pressure sensors. *Sensors* **2010**, *10*, 9541–9563. [CrossRef] [PubMed]
14. Lim, H.C.; Schulkin, B.; Pulickal, M.J.; Liu, S.; Petrova, R.; Thomas, G.; Wagner, S.; Sidhu, K.; Federici, J.F. Flexible membrane pressure sensor. *Sens. Actuators A Phys.* **2005**, *119*, 332–335. [CrossRef]
15. Ahmed, M.; Butler, D.P.; Celik-Butler, Z. Mems absolute pressure sensor on a flexible substrate. In Proceedings of the 2012 IEEE 25th International Conference on Micro Electro Mechanical Systems (MEMS), Paris, France, 29 January–2 February 2012; pp. 575–578.
16. Zhou, L.; Jung, S.; Brandon, E.; Jackson, T.N. Flexible substrate micro-crystalline silicon and gated amorphous silicon strain sensors. *Electron Devices IEEE Trans.* **2006**, *53*, 380–385. [CrossRef]
17. Wang, Y.-H.; Lee, C.-Y.; Chiang, C.-M. A mems-based air flow sensor with a free-standing micro-cantilever structure. *Sensors* **2007**, *7*, 2389–2401. [CrossRef]
18. Greenwood, J.C. Silicon in mechanical sensors. *J. Phys. E Sci. Instrum.* **1988**, *21*, 1114–1128. [CrossRef]
19. Eaton, W.P.; Smith, J.H. Micromachined pressure sensors: Review and recent developments. *Smart Mater. Struct.* **1997**, *6*, 530–539. [CrossRef]
20. Ko, W.H.; Wang, Q. Touch mode capacitive pressure sensors. *Sens. Actuators A Phys.* **1999**, *75*, 242–251. [CrossRef]
21. Barlian, A.A.; Park, W.-T.; Mallon, J.R.; Rastegar, A.J.; Pruitt, B.L. Review: Semiconductor piezoresistance for microsystems. *IEEE Proc.* **2009**, *97*, 513–552. [CrossRef] [PubMed]
22. Hopcroft, M.A.; Nix, W.D.; Kenny, T.W. What is the young's modulus of silicon? *Microelectromech. Syst. J.* **2010**, *19*, 229–238. [CrossRef]
23. Tian, B.; Zhao, Y.; Jiang, Z.; Zhang, L.; Liao, N.; Liu, Y.; Meng, C. Fabrication and structural design of micro pressure sensors for tire pressure measurement systems (TPMS). *Sensors* **2009**, *9*, 1382–1393. [CrossRef] [PubMed]
24. Kanda, Y.; Yasukawa, A. Optimum design considerations for silicon piezoresistive pressure sensors. *Sens. Actuators A Phys.* **1997**, *62*, 539–542. [CrossRef]

25. Peng, C.-T.; Lin, J.-C.; Lin, C.-T.; Chiang, K.-N. Performance and package effect of a novel piezoresistive
     pressure sensor fabricated by front-side etching technology. *Sens. Actuators A Phys.* **2005**, *119*, 28–37.
     [CrossRef]
26. Chiang, C.-F.; Graham, A.B.; Lee, B.J.; Chae, H.A.; Ng, E.J.; O'Brien, G.J.; Kenny, T.W. Resonant pressure
     sensor with on-chip temperature and strain sensors for error correction. In Proceedings of the 2013 IEEE
     26th International Conference on Micro Electro Mechanical Systems (MEMS), Taipei, Taiwan, 20–24 January
     2013; pp. 45–48.

13

Wireless Sensors Grouping Proofs for Medical Care and Ambient Assisted-Living Deployment

Denis Trček

Academic Editor: Vittorio M. N. Passaro

Faculty of Computer and Information Science, University of Ljubljana, Večna pot 113, Ljubljana 1000, Slovenia; denis.trcek@fri.uni-lj.si

Abstract: Internet of Things (IoT) devices are rapidly penetrating e-health and assisted living domains, and an increasing proportion among them goes on the account of computationally-weak devices, where security and privacy provisioning alone are demanding tasks, not to mention grouping proofs. This paper, therefore, gives an extensive analysis of such proofs and states lessons learnt to avoid possible pitfalls in future designs. It sticks with prudent engineering techniques in this field and deploys in a novel way the so called non-deterministic principle to provide not only grouping proofs, but (among other) also privacy. The developed solution is analyzed by means of a tangible metric and it is shown to be lightweight, and formally for security.

Keywords: wireless networks; internet of things; health care; ambient assisted living; PPDR networks; RFID; lightweight protocols; security; Yoking proofs

1. Introduction

Rapid proliferation of pervasive computing objects, like sensors and RFIDs, significantly changes not only the applications landscape, but also increases concerns related to security, privacy and even safety and this especially holds true in medical settings. The population of this kind of devices is quite heterogeneous. Despite this diversity, a notable common property is limited power supply, while a large part of them also lacks computing resources. A straightforward consequence is a requirement for lightweight cryptographic protocols. Knowing that cryptographic protocols design, alone, is a tricky issue, adding stringent computing resources and power limitations results in additional complexity. Knowing further that these devices are becoming rapidly adopted in medical settings, appropriate solutions are of utmost importance (for those interested, an extended survey that covers IoT security and energy issues can be found in [1]).

This research paper addresses an important, frequently considered problem in medical settings where a cluster of computationally-weak devices (e.g, RFIDs) has to act in orchestration. Such clusters should act in a way where it can be proved that the responses are obtained simultaneously. Traditional application areas are drugs administration, but clinical experiments are taking place for tracing patient's physiological signals from a wireless body area networks while she moves around hospital's premises as shown in Figure 1 (an early such can be found in [2], while more recent on in [3]). It should be added that the application areas are by no means limited to medical domain—they range from industrial assembly lines to logistics and PPDR (public protection and disaster relief) settings.

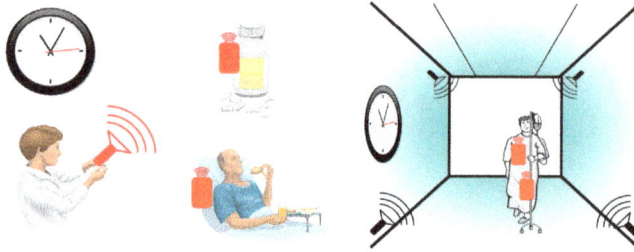

Figure 1. Two application scenarios for Yoking proofs (drugs administration and ubiquitous patients' physiological functions monitoring in clinics or ambient assisted living).

The strategy in this paper is to incrementally build upon protocols that have already been subject to scientific scrutiny. The reasoning is as follows: knowing that design of a watertight cryptographic protocol can be an elusive endeavor, it makes sense to avoid designing a new solution from scratch to minimize the possibility that the developed solution would turn out to be just another Pandora's Box [4]. By doing so this paper exposes important issues that seem to be often overlooked recently and which were already stated by Abadi and Needham in the 1990s—these are prudent engineering principles [5]. Last but not least, formal verification is a must for proving authentication and confidentiality (privacy) properties of developed Yoking proof schemes, which is the case with the developed solution in this paper.

The paper is structured as follows. There is an overview of the field in the second section. In the third section it is followed by an analysis of solutions that present the basis for new protocols, which are given in the same section. In the fourth section these are analyzed from two perspectives: authentication and privacy provisioning and lightweight properties perspective. There are conclusions in the fifth section, while the paper ends up with acknowledgements, references, appendix and author's *vita*.

2. Overview of the Field

Proofs of simultaneous presence of RFIDs were introduced by Juels in 2004, when he coined the term Yoking proofs [6]. These are intended to cover scenarios where proofs of simultaneous scanning of RFIDs are needed. For their analysis and design presented in this paper the following notation will be used:

- A, B, C, and X denote entities (tags' identifiers), V denotes a verifier and R denotes a reader;
- r denotes a random value (also a nonce), while its subscript denotes an entity that has generated it (e.g., r_A denotes a random value generated by A);
- $s_{A,B}$ and $x_{A,B}$ denote a secret key x or secret s that is shared between entities A and B, while x_A and x_B denote A's and B's own secrets (known also only to a server);
- $MAC_s[m]/MAC_x[m]$ denotes a cryptographic (hashed) message authentication code obtained by using a secret key x or s that is , e.g., appended to a message m before hashing;
- P_{AB} denotes a proof that entities (objects) A and B were scanned simultaneously.

Now starting with the first solutions proposed by Juels—their advantage was clear and minimalistic design, which especially holds true for the lightweight variant that is based on MACs—it is given in Figure 2. This work has gained interest and soon research followed that found attacks against it. The first one was done by Saito and Sakurai where they proposed timestamps to cure the discovered weakness [7]. However, this improved scheme was also vulnerable to reply attacks, as discovered by Piramuthu [8]. If an attacker submits a future timestamp and obtains the response from the first tag he can use it later on when the corresponding time becomes true. Then the attacker queries the second tag and completes the protocol ending up with a proof that both tags have been read simultaneously, although this has not been the case.

Therefore Piramuthu has proposed an improvement that is given in Figure 2. According to Piramuthu, this solution should also provide privacy.

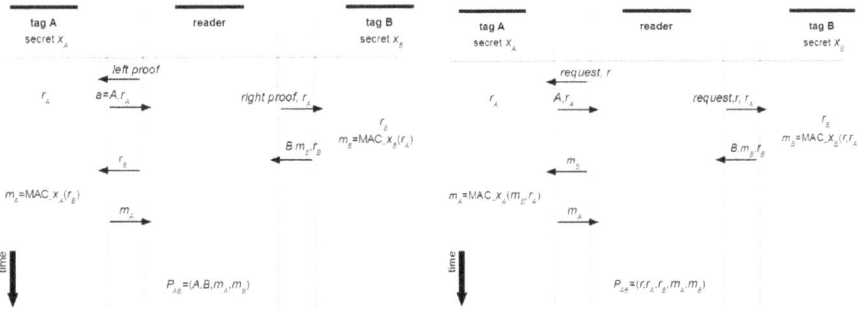

Figure 2. Juels' Yoking proof protocol (**Left**), and Piramuthu's extension of Juels' protocol (**Right**).

There are many unclear issues with Piramuthu's version. It is not clear how values for r_A and r_B are produced. If r from a reader serves as an input, then being interrogated by the same "r" tags will produce the same r_A, which gives good grounds to attacks because of not adhering to prudent security protocol engineering principles [5]. Next, how can a prover find A's and B's identities on the basis of received (r, r_A, r_B, m_X, m_B)? Further, Piamuthu's improvement is claimed to be vulnerable and the attack is shown in Figure 2 [9]. However, this procedure is actually not revealing a weakness in Piramuthus's protocol. If following exactly the attack as described, this attack cannot be successful. Actually, P_{AB} has to be verifiable in its entirety. Thus, a verifier can find out that the key for verification of m_X differs from that of A. If authors wanted to claim that the protocol provides proof P_{XB}, then also this claim would be problematic. There can be a time delay between interrogation of B and X, but the verification of will still be fine. However, authors in [9] that there is no privacy identifiers of tags A and B are sent in plaintext. Further, they present a new scheme called Clumping proof, where they include, explicitly, the verifier in the proof-generating phase. This entity is first mutually authenticated with a reader, and afterward a timestamp TS is exchanged. As opposed to Piramuthu's scheme, this timestamp is encrypted by the verifier. This encrypted value t of timestamp, which is obtained by a keyed hash function H (i.e., $t = H_V(TS)$), is divided into its upper part (MSB) and its lower part (LSB)—the MSB part is exchanged with tag A and the LSB part with tag B. Further, changeable identifiers are introduced that act as pseudonyms (a special kind of function called Nun is used for this purpose). Further, counters in tags are introduced to make explicit to which run of a protocol particular messages belong to—the whole scheme is presented in Figure 3.

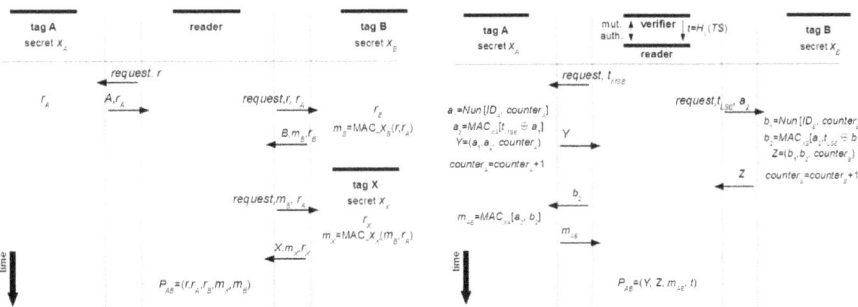

Figure 3. An attack on Piramuthu's protocol by Peris-Lopez *et al.* (**Left**) and Clumping proof scheme (**Right**).

The Clumping proof scheme does introduce some improvements, one of them being use of counters and the other one being (almost strict) chaining of all calculations. Nevertheless, this scheme needs an improvement—the hashed timestamp is an inherent point of attack. An attacker can shuffle a bit (or some bits) of the timestamp exchanged at the beginning between the verifier and the reader, and this cannot be noticed until the verifier checks the proof. To cure the situation, the verifier and the reader should share a common secret and use it to obtain a hashed message integrity code that is appended to the exchanged hashed timestamp. This timestamp has to be verified by the reader before the hashed LSB and MSB parts are transmitted to tags. By the same token ideally, tags should perform such integrity checks as well.

Extending further the above line of reasoning leads to very interesting consequences that have not been discussed in the literature so far. Despite the above-described attack, and assuming that the subsequent exchanges of messages are not attacked, the calculations following the attacked time stamp exchange are still providing a usable output. This output does not provide a proof that the tags were scanned synchronously at a certain absolute time, but a proof that the tags were scanned simultaneously at some (undefined) temporal point. Further, from the tags' point of view timestamps are semantically equivalent to nonces and random values. While freshness of nonces and random values can be verified to some extent by RFIDs (see, e.g., [4]), the verification of timestamps (*i.e.*, absolute time values) should not be assumed because such checks are very demanding for RFIDs. Adding a fact that authentication of readers by tags is also very demanding, the provisioning of grouping proofs with exact time is dependent on (honest) readers. Interestingly, authors require an on-line involvement of a verifier and its interaction with a reader. As a result, one actually ends up with a variant of on-line (authentication) protocol.

Lessons learned so far remain valid for research that followed the above described designs. In Cho *et al.*, presented a variant of Piramuthu's protocol [10]. Later, protocols that support anonymization appeared including Burmester *et al.* [11] and Chien and Liu [12]. In this latter paper authors state that malfunctioning or malicious tags can lead to calculation of useless proofs and consequently, denial of service attacks. Therefore on-line authentication schemes should be deployed and they propose one. However, the same authors later developed another off-line grouping verification scheme [13]. Probably the reason is that on-line grouping proofs are better replaced by on-line authentication schemes where proofs of simultaneous reading are formed by trusted back-ends.

All these protocols are analyzed for their weaknesses and computational costs in [14] (interestingly, Yoking proofs are all but simple even the recent proposal by [15] has been found vulnerable [16]). On this basis guidelines for attacks resistant grouping protocols design are given that can be summarized as follows. The basic design issue is severely restricted power supply (so tags should compute just, e.g., pseudo-random message authentication codes). Next, except for the first one, every input to a given tag should be a derivation of computations that can be only carried out by fellow tags participating in the proof. Next, if possible, tags should use group identifiers and group keys to prove membership to a group and should check each other's computations to make sure that only group members participate in the proof. Next, time stamps should be used to thwart replay attacks. Next, performance issues should be considered (and some approximate metrics is provided for this purpose). Next, forward secrecy should be enabled, so when secret keys are disclosed all dependent past cryptographic calculations should not be endangered. Finally, privacy is of a prime concern, therefore dynamic anonymous identifiers have to be used.

Taking into account what has been stated so far the above guidelines have to be adjusted as follows. Grouping proofs can be off-line or on-line. With off-line proofs a reader is not able to verify the calculations, thus there is an inherent risk of producing false proofs (be it on the account of attacking tags, or attackers that interfere with exchanged messages, *etc.*), because such cases can be detected only during the verification phase. Next, use of group IDs and group keys is desired, but it limits applicability of grouping proofs, so such identifiers should not be generally assumed. Next, checks of previous calculations by fellow tags are desired, but due to severely restrained resources they should not be generally assumed (this requirement goes against the famous 5 cents production costs limit). Next, as timestamps have the same semantics to RFIDs as random values and nonces, their use adds security only in case of a trusted reader. Next, performance metrics should be as fine graded as possible. One such metric that nails down the protocol's requirements to the number of NAND gates, which reflects technological reality, while at the same time it is transformable to power/energy consumption, is described in [4]. Finally, when designing privacy protocols, such ones should be the basis that have been subject to scientific scrutiny.

3. New Solution for Orchestrated Security and Privacy

A seminal work in formal analysis of strength of privacy providing RFID protocols is given in [17] and among the strongest protocols in this domain is the protocol developed by Okhubo, Suzuki, and Kinoshita [18]. With this protocol this holds true for privacy provisioning, but its weakness is a practical deployment issue, because the protocol can be forced into de-synchronization with a database [4]. To prevent this kind of attack, and to preserve the strengths of Okhubo, Suzuki, and Kinoshita protocol, the non-deterministic approach, called ND-PEPS, will be used (its newer version that enables not only privacy, but also confidential exchange of, e.g., measured physiological quantities [4]). ND-PEPS is called non-deterministic because it deploys asymmetry in computational power between a tag and a reader (or verifier). More precisely, the tag computes randomized responses from a certain interval of possible responses (which is not computationally demanding), while the reader (which is verifier in our case) has to compute all possible responses and search for a match to identify the tag (which is computationally not hard for the verifier). The ND-PEPS protocol is shown in Figure 4 (a Dolev-Yao adversary model is assumed, tamper resistance of tags is assumed, and only the back-end server is trusted).

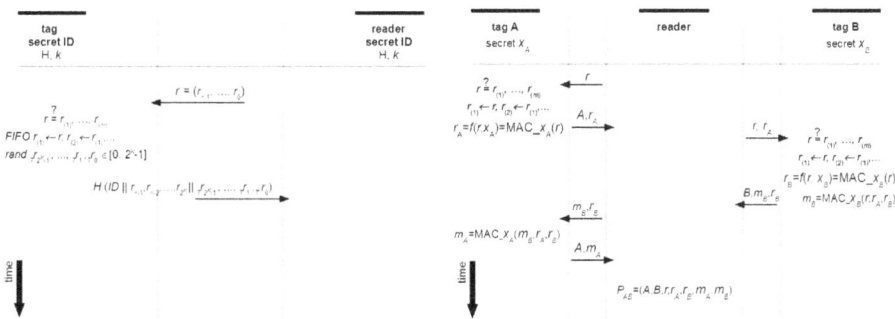

Figure 4. The ND-PEPS protocol (**Left**) and enhanced version of Juels/Piramuthu "Yoking proof" protocol (**Right**).

Now let a reader and a tag share a secret value *ID*, and also agree upon parameter k and a strong one-way hash function *H*. To authenticate the tag the reader sends to the tag a random query r. Upon receipt the tag compares the received r against previously received ones to prevent attacks with messages, which are not fresh. If this challenge is fresh, it is written in FIFO (first in, first out) memory and the oldest one is discarded. Afterward, the tag generates a random value Tr, replaces the last bits in the challenge with this value, and appends the result to the secret *ID*. This concatenated sequence is hashed and the output is sent to the reader. Upon receipt, the reader (*i.e.*, verifier) generates all possible values from the closed interval $[0, 2^{k-1}]$ for tags for which it keeps track in its database and obtains possible hashed sequences. Using these sequences and the received message the reader looks for a match—once it is found, the tag is authenticated.

3.1. Design of an Improved Grouping Proof Architecture

The basis for the solutions presented here is Juels' protocol, which despite its deficiencies has many nice properties. It has a clean design and semantics of its steps, it sticks with a simplistic architecture approach and it is lightweight. Likely because of these properties it was a good candidate for successful analysis and discovery of weaknesses by Piramuthu, who proposed its improvements (as discussed above and shown in Figure 2). This architecture will be taken as a basis for improved version that provides also privacy through anonymization. Further, a seminal work on prudent engineering principles for designing cryptographic protocols [5] will be taken into account to remedy accordingly Piramuthu's protocol weak points that could open doors to attacks.

Starting with the challenge r that is sent in the first step of the protocol run, its value makes no sense if not used by tag A. However it should play a crucial role as a value around which all messages are 'chained'. Further, value r_A should depend on the challenge r and secret x_A—the same chaining logic holds true for value r_B, m_B, and m_A. Such structure of messages supports consistent semantic interpretation, while their logical linking provides a provably linked sequence that belongs to one particular run of the protocol (the whole protocol is given in Figure 4). Each tag checks the freshness of r (of course, to a limited extent) by comparing it with a sequence of previous challenges stored in a first-in-first-out (FIFO) memory. Note that the above architecture is a conceptual basis that serves for clarification of the derivation of the final (optimized and secure) solution, which will be presented in the next subsection.

3.2. A New Grouping Proof with Privacy Provisioning

The protocol given in Figure 4 can be relatively easily upgraded in such a way that not only Yoking proofs are enabled, but also privacy is achieved. This is done by applying the ND-PEPS principle and the obtained architecture is presented in Figure 5.

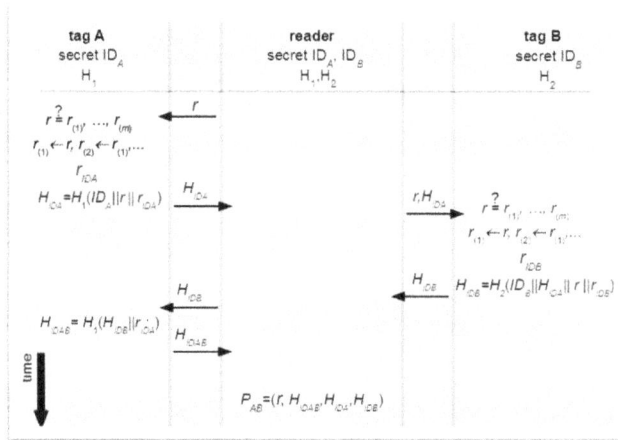

Figure 5. Yoking proof based grouping scheme enhanced by ND-PEPS.

In the first step a challenge r is sent and checked by tag A for freshness. If r is fresh, it enters FIFO and the oldest r is discarded. Next, the tag calculates a random r_{IDA} and concatenates it with its ID and r. The calculated r_{IDA} is from a pre-defined interval, so upon receipt, the verifier generates all values from this interval using all tags' IDs that are kept in its database, and obtains all possible hashed sequences. These sequences serve to find a match with the received part of P_{AB} to identify tag A. Similarly, using the rest of exchanged messages and taking into account their cascading, the verifier can analyze the proof P_{AB}, while the identities of A and B remain hidden to an outside observer (and even to the reader, if it is not the verifier).

4. Discussion

The developed solution as this is the case with the original Yoking proofs is intended for off-line grouping proofs (as discussed earlier, on-line grouping proofs are better replaced by on-line authentication schemes where proofs of simultaneous reading can be formed by back-end systems). Further, in order to preserve its general applicability (without a need for RFIDs that use group keys, *etc.*) the improved protocol requires a minimal number of secrets to be stored on RFIDs in principle, only tags' *ID*s. However, as also suggested in the above text (the attack on time-stamp with Peres-Lopez protocol), integrity provisioning of exchanged messages should be deployed. This would require an additional shared secret between a tag and a reader (or back-end system) and this secret would be used to derive hashed message integrity codes, MICs. Such MICs could be verified also by tags, not only readers (these checks are intentionally not included in Figure 5 to preserve clarity). Further, responses from readers are constantly changing in order to provide privacy through anonymity and untraceability—only authenticated readers (or back-end systems) will be able to find a match. Therefore, not only authentication of tags, but also (indirect) authentication of readers (or back-end systems) is achieved. Further, replies are prevented, because the exchanged cryptograms are logically linked to one session and to a particular group of participating RFIDs. As informal analyses are often subject to mistakes, the new protocol from Figure 5 has been thoroughly formally analyzed (see the appendix).

However, simplicity, clarity, and elegance of the protocol come at a price. First, invalid proofs cannot be prevented *per se*, although it is questionable if they can be achieved at all without on-line communications (these falsified proofs may go on the account of malfunctioning or rogue tags, or malicious readers). Next, the scheme is not providing an absolute time of when the tags group was scanned, but only a proof that such scanning has taken place at some time in the past (again, this seems to be an inherent problem for off-line grouping proofs). However, if the reader can be operated in a reliable manner (e.g., being a secure tablet operated by a trusted doctor with an access to a timing base), this functionality is provided. Further, replay attacks are prevented, as well as one-time proof is obtained to the degree enabled by that technological reality on RFID chips and, e.g., their pseudo-number generators.

Getting to computational costs—a large part of IoT devices lacks computing power and has the following typical structure: communications part, processing gates, memory gates and optionally one or more sensors. As communication is on RF such devices obviously need receiver and transmitter parts—these parts are, of course, excluded from calculation because they are needed regardless of an implemented a protocol. For the rest of the gates, computing resources are varying and these should be quantified.

One appropriate metric is described in [13]. It is based on technological reality of RFIDs production, where NAND gates are used. The total number of NAND gates required for a protocol gives the quantitative value about how lightweight a protocol is. This total is obtained by summing the gates needed for processing (*i.e.*, to implement protocol by using appropriate Boolean functions), the gates needed for storage (of identifiers, secret values and their calculations), and the gates that are equivalent to storage gates that would be needed to transfer bits, which are exchanged between a reader and a tag. Further, cryptographic protocols are built by using certain building blocks. Starting with memory cells, these are D cells and storing one bit with such cells requires five NAND gates [19]. To perform addition, eleven gates are needed for one bit full-adder, while addition modulo 2^n requires n-times eleven NAND gates, and bitwise XOR requires four gates for each pair of bits [19]. Pseudorandom number generators can be implemented with shift registers and a shift register with four bits requires approximately 60 NAND gates, with approximately eight bits, 120 NAND gates, and so on [20] (using digital circuit artifacts quality random number generators can be built with approximately 300 NAND gates [21], while a more promising solution suited for EPCGen2 tags is given in [22]). Symmetric encryption can be done with light DES that requires approximately 1800 gates [23], while AES requires approximately 3000 gates [24]. Strong one-way hash functions are generally expensive—for example, SHA-3 can be implemented with approximately 2500 gates [25] (unfortunately, it seems SHA-3 has a built-in side channel [26]). Recent advancements are slowly making realistic also deployment of asymmetric encryption, so, for example, a 1024 bit public key scheme can be implemented with approximately 4600 logic gate equivalents [27]. Finally, as stated in [4], "For a protocol to qualify as being lightweight, the typical five-cent production cost limit has to be taken into account, which means that today one could count with 5000 to 8000 logic gates being dedicated to security and privacy".

Let us analyze the part for tag A, which is more demanding than that for tag B. The tag has to store its ID and this requires 96×5 NAND gates. Further, it has to compute a strong one-way hash function by using DESL, with approximately 1800 logic gates needed. Further, a pseudo-random value is required, and assuming that its length is eight bits, 120 NAND gates are needed. To prevent exhaustive challenges attack and to ensure their freshness, 64 bits for challenges are assumed and twenty of them are stored in the FIFO, which results in $64 \times 5 \times 20$ storage cells (to extend the range of stored challenges, only the first k-bits of each challenge can be stored in FIFO). The total number of gates for storage and processing is therefore approx. 2920 gates (the gates needed for comparing the new challenges with the stored ones have been neglected). Now the cost of communications has to be taken into account. Assuming that hashed outputs are also 64 bits long, the related cost is $64 \times 5 \times 4$ NAND gates (four 64 bit-long messages are exchanged between tag A and a reader). Therefore the total estimated cost of the protocol is 4180 logic gates and the obtained solution is lightweight. Even when increasing its strength by extending its challenges and hashed values to 128 bits, this would keep its cost within the boundaries of lightweight protocols.

Now, how to relate the number of NAND gates to power consumption? A NAND gate in CMOS technology consists of four transistors. Each of them is subject to static and dynamic power consumption. The first part is bound by leakage current, while multiplying it by supply voltage gives static power consumption. Getting to dynamic power consumption, it consists of transient power consumption and capacitive-load power consumption. As to the transient power consumption (for a single bit switching), this is a product of a dynamic power-dissipation capacitance, square of supply voltage and input signal frequency [28]. As to the capacitive-load power consumption, this can be usually neglected compared to transient power consumption [28].

And finally, the properties of the new scheme are summarized in Table 1.

Table 1. Summary of security related properties of the new solution (*on-line mode depends on reader's real-time access to a back end system).

Authentication	Privacy	Data-Base Desync Prevention	Tracking Prevention	On-Line Mode	Replay Prevention	Forge Proof	Lightweight
yes	yes	yes	yes	yes*	(yes)	yes	yes

5. Conclusions

Numerous areas of our lives are becoming increasingly dependent on IoT structures that range from NFCs, RFIDs, RFIDs with sensors, and bare sensors on one side of the spectrum to capable sensor (and actuator) motes' networks on the other side of the spectrum. One common characteristic of these structures are stringent power (energy) requirements, while a large part of this digital ecosystem also lacks computing resources. Further, the need for grouping proofs for such structures is growing. One such example are proofs of simultaneous readouts from wireless medical sensors body area networks that are required together with privacy provisioning, and another is controlled drugs administration. To improve economic feasibility of such networks and to ensure a minimal interference with the patients' activities, the deployed devices have to be small, which means that weak devices will be deployed, and that they will be passively powered. Therefore, lightweight grouping proofs with privacy provisioning for such structures are high on the agenda.

However, history is teaching us that the design of secure grouping proofs is an elusive endeavor. This paper, therefore, uses an approach where already-developed solutions (which have been subject to public scrutiny) are taken as a basis. Next, it improves their design to compensate the discovered weaknesses. As a result, a new grouping proof protocol is presented that is based on elegant Yoking proof scheme. By applying a so called non-deterministic principle it is further enhanced to provide privacy. Next, it is analyzed from security point of view, and from the computing requirements point of view. By using a tangible metric it is proved that the presented solution is lightweight, while the appendix contains a formal proof of its security.

Summing up, the number of IoT objects (like those in medical and assisted living settings) is expected to exceed other kinds of devices connected to the Internet. Therefore, the related security research area will be gaining importance. Clearly, as there is no size-fits-all solution, emerging areas of applications will call for new solutions like those described in [29], but grouping proofs can already be identified as being one of them.

Acknowledgments: Authors acknowledge the financial support of the Slovenian Research Agency through research program Pervasive computing (P2-0359). Authors also acknowledge the financial support of the EU through FP 7 project SALUS—Security and interoperability in next generation PPDR communication infrastructure (contract #313296). Finally, the non-financial benefits of collaboration in COST IC 1403 (Cryptanalysis of ubiquitous computing systems) and COST IC 1303 (Algorithms, Architectures and Platforms for Enhanced Living Environments) projects should be mentioned.

Appendix

This appendix gives a formal specification and verification of the scheme from Section 3.2. It is needed for verification with AVISPA model checkers, which use the widely used and most universal Dolev-Yao adversary model [30] (thus specific attacks like distance-bounding attack are excluded—for their deeper analysis a reader is referred to [31]). The specification is in HLPSL and the details about this language can be found in [32] (also without familiarity with HLPSL, its specifications can be mainly understood by readers who have some knowledge of formal verification techniques).

```
role role_A(A:agent,B:agent,R:agent,H1:hash_func,IDa:text,SND,RCV:channel(dy))
played_by A
def=
    local
        State:nat,Ridb:text,Rida:text,IDb:text,Rand:text,H2:hash_func
    init
        State := 0
    transition
        1. State=0 /\ RCV(Rand') =|> State':=1 /\ Rida':=new() /\ secret(IDa',sec_3,A,R) \ SND(H1(IDa.
Rand'.Rida'))
        5. State=1 /\ RCV(H2(IDb'.H1(IDa.Rand.Rida).Rand.Ridb')) =|> State':=2 /\ secret(IDa',sec_3,A,R)
/\ secret(IDb',sec_4,B,R) /\ SND(H1(Rida.H2(IDb'.H1(IDa.Rand.Rida).Rand.Ridb')))
end role
role role_B(A:agent,B:agent,R:agent,H2:hash_func,IDb:text,SND,RCV:channel(dy))
played_by B
def=
    local
        State:nat,Ridb:text,Rida:text,IDa:text,H1:hash_func,Rand:text
    init
        State := 0
    transition
        3. State=0 /\ RCV(Rand'.H1(IDa'.Rand'.Rida')) =|> State':=1 /\ secret(IDa',sec_3,A,R) /\
Ridb':=new() /\ secret(IDb',sec_4,B,R) /\ SND(H2(IDb.H1(IDa'.Rand'.Rida').Rand'.Ridb'))
```

end role
role role_R(A:agent,B:agent,R:agent,H1:hash_func,H2:hash_func,IDa:text,IDb:text,SND,RCV:channel(dy))
played_by R
def=
    local
        State:nat,Ridb:text,Rida:text,Rand:text
    init
        State := 0
    transition
        1. State=0 /\ RCV(start) = | > State':=1 /\ Rand':=new() /\ SND(Rand')
        2. State=1 /\ RCV(H1(IDa.Rand.Rida')) = | > State':=2 /\ secret(IDa',sec_3,A,R) /\ SND(Rand.
H1(IDa.Rand.Rida'))
        4. State=2 /\ RCV(H2(IDb.H1(IDa.Rand.Rida).Rand.Ridb')) = | > State':=3 / secret(IDa',sec_3,
A,R) /\ secret(IDb',sec_4,B,R) /\ SND(H2(IDb.H1(IDa.Rand.Rida).Rand.Ridb'))
        6. State=3 /\ RCV(H1(Rida.H2(IDb.H1(IDa.Rand.Rida).Rand.Ridb))) = | > State':=4 /\ secret
(IDa',sec_3,A,R) /\ secret(IDb',sec_4,B,R)
end role
role session1(A:agent,B:agent,R:agent,H1:hash_func,H2:hash_func,IDa:text,IDb:text)
def=
    local
        SND3,RCV3,SND2,RCV2,SND1,RCV1:channel(dy)
    composition
        role_R(A,B,R,H1,H2,IDa,IDb,SND3,RCV3) /\ role_B(A,B,R,H2,IDb,SND2,RCV2) /\ role_A
(A,B,R,H1,IDa,SND1,RCV1)
end role
role environment()
def=
    const

    hash_0:function,ida:text,h1:hash_func,tagb:agent,taga:agent,reader:agent,h2:hash_func,idb:text,
auth_1:protocol_id,auth_2:protocol_id,sec_3:protocol_id,sec_4:protocol_id
    intruder_knowledge = reader,h1,h2
    composition
        session1(taga,tagb,reader,h1,h2,ida,idb)
end role
goal
    authentication_on auth_1
    authentication_on auth_2
    secrecy_of sec_3
    secrecy_of sec_4
end goal
environment()

References

1.   Granjal, J.; Monteiro, E.; Silva, J.S. Security in the integration of low-power Wireless Sensor Networks with the Internet: A survey. *Ad. Hoc. Netw.* **2015**, *24*, 264–287. [CrossRef]
2.   Najera, P.; Lopez, J.; Roman, R. Real-time location and inpatient care systems based on passive RFID. *J. Netw. Comput. Appl.* **2011**, *34*, 980–987. [CrossRef]
3.   Yang, M.T.; Huang, S.Y. Appearance-Based Multimodal Human Tracking and Identification for Healthcare in the Digital Home. *Sensors* **2014**, *14*, 14253–14277. [CrossRef] [PubMed]

4. Trček, D.; Brodnik, A. Hard and soft security provisioning for computationally weak pervasive computing systems in e-health. *IEEE Wirel. Commun.* **2013**, *20*, 22–29. [CrossRef]

5. Abadi, M.; Needham, R. Prudent Engineering Practice for Cryptographic Protocols. *IEEE Trans. Softw. Eng.* **1996**, *22*, 6–15. [CrossRef]

6. Juels, A. "Yoking-Proofs" for RFID Tags. In Proceedings of the Second IEEE Annual Conference on Pervasive Computing and Communications Workshops, Orlando, FL, USA, 14–17 March 2004; pp. 138–142.

7. Saito, J.; Sakurai, K. Grouping Proof for RFID Tags. In Proceedings of the 19th Internation Conference on Advanced Information Networking and Applications (AINA), Tamkang, Taiwan, 28–30 March 2005; p. 621.

8. Piramuthu, S. Protocols for RFID tag/reader authentication. *Decis. Support Syst.* **2007**, *43*, 897–914. [CrossRef]

9. Peris-Lopez, P.; Hernandez-Castro, J.C.; Estevez-Tapiador, J.M.; Ribagorda, A. Solving the Simultaneous Scanning Problem Anonymously: Clumping Proofs for RFID Tags. In Proceedings of the Security, Privacy and Trust in Pervasive and Ubiquitous Computing (SECPerU), Istanbul, Turkey, 19 July 2007; pp. 55–60.

10. Cho, J.-S.; Yeo, S.-S.; Hwang, S.; Rhee, S.-Y.; Kim, S.K. Enhanced yoking proof protocols for RFID tags and tag groups. In Proceedings of the International Conference on Advanced Information Networking and Applications Workshops (AINAW), Okinawa, Japan, 25–28 March 2008; pp. 1591–1596.

11. Burmester, M.; de Medeiros, B.; Motta, R. Provably secure grouping-proofs for RFID Tags. In Proceedings of the 8th Smart Card Research and Advanced Applications (CARDIS), LNCS, London, UK, 8–11 September 2008; pp. 176–190.

12. Chien, H.-Y.; Liu, S.-B. Tree-based RFID yoking proof. In Proceedings of the Conference on Networks Security, Wireless Communications and Trusted Computing (NSWCT), Hubei, China, 25–26 April 2009; pp. 550–553.

13. Chien, H.-Y.; Yang, C.-C.; Wu, T.-C.; Lee, C.-F. Two RFID-based solutions to enhance inpatient medication safety. *J. Med. Syst.* **2010**, *35*, 369–375. [CrossRef] [PubMed]

14. Pedro, P.L.; Agustin, O.; Julio, C.; van der Lubbe, J.C.A. Flaws on RFID Grouping-proofs Guidelines for Future Sound Protocols. *J. Netw. Comput. Appl.* **2010**, *34*, 833–845.

15. Peris-Lopez, P.; Orfila, A.; Mitrokotsa, A.; van der Lubbe, J.C.A. A comprehensive RFID solution to enhance inpatient medication safety. *Inter. J. Med. Inform.* **2011**, *80*, 13–24. [CrossRef] [PubMed]

16. Safkhani, M.; Bagheri, N.; Naderi, M. A note on the security of IS-RFID, an inpatient medication safety. *Inter. J. Med. Inform.* **2014**, *83*, 82–85. [CrossRef] [PubMed]

17. Avoine, D. *Radio Frequency Identification: Adversary Model and Attacks on Existing Protocols*; Technical Report LASEC-REPORT-2005-001; Swiss Federal Institute of Technology in Lausanne, School of Computer and Communication Sciences: Lausanne, Switzerland; September; 2005.

18. Ohkubo, M.; Suzuki, K.; Kinoshita, S. Cryptographic approach to "privacy-friendly" tags. *RFID Privacy Workshop* **2005**, *82*, 1–9.

19. Vodovnik, L.; Rebersek, S. *Digital Circuits*; Faculty of Electrical Engineering: Ljubljana, Slovenia, 1986.

20. Horowitz, P.; Hill, W. *The Art of Electronics*; Cambridge University Press: New York, NY, USA, 1989.

21. Epstein, M.; Hars, L.; Krasinski, B.; Rosner, M.; Zheng, H. Design and Implementation of a True Random Number Generator Based on Digital Circuit Artifacts. *Lect. Notes Comput. Sci.* **2003**, *2779*, 152–165.

22. Peinado, A.M.; Amparo, J.F.-S. EPCGen2 Pseudorandom Number Generators: Analysis of J3Gen. *Sensors* **2014**, *14*, 6500–6515. [CrossRef] [PubMed]

23. Poschmann, G.; Leander, K.; Schramm, C.P. New Light-Weight Crypto Algorithms for RFID. In Proceedings of the IEEE International Symposium on Circuits and Systems (ISCAS), New Orleans, LA, USA, 27–30 May 2007; pp. 1843–1846.

24. Feldhofer, M.; Dominikus, S.; Wolkerstorfer, J. Strong Authentication for RFID Systems Using the AES Algorithm. In Proceedings of the CHES, LNCS 3156, Boston, US, 11–13 August 2004; pp. 357–370.

25. Kavun, E.B.; Yalcin, T. A Lightweight Implementation of Keccak Hash Function for Radio-Frequency Identification Applications. *LNCS* **2010**, 258–269.

26. Newman, L.H. Can you trust NIST? IEEE Spectrum. Available online: http://spectrum.ieee.org/telecom/security/can-you-trust-nist (accessed on 16 December 2015).

27. Oren, Y.; Feldhofer, M. A Low-Resource Public-Key Identification Scheme for RFID Tags and Sensor Nodes. In Proceedings of the 2nd ACM Conference on Wireless Network Security (WiSec), Zurich, Switzerland, 16–19 March 2009; pp. 59–68.

28. Sarwar, A. *CMOS Power Consumption and Cpd Calculation, TI SCAA035B document*; Texas Instruments: Dallas, TX, USA, 1997.
29. Stelte, B.; Dreo, G.R. Thwarting attacks on ZigBee—Removal of the KillerBee stinger. In Proceedings of the 9th International Conference on Network and Service Management (CNSM), Zurich, Switzerland, 14–18 October 2013; pp. 219–226.
30. AVISPA Team, AVISPA User Manual, v 1.1. Available online: http://www.avispa-project.org/package/user-manual.pdf (accessed on 26 December 2015).
31. Hancke, G.P. Design of a secure distance-bounding channel for RFID. *J. Netw. Comput. Appl.* **2011**, *34*, 877–887. [CrossRef]
32. AVISPA Team, The High Level Protocol specification Language, deliverable 2.1. Available online: http://www.avispa-project.org/delivs/2.1/d2-1.pdf (accessed on 26 December 2015).

14

Sensor Fusion Based Model for Collision Free Mobile Robot Navigation

Marwah Almasri *, Khaled Elleithy * and Abrar Alajlan *

Academic Editors: Lianqing Liu, Ning Xi, Wen Jung Li, Xin Zhao and Yajing Shen

Computer Science and Engineering Department, University of Bridgeport, 126 Park Ave, Bridgeport, CT 06604, USA

*   Correspondence: maalmasr@my.bridgeport.edu (M.A.); elleithy@bridgeport.edu (K.E.); aalajlan@my.bridgeport.edu (A.A.)

Abstract: Autonomous mobile robots have become a very popular and interesting topic in the last decade. Each of them are equipped with various types of sensors such as GPS, camera, infrared and ultrasonic sensors. These sensors are used to observe the surrounding environment. However, these sensors sometimes fail and have inaccurate readings. Therefore, the integration of sensor fusion will help to solve this dilemma and enhance the overall performance. This paper presents a collision free mobile robot navigation based on the fuzzy logic fusion model. Eight distance sensors and a range finder camera are used for the collision avoidance approach where three ground sensors are used for the line or path following approach. The fuzzy system is composed of nine inputs which are the eight distance sensors and the camera, two outputs which are the left and right velocities of the mobile robot's wheels, and 24 fuzzy rules for the robot's movement. Webots Pro simulator is used for modeling the environment and the robot. The proposed methodology, which includes the collision avoidance based on fuzzy logic fusion model and line following robot, has been implemented and tested through simulation and real time experiments. Various scenarios have been presented with static and dynamic obstacles using one robot and two robots while avoiding obstacles in different shapes and sizes.

Keywords: autonomous mobile robots; collision avoidance; path following; fusion; fuzzy logic

1. Introduction

The area of autonomous mobile robots has gained an increasing interest in the last decade. Autonomous mobile robots are robots that can navigate freely without human involvement. Due to the increased demand of this type of robots, various techniques and algorithms are developed. Most of them are focused on navigating the robot in collision-free trajectories with the controlling of the robot's speed and direction. The robot can be mounted by different kinds of sensors in order to observe the surrounding environment and thus steer the robot accordingly. However, many factors affect the reliability and efficiency of these sensors. The integration of multi-sensor fusion systems can overcome this problem by combining inputs coming from different types of sensors, hence have more reliable and complete outputs. This plays a key role in building a more efficient autonomous mobile robotic system.

There are many sensor fusion techniques that have been proven to be effective and beneficial, especially in detecting and avoiding obstacles as well as path planning of the mobile robot. Fuzzy logic, neural network, neuro-fuzzy, and genetic algorithms are examples of well-known fusion techniques that help in moving the robot from the starting point to the target without colliding with any obstacles along its path.

Obstacles detected can be moving or static objects in known or unknown environments. In addition, the path planning behavior can be categorized as global path planning where the environment is entirely known in advance, or local path planning where the environment is partly known or not known at all. The latter case is called dynamic collision avoidance [1].

For the purpose of collision avoidance and path following approaches, different types of sensors such as camera, infrared sensor, ultrasonic sensor, and GPS can detect different aspects of the environment. Each sensor has its own capability and accuracy, whereas integrating multiple sensors enhances the overall performance and detection of obstacles. Many researchers have used sensor fusion to fuse data from various types of sensors, which improved the decision making process of routing the mobile robot. A hybrid mechanism was introduced by [2] which uses the neuro-fuzzy controller for collision avoidance and path planning behavior for mobile robots in an unknown environment. Moreover, an adaptive neuro-fuzzy inference system (ANFIS) was applied for an autonomous ground vehicle (AGV) to safely reach the target while avoiding obstacles by using four ANFIS controllers [3]. Another sensor fusion based on Unscented Kalman Filter (UKF) was used for mobile robots' localization problems. Accelerometers, encoders, and gyroscopes were used to obtain data for the fusion algorithm. The proposed work was tested experimentally and was successfully capable of tracking the motion of the robot [4]. In [5], Teleoperated Autonomous Vehicle (TAV) was designed with collision avoidance and path following techniques to discover the environment. TAV includes GPS, infrared sensors, and the camera. Behavior based architecture is proposed which consist of obstacle avoidance module (OAM), Line Flowing Module (LFM), Line Entering Module (LEM), Line Leaving Module (LLM) and U-Turn Module (UTM). Sensor fusion based on Fuzzy logic was used for collision avoidance where neural network fusion was used for the line following approach.

This paper focuses on the integration of multisensory information from range finder camera and infrared sensors using Fuzzy logic fusion system for collision avoidance and line follower mobile robots. The proposed methodology develops membership functions for inputs and outputs and designs fuzzy rules based on these inputs and outputs.

The rest of the paper is organized as follows: Section 2 presents the related work. Section 3 demonstrates the proposed methodology based on fuzzy logic system. Section 4 shows the simulation and real time implementations. Section 5 discusses the results in details. Section 6 concludes the paper.

2. Related Work

Many obstacle detection, obstacle avoidance, path planning techniques have been proposed in the field of autonomous robotic systems. This section presents some of these techniques with the collaboration of sensor fusion to obtain best results.

Chen and Richardson proposed a collision avoidance mechanism for mobile robot navigating in unknown environments based on a dynamic recurrent neuro-fuzzy system (DRNFS). In this technique, a short memory is used that is capable of memorizing the past and the current information for a more reliable behavior. The ordered derivative algorithm is implemented for updating the DRNFS parameters [6]. Another collision avoidance approach for mobile robots was proposed by [7], which is based on multi sensor fusion technology. With the use of ultrasonic sensors and infrared distance sensors, a 180° rolling window was established in front of the robot. The robot's design has mostly focused on four main layers as follows: energy layer, driver layer, sensor layer, and, finally, the master layer [7].

In addition, a collision avoidance algorithm for a network-based autonomous robot was discussed in [8]. The algorithm is based on the Vector Field Histogram (VFH) algorithm with the consideration of the network's delay. The system consists of sensors, actuators, and the VFH controller. Kalman filter fusion is applied for the robot's localization in order to compensate for the delay between the odometry and environmental sensor readings [8].

The Kalman filtering fusion technique for multiple sensors has been applied in [9]. The Kalman filter is used for predicting the position and distance to the obstacle or wall using three infrared range finder sensors. Authors claimed that this technique is mostly helpful in robots' localization, automatic robots' parking, and collision avoidance [9].

Furthermore, in [10], a path control for mobile robots based on sensor fusion is presented where the deliberative/reactive hybrid architecture is used for handling the mobile robot motion and path control. The sensor fusion technology helps the robot to reach the target point successfully [10]. Another multi sensor fusion system was designed in [11]. This system was mainly for navigating coal mine rescue robots. It used various types of sensors such as infrared and ultrasonic sensors with digital signal processing. The multi-sensor data fusion system helped in decreasing errors caused by the blind zone of ultrasonic sensors [11].

Moreover, a transferable belief model (TBM) was applied in mobile robot for the purpose of a collision-free path planning navigation in a dynamic environment which contains both static and moving objects. TBM was used for building the fusion system. In addition, a new path planning mechanism has been proposed based on TBM. The main benefit of designing such mechanisms is the recognition of the obstacle's type whether it is dynamic or static without the need of any previous information [12].

In [13], the authors developed a switching path-planning control scheme that helped in avoiding obstacles for a mobile robot while reaching its target. In this scheme, a motion tracking mode, obstacle avoidance mode, and self-rotation mode were designed without the need of any previous environmental information [13].

Another multi-sensor particle filter fusion based algorithm for mobile robot localization was proposed by [14]. The algorithm was able to fuse data coming from various types of sensors. Authors also proposed an easy and fast deployment mechanism of the proposed system. A laser range-finder, a WiFi system, many external cameras, and a magnetic compass along with a probabilistic and mapping strategy were used to validate the work proposed [14].

In [15], a novel multi-sensor data fusion methodology for autonomous mobile robots in unknown environments was designed. The flood fill algorithms and fuzzy algorithms were used for the robot's path planning, whereas Principal Component Analysis (PCA) was used for object detection. Multiple sensor data were fused using Kalman Filter fusion technique from infrared sensor, ultrasonic sensor, camera, and accelerometer. The proposed technique has successfully reduced the time and energy consumption.

3. Proposed Methodology

This section presents the proposed methodology for mobile robot collision free navigation with the integration of the fuzzy logic fusion technique. The mobile robot is equipped with distance sensors, ground sensors, camera, and GPS. Distance sensors which are infrared sensors, and the camera are used for collision avoidance behavior where the ground sensors are used for path follower behavior. GPS is used to get the robot's position. The goal of the proposed technique is as follows:

- The capability of the mobile robot to avoid obstacles along its path;
- The integration of sensor fusion using fuzzy logic rules based on sensor inputs and defined membership functions;
- The capability of the mobile robot to follow a predetermined path;
- The performance of the mobile robot when programmed with the fuzzy logic sets and rules.

3.1. Robot and Environment Modeling

Webots Pro simulator is used to model the robot and the environment. Webots Pro is a Graphical User Interface (GUI) which creates an environment that is suitable for mobile robot simulation. It also allows creating obstacles in different shapes and sizes. The mobile robot used in Webots Pro simulator is called E-puck robot which is equipped with a large choice of sensors and actuators such as camera, infrared sensors, GPS, and LED sensors [16].

The environment in this paper is modeled with a white floor that has a black line in order for the robot to follow it. It also has solid obstacles where the robot should avoid them. The environment in

Webots Pro is called "world." A world file can be built using a new project directory. Each project file composed of four main windows which are: the Scene tree which represents a hierarchical view of the world, the 3D window that demonstrates the 3D simulation, the Text editor that has the source code (Controller), and the Console that shows outputs and compilation [16].

The two differential wheel robot (E-puck robot) that is used in this paper is equipped with eight infrared sensors (distance sensors), a camera, and three ground sensors which are also infrared sensors. The eight distance sensors are used to detect obstacles. Each distance sensor has a range of 0 to 2000 where 0 is the initial value of the distance sensor, which means there is no obstacle detected. As the mobile robot approaches the obstacle, its value is increased accordingly. When an obstacle is detected, the distance sensor value will be 1000 or more depending on the distance between the sensor and the obstacle. The camera sensor that is used in this work is a range finder type of camera which allows obtaining distance in meters between the camera and the obstacle from the OpenGL context of the camera. Finally, the three ground sensors are located in front of the e-puck robot where all of them are pointing directly to the ground. These sensors are used to follow the black line drawn on the floor. Figure 1 shows the E-puck robot top view with different types of sensors.

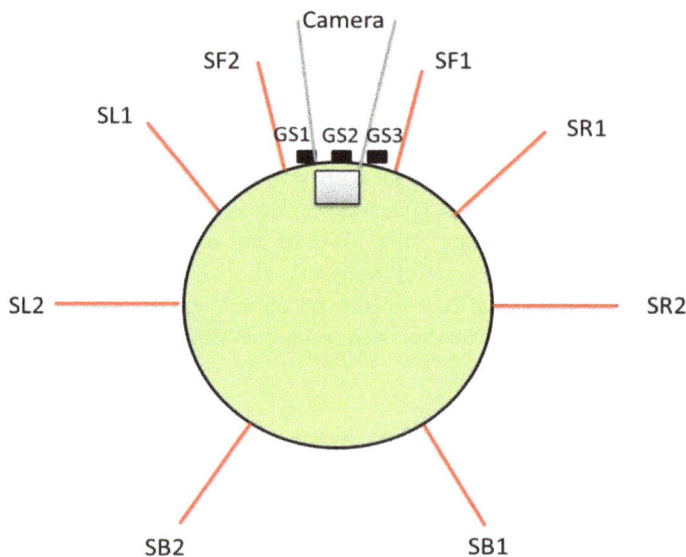

Figure 1. The top view of E-puck robot with various types of sensors.

3.2. Design of the Fusion Model

Multisensory fusion model is designed for better obstacle detection and avoidance by fusing eight distance sensors and the range finder camera. The fusion model is based on Fuzzy Logic fusion technique using MATLAB software. A fuzzy logic system (FLS) is composed of four main parts which are: fuzzifier, rules, inference engine, and defuzzifier. The block diagram of FLS is shown in Figure 2.

The fuzzification stage is the process of converting a set of inputs to fuzzy sets based on defined fuzzy variables and membership functions. According to a set of rules, the inference is made. Finally, at the defuzzification stage, membership functions are used to map every fuzzy output to a crisp output.

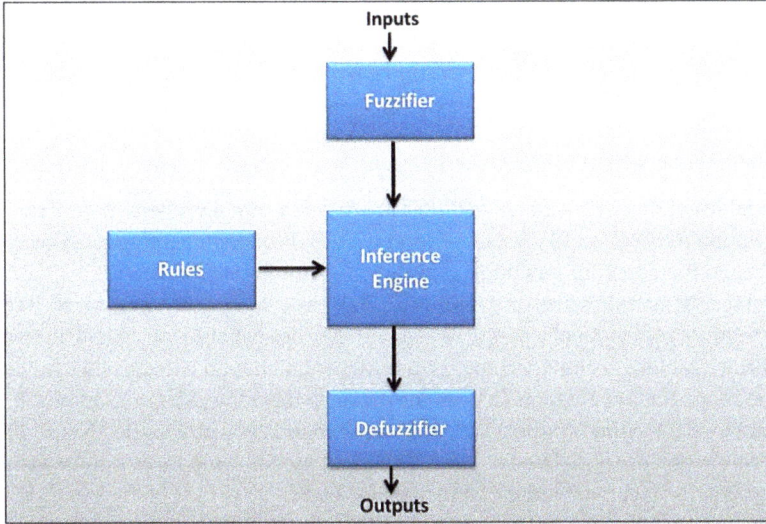

Figure 2. Block diagram of the Fuzzy Logic System.

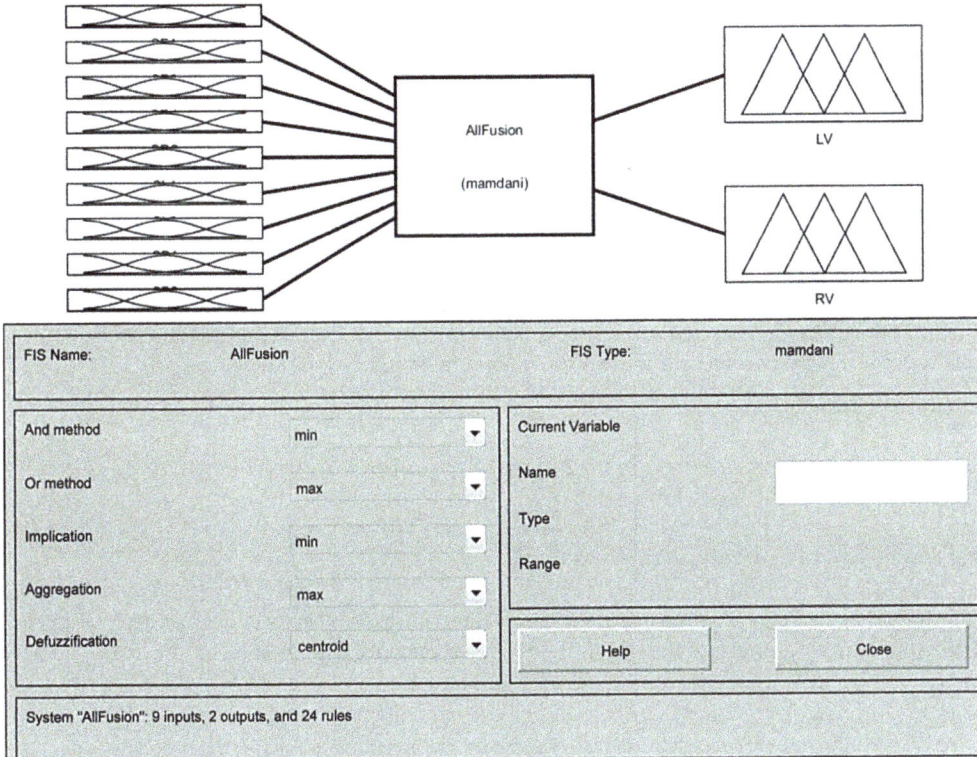

Figure 3. Fuzzy Inference System (FIS) with inputs and outputs.

3.2.1. Fuzzy Sets of the Input and Output

There are nine inputs to the fuzzy logic system and two outputs. The inputs are basically the values of eight distance sensors donated as SF1, SF2, SR1, SR2, SL1, SL2, SB1, and SB2. These sensors measure the amount of light in a range of 0 to 2000 where the threshold is set to 1000 for detected obstacle. The ninth input is the range finder camera value that measures the distance to an obstacle. Two outputs are generated left velocity (LV) and right velocity (RV). Figure 3 shows the Mamdani System using Fuzzy Inference System (FIS) with nine inputs and two outputs.

3.2.2. Membership Functions of the Input and Output

Input variables of distance sensors readings are divided into membership functions which are Obstacle Not Found (OBSNF), and Obstacle Found (OBSF). Both membership functions are a type of trapezoidal-shaped membership function.

The range for the distance sensor values is [0, 2000] and the threshold is set to 1000 where the value of 1000 or more means an obstacle is found and the robot should avoid it. The input variables of the range finder camera are divided into two trapezoidal-shaped membership functions "Near" and "Far". The range finder camera measures the distance from the camera to an obstacle in meters. The overall range of the camera input is [0, 1] where 0.1 m is considered as "Near" distance, and collision behavior avoidance should be applied. The input membership functions for distance sensors and the camera are displayed in Figures 4 and 5 respectively.

Let us assume that x is the sensor value and R is the range of all sensors values where $x \in R$. The trapezoidal-shaped membership function based on four scalar parameters i, j, k, and l, can be expressed as in Equation (1).

$$\mu_{trap}(x:i,j,k,l) = max(min\left(\frac{x-i}{j-i}, 1, \frac{l-x}{l-k}, 0\right)) \tag{1}$$

The output variables of left and right velocities of the mobile robot (LV and RV) are divided into two membership functions negative velocity "NEG_V" and positive velocity "POS_V". The effect and the action of these two memberships on the differential wheels of the robot are summarized as follows:

- If both LV and RV speeds are set to POS_V, then the robot will move forward;
- If LV is set to POS_V and RV is set to NEG_V, then the robot will turn right;
- If LV is set to NEG_V and RV is set to POS_V, then the robot will turn left.

The "NEG_V" is a Z-shaped membership function. This function is represented in Equation (2) where u and q are two parameters of the most left and most right of the slope.

$$\mu_z(x) = \begin{cases} 1, & x \leq u \\ 1 - 2\left(\frac{x-u}{q-u}\right)^2, & u \leq x \leq \frac{u+q}{2} \\ 2\left(\frac{x-q}{q-u}\right)^2, & \frac{u+q}{2} \leq x \leq q \\ 0, & x \geq q \end{cases} \tag{2}$$

In addition, the "POS_V" is an S-shaped membership function where y1 and y2 are two parameters of the leftmost and rightmost of the slope. The S-shaped membership function can be expressed as in Equation (3). Figure 6 shows the output membership functions.

$$\mu_s(x) = \begin{cases} 1, & x \leq y1 \\ 2\left(\frac{x-y1}{y2-y1}\right)^2, & y1 \leq x \leq \frac{y1+y2}{2} \\ 1 - 2\left(\frac{x-y2}{y2-y1}\right)^2, & \frac{y1+y2}{2} \leq x \leq y2 \\ 0, & x \geq y2 \end{cases} \tag{3}$$

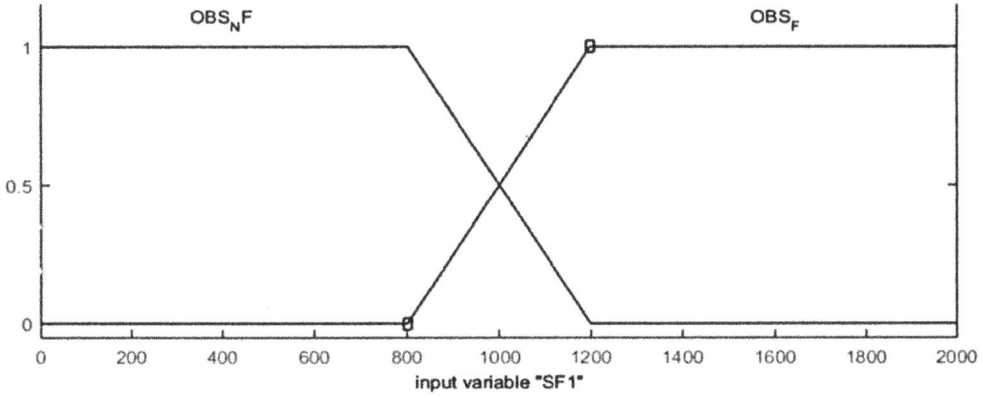

Figure 4. Input membership functions for the distance sensors.

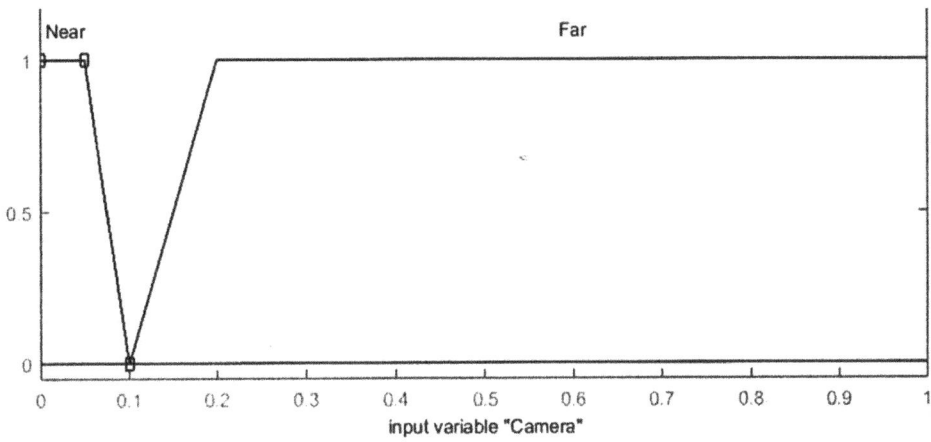

Figure 5. Input membership functions for the camera.

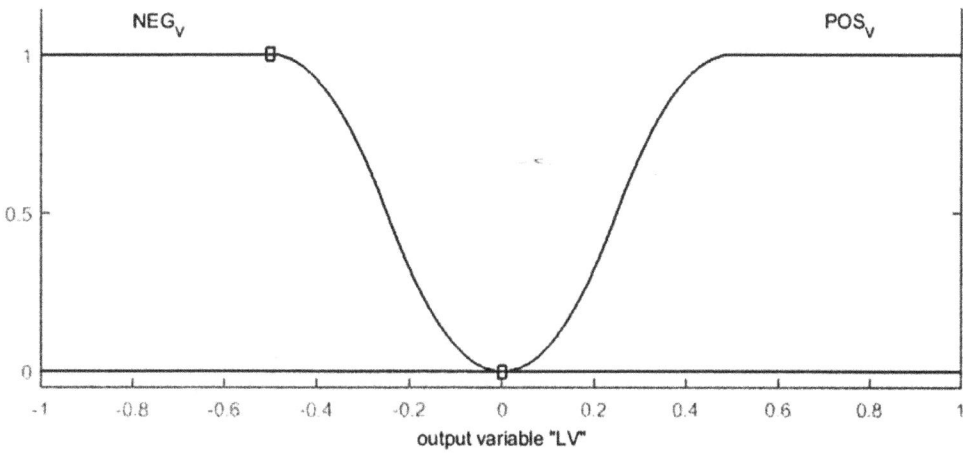

Figure 6. Output membership functions.

3.2.3. Designing Fuzzy Rules

Based on the membership functions of the fuzzy set and inputs and outputs, rules are defined. There are 24 rules for collision avoidance of the mobile robot.. We can use AND or OR operations for connecting membership values where the fuzzy AND is the minimum of two or more membership values and OR is the maximum of two or more membership values. Let $\mu\gamma$ and $\mu\delta$ be two membership values, then the fuzzy AND and fuzzy OR are described as in Equations (4) and (5), respectively. In addition, Table 1 lists all the rules with the fuzzy AND operator that express the movement behavior of the mobile robot.

$$\mu_\gamma \text{ AND } \mu_\delta = \text{ min } \left(\mu_\gamma, \mu_\delta\right) \tag{4}$$

$$\mu_\gamma \text{ OR } \mu_\delta = \text{ max } \left(\mu_\gamma, \mu_\delta\right) \tag{5}$$

3.2.4. Defuzzification

The last step of designing the fuzzy logic fusion system is the defuzzification process where outputs are generated based on fuzzy rules, membership values, and a set of inputs. The method used for defuzzification is the Centroid method.

Moreover, the fuzzy logic fusion model was designed for preventing the mobile robot from colliding with any obstacles while following the line. The fusion model composed of nine inputs, two outputs, and 24 rules. Figure 7 demonstrates the proposed methodology. As shown in Figure 7, the initialization of the robot and its sensors is the first step. After that, the distance sensors and camera values are fed into the fuzzy logic fusion system for obstacle detection and distance measurements. If an obstacle is found, the mobile robot will adjust its speed for turning left or right based on the position of the obstacle. The decision is made based on defined fuzzy rules. After avoiding the obstacle, the mobile robot should continue following the line by obtaining ground sensor values and finally adjust its speed accordingly. On the other hand, if there is no obstacle detected, the mobile robot should follow the line while it checks for obstacles to avoid at each time step.

Table 1. Fuzzy logic rules.

No	SF1	SF2	SR1	SR2	SL1	SL2	SB1	SB2	LV	RV
1	OBS_F	OBS_F	OBS_NF	OBS_NF	OBS_NF	OBS_NF	OBS_NF	OBS_NF	NEG_V	POS_V
2	OBS_F	OBS_NF	OBS_NF	OBS_NF	OBS_NF	OBS_NF	OBS_NF	OBS_NF	NEG_V	POS_V
3	OBS_NF	OBS_F	OBS_NF	OBS_NF	OBS_NF	OBS_NF	OBS_NF	OBS_NF	NEG_V	POS_V
4	OBS_NF	OBS_NF	OBS_NF	OBS_NF	OBS_F	OBS_F	OBS_NF	OBS_NF	POS_V	NEG_V
5	OBS_NF	OBS_NF	OBS_NF	OBS_NF	OBS_F	OBS_NF	OBS_NF	OBS_NF	POS_V	NEG_V
6	OBS_NF	OBS_NF	OBS_NF	OBS_NF	OBS_NF	OBS_F	OBS_NF	OBS_NF	POS_V	NEG_V
7	OBS_NF	OBS_NF	OBS_F	OBS_F	OBS_NF	OBS_NF	OBS_NF	OBS_NF	NEG_V	POS_V
8	OBS_NF	OBS_NF	OBS_F	OBS_NF	OBS_NF	OBS_NF	OBS_NF	OBS_NF	NEG_V	POS_V
9	OBS_NF	OBS_NF	OBS_NF	OBS_F	OBS_NF	OBS_NF	OBS_NF	OBS_NF	NEG_V	POS_V
10	OBS_NF	OBS_NF	OBS_NF	OBS_NF	OBS_NF	OBS_NF	OBS_F	OBS_F	POS_V	POS_V
11	OBS_NF	OBS_NF	OBS_NF	OBS_NF	OBS_NF	OBS_NF	OBS_F	OBS_NF	POS_V	POS_V
12	OBS_NF	OBS_NF	OBS_NF	OBS_NF	OBS_NF	OBS_NF	OBS_NF	OBS_F	POS_V	POS_V
13	OBS_F	OBS_NF	OBS_NF	OBS_NF	OBS_F	OBS_F	OBS_NF	OBS_NF	POS_V	NEG_V
14	OBS_F	OBS_NF	OBS_NF	OBS_NF	OBS_F	OBS_NF	OBS_NF	OBS_NF	POS_V	NEG_V
15	OBS_F	OBS_NF	OBS_NF	OBS_NF	OBS_NF	OBS_F	OBS_NF	OBS_NF	POS_V	NEG_V
16	OBS_F	OBS_NF	OBS_F	OBS_F	OBS_NF	OBS_NF	OBS_NF	OBS_NF	NEG_V	POS_V
17	OBS_F	OBS_NF	OBS_F	OBS_NF	OBS_NF	OBS_NF	OBS_NF	OBS_NF	NEG_V	POS_V
18	OBS_F	OBS_NF	OBS_NF	OBS_F	OBS_NF	OBS_NF	OBS_NF	OBS_NF	NEG_V	POS_V
19	OBS_NF	OBS_F	OBS_NF	OBS_NF	OBS_F	OBS_F	OBS_NF	OBS_NF	POS_V	NEG_V
20	OBS_NF	OBS_F	OBS_NF	OBS_NF	OBS_F	OBS_NF	OBS_NF	OBS_NF	POS_V	NEG_V
21	OBS_NF	OBS_F	OBS_NF	OBS_NF	OBS_NF	OBS_F	OBS_NF	OBS_NF	POS_V	NEG_V
22	OBS_NF	OBS_F	OBS_F	OBS_F	OBS_NF	OBS_NF	OBS_NF	OBS_NF	NEG_V	POS_V
23	OBS_NF	OBS_F	OBS_F	OBS_NF	OBS_NF	OBS_NF	OBS_NF	OBS_NF	NEG_V	POS_V
24	OBS_NF	OBS_F	OBS_NF	OBS_F	OBS_NF	OBS_NF	OBS_NF	OBS_NF	NEG_V	POS_V

4. Simulation and Real Time Implementation for Mobile Robot Navigation

The environment and the robot are modeled using the Webots Pro simulator for mobile robot collision free navigation. The e-puck used has eight distance sensors which are infrared sensors, camera, three ground sensors, and GPS. The e-puck first senses the environment for possible collisions by using the distance sensors and the range finder camera readings. If there is no obstacle detected, the e-puck follows a black line drawn on a white surface. Snapshots of the simulation and real time experiment for one robot detecting and avoiding an obstacle while following the line are depicted in Figures 8 and 9 respectively. Both figures show the environment with one mobile robot moving forward until it detects an obstacle. After the detection of the obstacle, all readings are fed into the proposed fuzzy logic fusion model and, based on the defined fuzzy rules, the e-puck will turn accordingly by adjusting the left and right wheels velocities. After that, the e-puck will continue moving forward and follow the line.

In addition, a more complex environment with various obstacles in different shapes and sizes has been modeled and tested through simulation and real time experiments. Figures 10 and 11 present snapshots of the simulation and real time experiments for two robots following a black line and avoiding different types of obstacles. As shown in these figures, both robots face and detect each other successfully. Each robot tries to avoid the other by adjusting its speeds and returns to follow the line. These robots are considered as dynamic obstacles to each other.

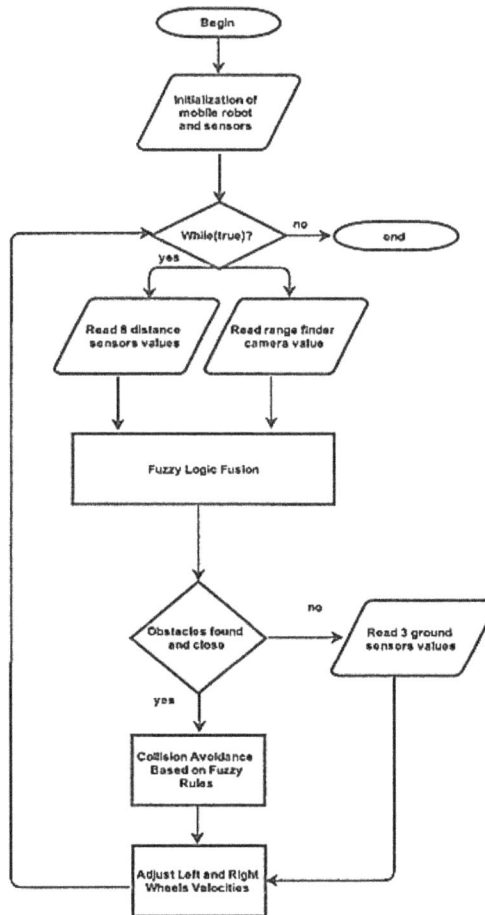

Figure 7. Flowchart of the proposed methodology.

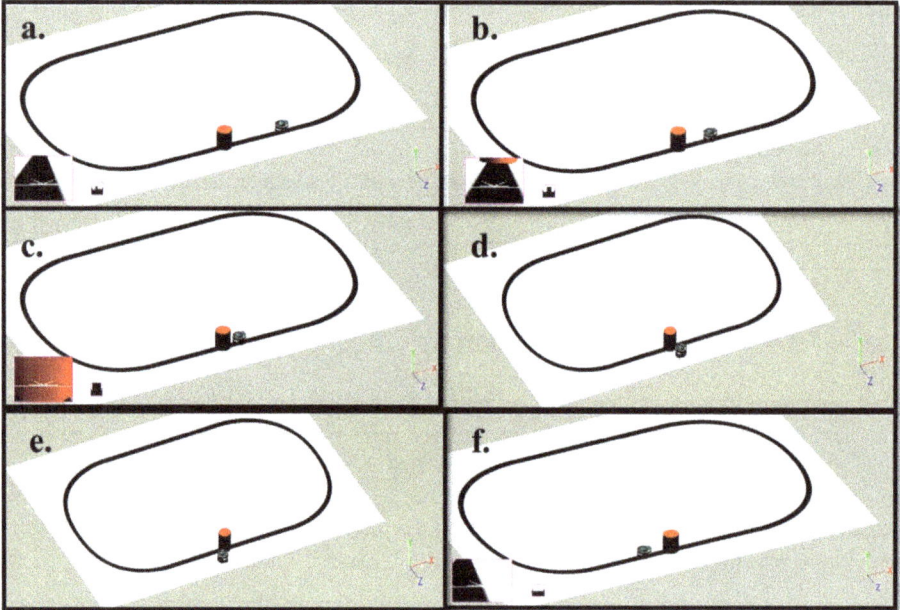

Figure 8. Simulation snapshots of one robot and one obstacle. (**a**) The beginning of the simulation; (**b**) The robot moves forward; (**c**) The robot detects an obstacle on its way; (**d**) The robot avoids the obstacle and turns left; (**e**) The robot tries to find the path again; (**f**) The robot continues following the line.

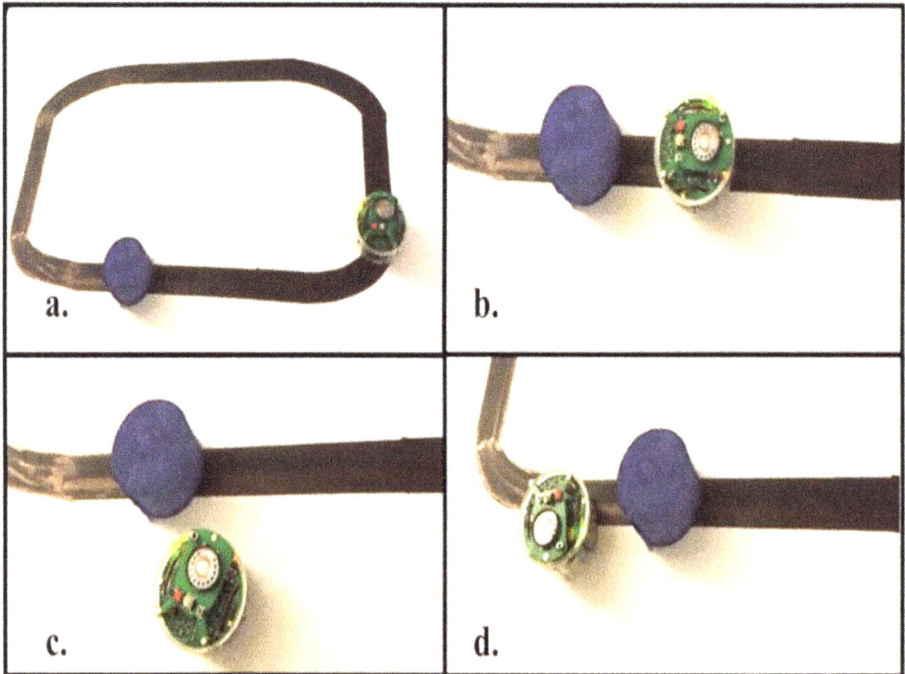

Figure 9. Snapshots of the real time experiment of one robot and one obstacle. (**a**) The beginning of the real time experiment; (**b**) The robot detects an obstacle on its way; (**c**) The robot avoids the obstacle and turns left; (**d**) The robot finds its path again and continues following it.

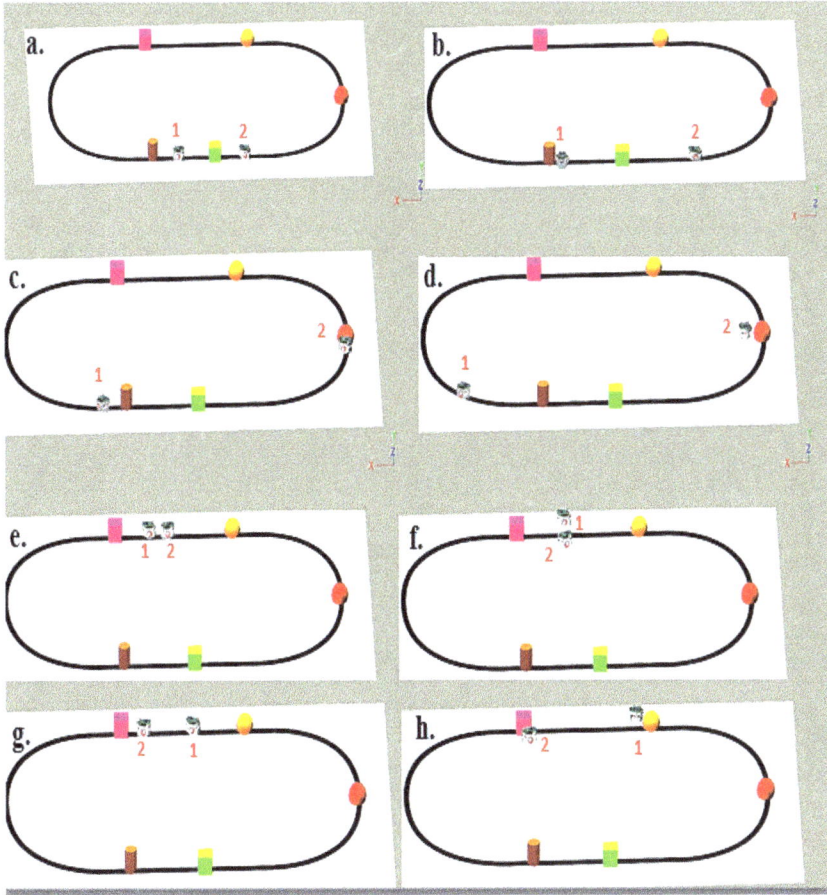

Figure 10. Simulation snapshots for multiple robots and obstacles. (**a**) Both robots at the beginning of the simulation; (**b**) Robot 1 detetcts an obstacle and Robot 2 moves forward; (**c**) Robot 1returns to the line and Robot 2 detects an obstacle; (**d**) Robot 1 continues following the line, and Robot 2 avoids the obstacle and moves around it; (**e**) Robot 1 and 2 detetc each other as dynamic obstacles; (**f**) Both robots avoid each other; (**g**) Both robots have successfully avoided each other and returned to the line; (**h**) Both robots detect other obstacles again.

5. Results and Investigation of the Proposed Model

5.1. Data Collection and Analysis

This section presents the sensors values obtained from the simulation at different time steps. Three different scenarios have been presented. The first one is a simple environment containing one obstacle and one robot, whereas the second one is a more complex environment that has more static and dynamic obstacles and two robots. The third scenario has more cluttered obstacles, which makes it a more challenging environment for the mobile robot navigation.

5.1.1. First Scenario

Tables 2 and 3 show the sensors values before and after applying the fuzzy logic fusion model in a simple environment at three different simulation times. At T1, the robot is far away from the obstacle; at T2, the robot is very close to the obstacle, and, at T3, the robot has passed the obstacle successfully. When distance sensor values are below the threshold (1000), it means that there is no obstacle detected.

However, when it goes above the threshold, it means that there is an obstacle and the robot needs to adjust its movement. As shown in Table 2, the front distance sensor SF1 has a value of 1159.18, which is higher than the threshold value at T2 and occurs before applying the fuzzy logic fusion method.

In addition, both front distance sensors SF1 and SF2 have higher values than the threshold, which are 1127.19 and1077.76, respectively, at T2 where the fuzzy logic technique is applied. Again, this means that there is an obstacle detected.

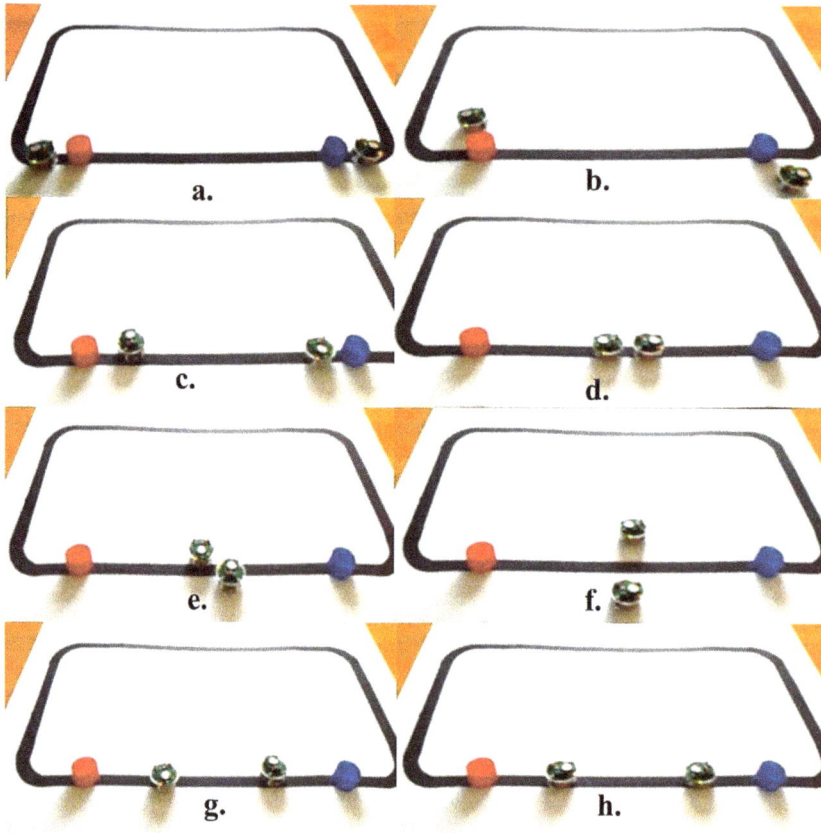

Figure 11. Real time experiment for multiple robots and obstacles. (**a**) Both robots detect obstacles; (**b**) Both robots avoid the obstacles; (**c**) Both robots return to the line; (**d**) Both robots detect each other as dynamic obstacles; (**e**) Both robots avoid each other; (**f**) Both robots have successfully avoided each other; (**g**) Both robots return to the line; (**h**) Both robots continue following the line.

Table 2. Distance sensor values before and after implementing the fusion model.

Distance Sensor	Without Fuzzy Logic Fusion			With Fuzzy Logic Fusion		
	T1 = 5	T2 = 15	T3 = 41	T1 = 5	T2 = 22	T3 = 41
SF1	14.71	1159.18	64.05	11.73	1127.19	72.72
SF2	35.19	220.33	53.72	48.48	1077.76	54.80
SR1	59.53	24.41	40.20	45.69	35.14	42.40
SR2	31.55	26.77	4.68	31.54	28.00	28.47
SL1	23.78	36.03	59.58	21.23	58.54	31.90
SL2	59.49	18.59	33.88	13.40	38.10	72.51
SB1	19.99	91.65	14.64	22.21	56.23	59.42
SB2	46.62	13.29	5.08	66.13	30.13	47.91

Table 3. Distance measurements by camera before and after implementing the fusion model.

Range Finder Camera	Without Fuzzy Logic Fusion			With Fuzzy Logic Fusion		
	T1 = 5	T2 = 13	T3 = 41	T1 = 5	T2 = 22	T3 = 41
Distance to Obstacle in Meter	0.337	0.097	X	0.339	0.040	X

Table 4. Ground sensor values at different simulation times.

Simulation Time	Ground Sensor 1	Ground Sensor 2	Ground Sensor 3	Delta
5	287.83	256.58	330.61	42.77
10	271.00	250.32	327.13	56.12
13	323.11	307.06	353.36	30.25
41	266.31	290.37	277.68	11.36

Table 3 shows the distance to obstacles measured in meters by the range finder camera at various simulation times T1, T2, and T3. As represented in Table 3, once the robot approaches the obstacle, the distance between the robot and the obstacle is decreased. At T2, the distance between the robot and the obstacle, where the obstacle is firstly detected by the camera and before applying the fuzzy logic method is 0.097 m where it is only 0.040 m after applying the fusion model. The camera can measure the distance up to one meter ahead. At T3, the camera could not measure the distance to an obstacle because it did not find any obstacles within its range.

In addition, Table 4 shows the three ground sensors values used for the line following approach at different simulation times. It also shows the delta values which are the difference between the left and right ground sensors. Delta values are used to adjust the robot's left and right speeds to follow the line.

Finally, Table 5 shows the robot's position, orientation, and velocities. The position of the robot has been obtained through the GPS sensor. The position and orientation of the robot according to Webots global coordinates system. As shown in Table 5, when the robot detects an obstacle at a time 22 s of the simulation, its left and right velocities are adjusted. The negative left velocity and the positive right velocity means that the robot is turning at the left direction to avoid the obstacle. At a time of 38 s, the robot has avoided the obstacle and turned right to continue following the line.

Table 5. The position, orientation, and velocities of the robot in a simple environment.

Simulation Time in Seconds	Position			Rotation Angle in Degree θ	Left Wheel Velocity	Right Wheel Velocity
	x	y	z			
5	0.34	0.05	1.24	−92.24	270.24	229.76
22	0.08	0.05	1.26	−165.58	−200.65	400.23
30	0.03	0.05	1.33	−100.87	200.65	189.44
38	−0.06	0.05	1.34	−36.04	305.34	−199.72
41	−0.14	0.05	1.24	−101.62	213.16	286.84

5.1.2. Second Scenario

This section demonstrates the proposed model in a more complex environment where it is composed of a number of obstacles in different sizes and shapes. Two robots are running in this scenario where both are avoiding obstacles and each other as well. Each robot considers the other as a dynamic obstacle. As presented in Figure 10, two robots (1 and 2) in opposite directions are following the line and overtaking obstacles. Figures 12–14 show all distance sensor readings, distance to obstacles in meters, and left and right velocities through the entire loop for the two robots, respectively. In Figure 13, the distances to obstacles are obtained by the range finder camera where sometimes the obstacle is either in a distance greater than one meter or it is outside the camera field of the view.

In later cases, the camera cannot measure the distances between the robots and the obstacles, which explains the gabs in Figure 13a,b.

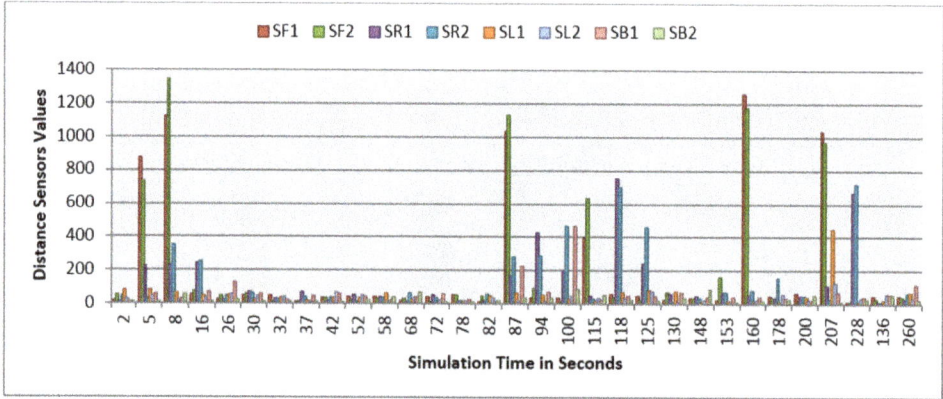

(a)

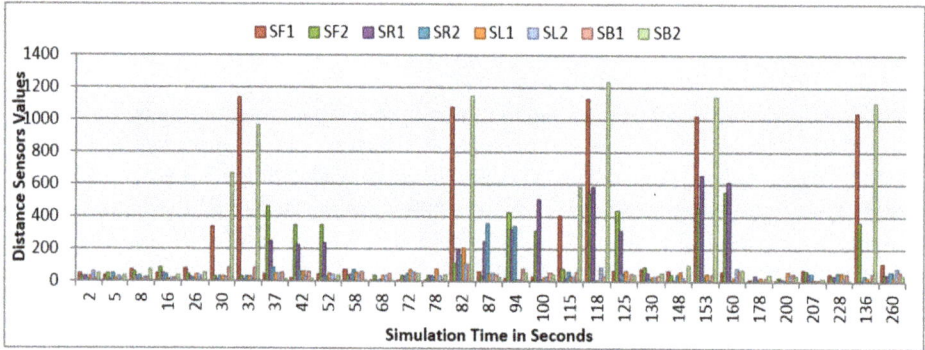

(b)

Figure 12. Distance sensor values for both robots at different simulation times. (**a**) Distance sensors readings at different simulation times for Robot 1; (**b**) Distance sensor readings at different simulation times for Robot 2.

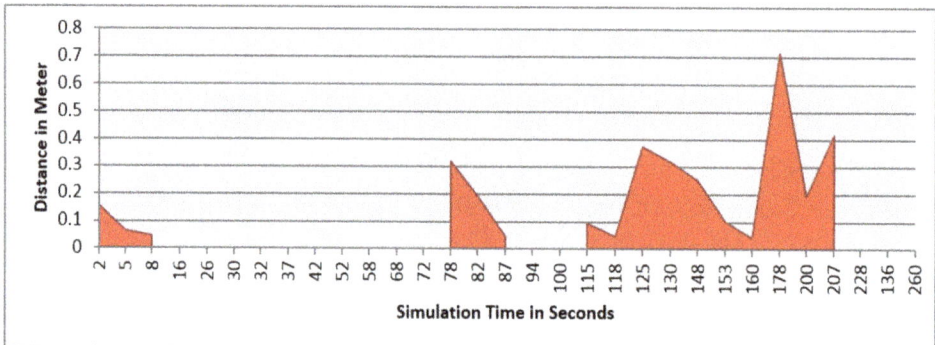

(a)

Figure 13. *Cont.*

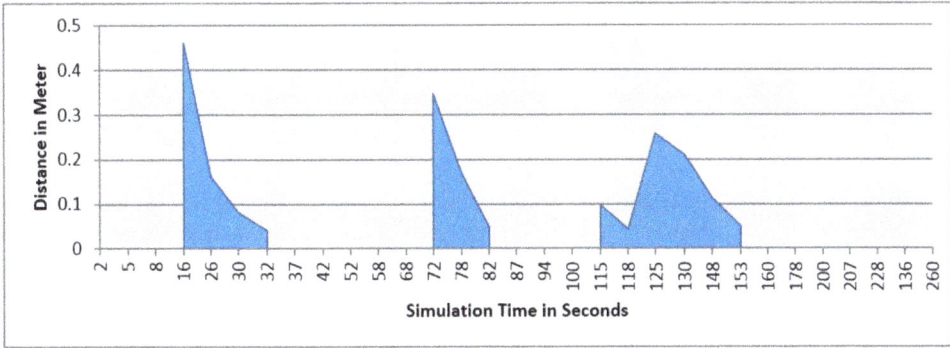

(b)

Figure 13. Distance to obstacles for both robots. (a) Distance to obstacles in meters for Robot 1; (b) Distance to obstacles in meters for Robot 2.

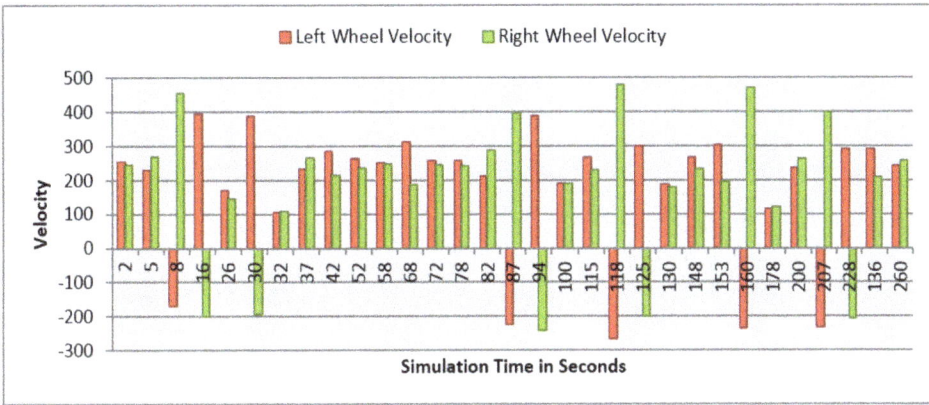

(a)

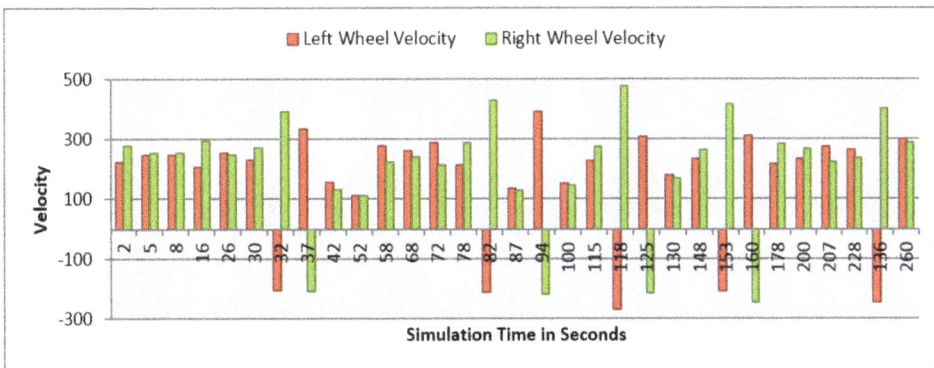

(b)

Figure 14. Left and right velocities for both robots at different simulation times. (a) Left and right velocities for Robot 1; (b) Left and right velocities for Robot 2.

At the beginning of the simulation, the distance between robot 1 and the obstacle is 0.15 m as shown in Figure 13a. At a time of eight seconds, robot 1 detects an obstacle where both front distance sensors (SF1 and SF2) have values greater than the threshold value as presented in Figure 12a. The distance between the robot and the obstacle at a time of eight seconds has been decreased to 0.047 m as shown in Figure 13a. As a result, the robot will adjust its speed accordingly to avoid colliding with the obstacle. Figure 14a, shows the left and right velocities for robot 1. At a time of eight seconds, the left wheel velocity is a negative value (−168) and the right wheel velocity is a positive value (454), which indicates that robot 1 is turning left to avoid collision. At time 16 s, robot 1 is turning right around the obstacle where the left wheel velocity is 395 and the right wheel velocity is −199. After that, the robot will continue following the line until another obstacle is detected. The ground sensor values and traveled distance by left and right wheels for both robots during the entire loop are demonstrated in Figures 15 and 16 respectively.

Furthermore, at time 115 s, robots 1 and 2 face each other after avoiding a couple of obstacles successfully. At that time, the distance between both robots is approximately 0.096 m. At a time of 118 s, the distance between them reaches 0.044 m. Both robots turn in opposite directions to avoid collision. To illustrate, at this time, the speed of robot 2 has been adjusted as shown in Figure 14b. Both robots will get around each other and return to follow the line. The position and orientation of both robots according to Webots global coordinates system are presented in Table 6.

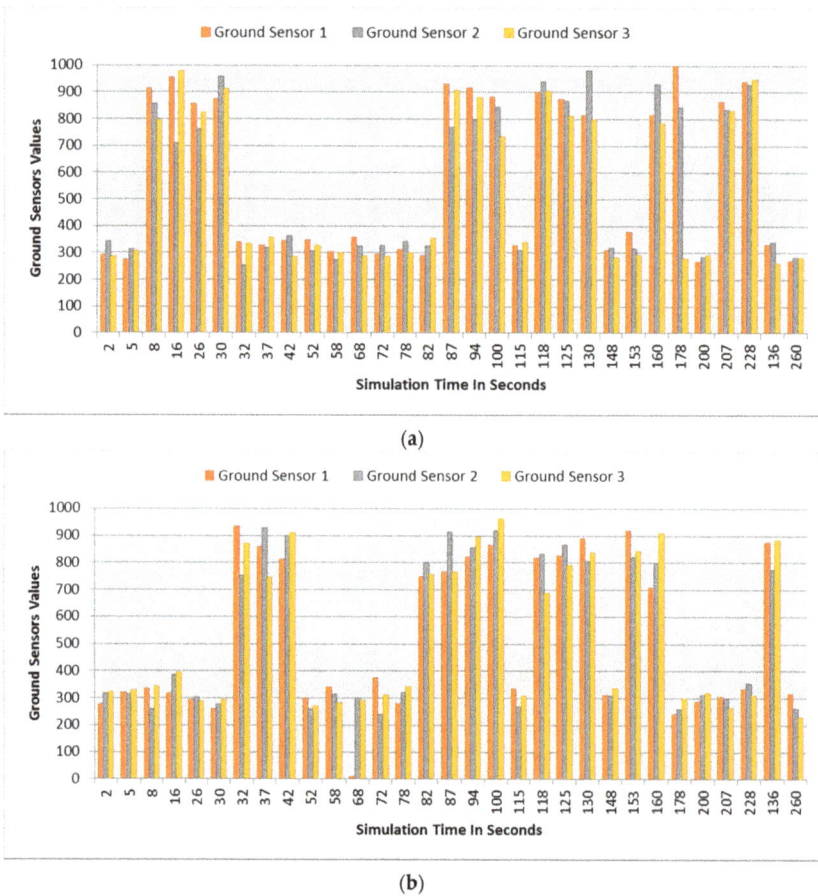

(a)

(b)

Figure 15. Ground sensors values for both robots at different simulation times. (a) Ground sensors values at different simulation times for robot 1; (b) Ground sensors values at different simulation times for robot 2.

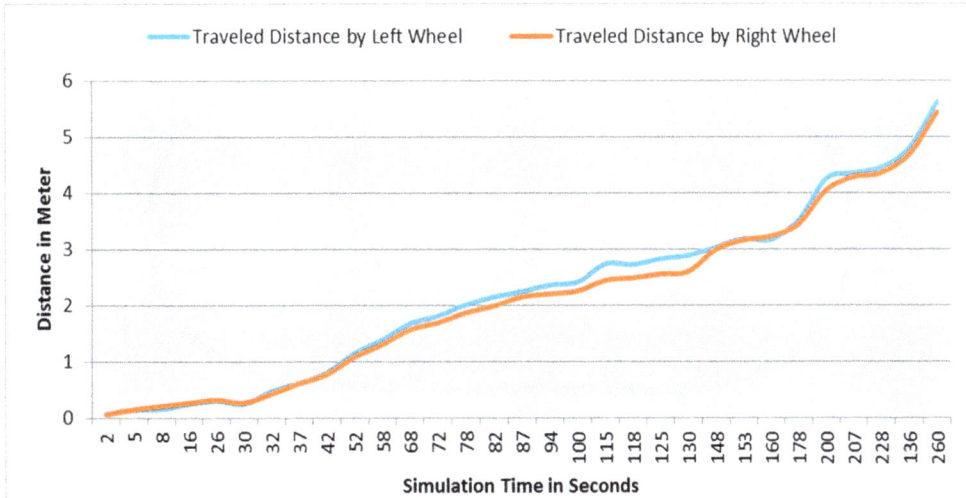

(a)

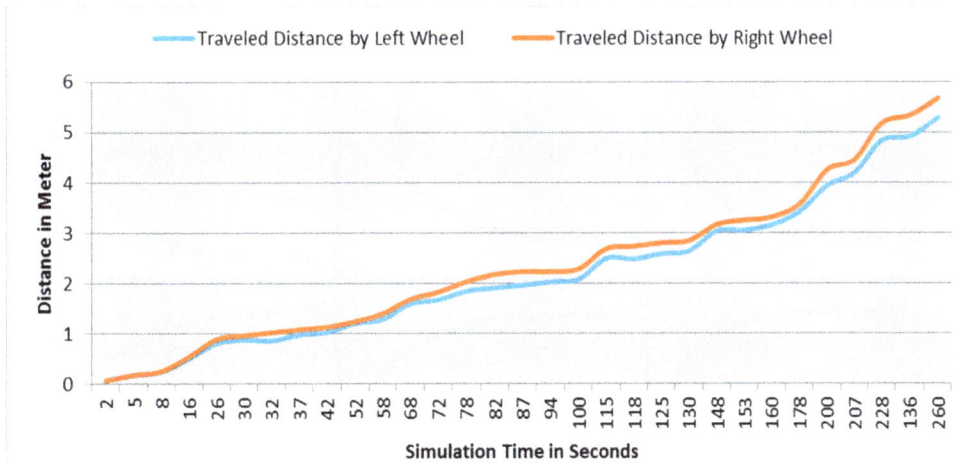

(b)

Figure 16. Traveled distance measured by left and right wheels for both robots. (**a**) Distance traveled by left and right wheels for robot 1; (**b**) Distance traveled by left and right wheels for robot 2.

5.1.3. Third Scenario

In this scenario, there are more cluttered obstacles in the environment where it is more challenging for the robot to avoid them. The robot needs to adjust its speed and orientation according to obstacles positions. Figures 17 and 18 represent the simulation of the mobile robot with many cluttered obstacles around.

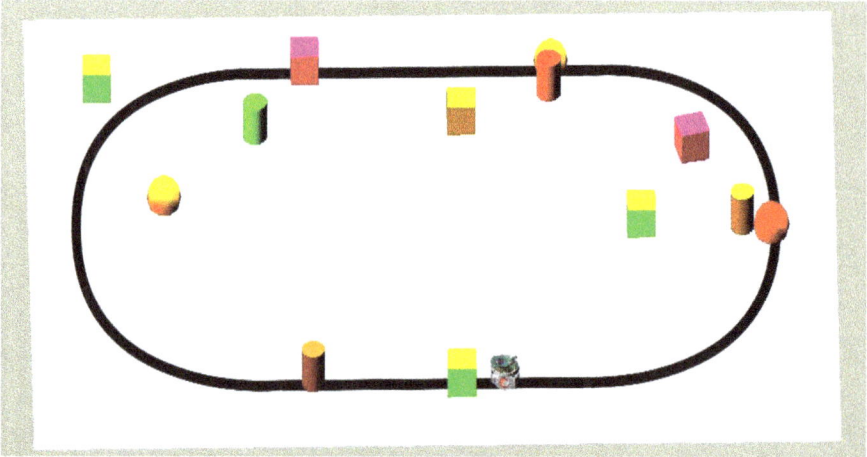

Figure 17. Simulation overview of the mobile robot and the environment.

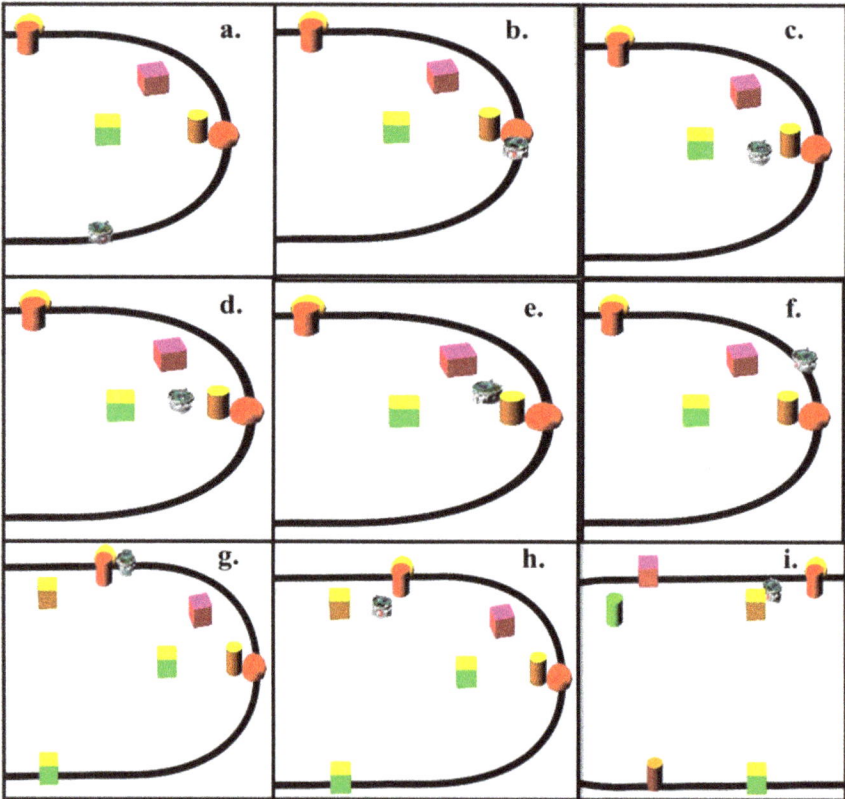

Figure 18. Simulation snapshots at various times. (**a**) The beginning of the simualtion; (**b**) The robot detects an obstacle and turns left; (**c**) The robot moves around the obstacle; (**d**) The robot between multiple scattered obstacles; (**e**) The robot turns right to avoid possible collision; (**f**) The robot returns to the line; (**g**) The robot detects another obstacle on its way; (**h**) The robot avoids obstacles; (**i**) The robot has successfully avoided obstacles.

Table 6. The position and orientation of both robots at different simulation times.

Simulation Time in Seconds	Robot 1				Robot 2			
	Position			Rotation Angle in Degree θ	Position			Rotation Angle in Degree θ
	x	y	z		x	y	z	
2	0.18	0.05	0.15	99.39	−0.20	0.05	0.15	−88.26
5	0.27	0.05	0.16	48.17	−0.31	0.05	0.15	−93.40
8	0.28	0.05	0.13	16.01	−0.40	0.05	0.15	−93.94
16	0.33	0.05	0.07	80.73	−0.62	0.05	0.23	−125.41
26	0.43	0.05	0.11	145.54	−0.79	0.05	0.52	−167.63
30	0.46	0.05	0.15	145.54	−0.79	0.05	0.59	110.00
32	0.47	0.05	0.16	145.54	−0.77	0.05	0.59	110.00
37	0.59	0.05	0.17	107.47	−0.72	0.05	0.62	174.72
42	0.73	0.05	0.23	127.69	−0.72	0.05	0.68	174.71
52	0.90	0.05	0.51	167.23	−0.78	0.05	0.76	−120.48
58	0.92	0.05	0.70	179.96	−0.81	0.05	0.83	158.72
68	0.87	0.05	0.99	−154.20	−0.71	0.05	1.10	142.67
72	0.80	0.05	1.10	−141.02	−0.61	0.05	1.19	121.69
78	0.65	0.05	1.21	−111.32	−0.44	0.05	1.25	95.44
82	0.54	0.05	1.24	−100.42	−0.34	0.05	1.23	4.69
87	0.41	0.05	1.26	−169.79	−0.34	0.05	1.17	4.70
94	0.39	0.05	1.34	−105.05	−0.27	0.05	1.13	69.45
100	0.31	0.05	1.36	−105.05	−0.21	0.05	1.14	134.26
115	0.14	0.05	1.25	261.32	0.05	0.05	1.25	89.07
118	0.13	0.05	1.28	−161.27	0.07	0.05	1.22	11.18
125	0.10	0.05	1.34	−96.53	0.10	0.05	1.16	75.91
130	0.04	0.05	1.34	−96.53	0.15	0.05	1.14	75.91
148	−0.01	0.05	1.24	84.02	0.28	0.05	1.24	265.34
153	−0.18	0.05	1.24	−101.41	0.36	0.05	1.22	4.43
160	−0.19	0.05	1.32	−179.93	0.37	0.05	1.15	69.20
178	−0.43	0.05	1.28	−56.51	0.46	0.05	1.25	84.13
200	−0.80	0.05	0.86	−9.46	0.88	0.05	0.94	18.56
207	−0.83	0.05	0.75	−81.51	0.91	0.05	0.78	6.13
228	−0.83	0.05	0.56	53.63	0.48	0.05	0.16	−71.84
136	−0.74	0.05	0.37	28.13	0.39	0.05	0.20	−170.91
260	−0.17	0.05	0.16	82.96	0.17	0.05	0.15	−94.52

Figure 19 demonstrates the distance sensor values, and Figure 20 shows the distance to obstacles obtained by the range finder camera at various simulation times. At the beginning of the simulation, the robot starts sensing the environment for possible obstacle detection. It also follows the predefined line using the ground sensors. As shown in Figure 18b, the robot turned left due to obstacle presence. At 32 s of the simulation time, SF1 (front distance sensor) reached a value of 1261, which indicates that there is an obstacle detected (Figure 19). In addition, the range finder camera has measured the distance to that obstacle which is 0.043 m as in Figure 20. At 37 s, the robot detects another obstacle on its right side and moves forward as in Figure 18c. Then, the robot gets stuck in between two obstacles and another obstacle in front of it. As shown in Figure 18d, the robot tries to travel in between both obstacles. At 50 s, the distance between the robot and the front obstacle is 0.21 m as in Figure 20. After that, the robot turns right to catch the line again as in Figure 18f. Figure 21 depicts the ground sensor values at different times, and Figure 22 shows the left and right wheels' velocities of the robot. At a time of 97 s, the robot detects another obstacle on its path and turns left as indicated in Figures 18g and 22. At 112 s, the robot detects an obstacle on its right side, which is very close to the first one and another obstacle at the front as in Figure 18h. Again, the robot tries to move in between both obstacles to recover its path as in Figure 18i. Table 7 summaries the robot's position and rotation angle in degrees at various simulation times.

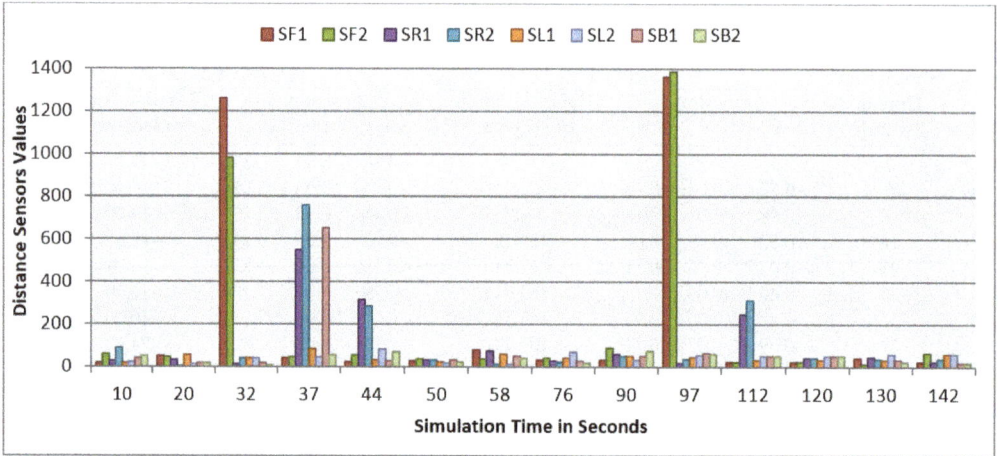

Figure 19. Distance sensor values at different simulation times.

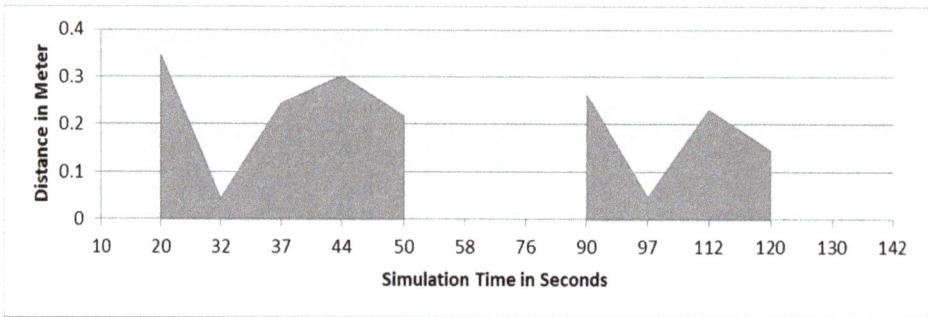

Figure 20. Distance to obstacles in meters at different simulation times.

5.2. Results and Discussion

The e-puck has successfully detected different types of obstacles (static and dynamic obstacles) with various shapes and sizes, and avoided them while it was following the line. Different scenarios have been presented with simple, complex, and challenging environments. Distance sensors and camera are used for obstacle detection and distance measurement. The distance sensor can only detect the obstacle when the robot is very close to the obstacle while the camera can detect it up to one meter ahead of the robot. Before applying the proposed fusion model, the distance sensors had detected the obstacle at a distance of 0.076 m from that obstacle to the robot while the camera has detected the obstacle at a distance of 0.097 m between the camera and the obstacle. On the other hand, after implementing the proposed fuzzy logic fusion methodology for collision avoidance behavior, the robot has detected the obstacle at a distance of 0.040 m. Detecting obstacles in a short distance range is very efficient and beneficial, especially in a dynamic environment where the robot quickly detects obstacles just gotten in its way. Figure 23 demonstrates the distance to obstacle measurements by using the distance sensor, or the camera, both with the integration of fusion model. As shown in Figure 23, fusing both sensors outweighs the performance of using each sensor separately.

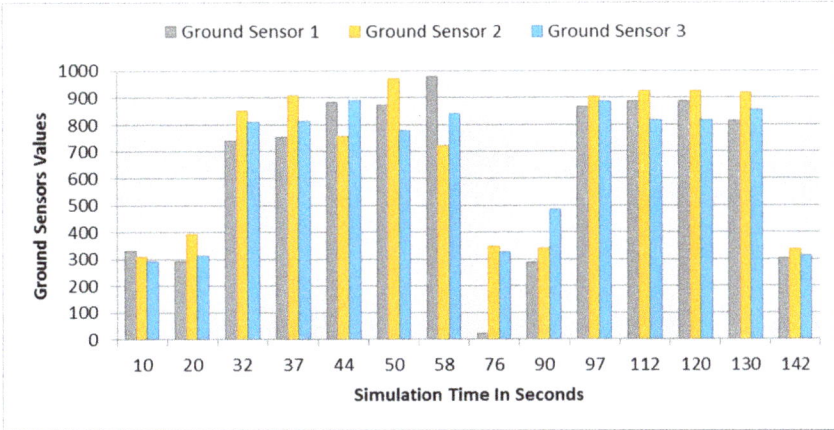

Figure 21. Ground sensor values at different simulation times.

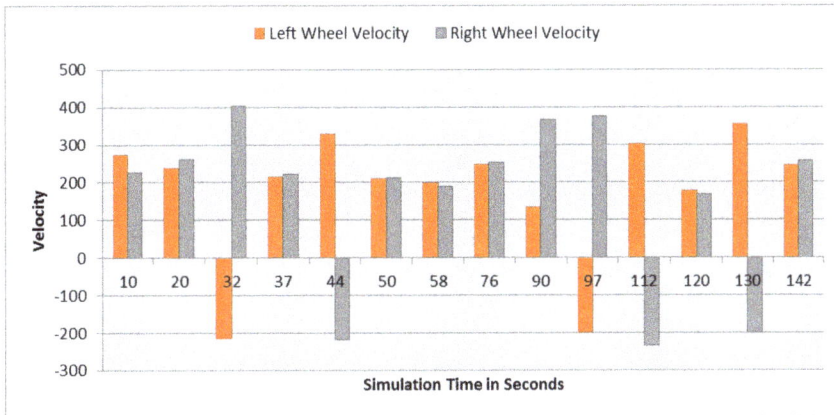

Figure 22. Left and right wheel velocities at different simulation times.

Furthermore, the distance traveled by the left and right robot's differential wheels is observed. As shown in Figure 24, the proposed fusion model has helped in reducing the distance traveled by the robot as opposed to each sensor separately, especially at the beginning of the simulation, which saves more energy, time, and computational load. In addition, an example of the proposed model using the MATLAB rule viewer is presented in Figure 25. In this figure, the sensor values of SF1, SF2, SR1, and SR2 which are the front and right distance sensors are higher than the set threshold. As a result, there are obstacles detected at the front and right sides of the robot's position. As shown in Figure 25, LV has a negative value and RV has a positive value, which means that the robot turns left due to the presence of obstacles at the front and right sides.

In addition, our approach aims at following the robot along a predefined path (black line on a white surface) while avoiding multiple obstacles on its way such as static, dynamic, and cluttered obstacles. Applying the fuzzy logic fusion has successfully reduced the distance traveled by the robot's wheels and minimized the distance between the robot and the obstacle detected as compared to a non-fuzzy logic approach, which is beneficially in a dynamic environment. Unlike other optimal planners such as Dijkstra, our approach does not focus on the shortest route and time taken to a specific target, the terrain characteristic, and energy of control actions.

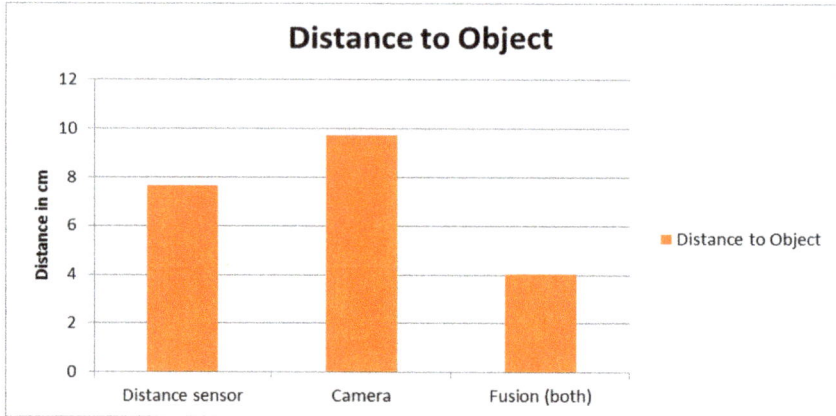

Figure 23. Distance measurements between the robot and the obstacle.

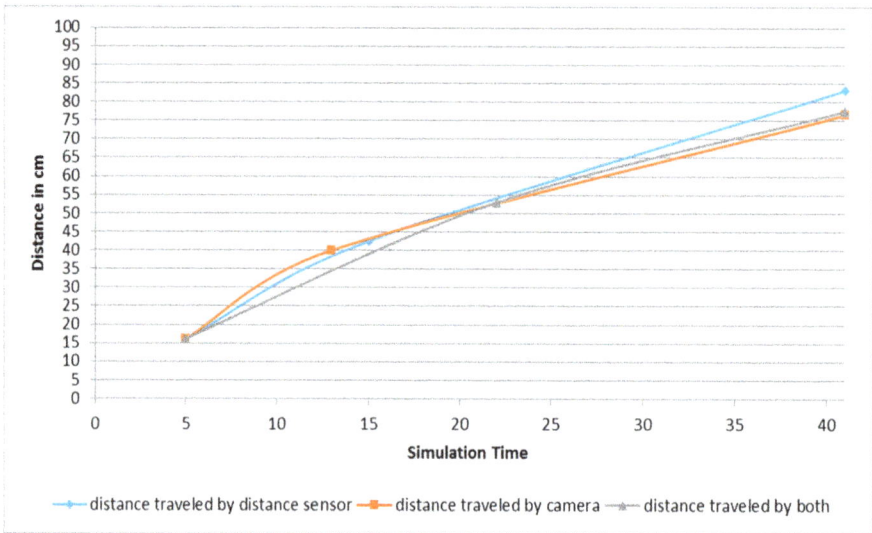

Figure 24. Average of distance traveled by the robot's differential wheels.

Table 7. Summaries of the robot's position and rotation angle at various simulation times.

Simulation Time in Seconds	Position			Rotation Angle in Degree θ
	x	y	z	
10s	−0.46	0.05	0.16	−101.09
20s	−0.72	0.05	0.33	−147.02
32s	−0.79	0.05	0.59	109.58
37s	−0.71	0.05	0.62	109.58
44s	−0.65	0.05	0.65	174.30
50s	−0.64	0.05	0.74	175.88
58s	−0.68	0.05	0.80	−119.31
76s	−0.78	0.05	0.93	171.94
90s	−0.52	0.05	1.23	107.53
97s	−0.34	0.05	1.23	9.67
112s	−0.27	0.05	1.07	74.42
120s	−0.19	0.05	1.05	92.64
130s	−0.11	0.05	1.14	139.23
142s	0.05	0.05	1.24	81.03

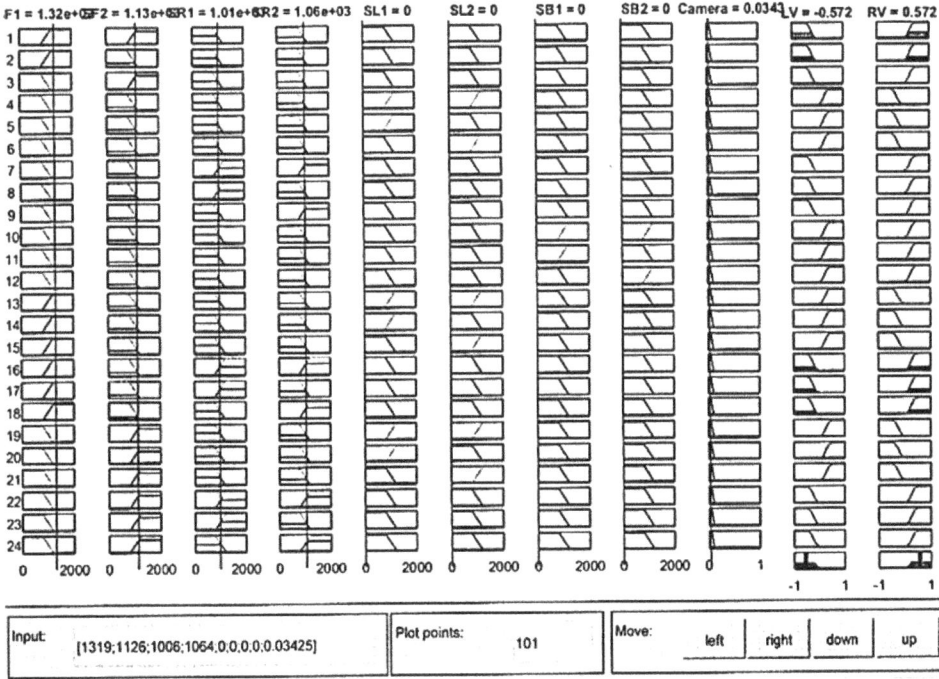

Figure 25. An example of the fusion model at MATLAB's rules viewer.

6. Conclusions

In this article, a multisensory fusion based model was proposed for collision avoidance and path following the mobile robot. Eight distance sensors and a range finder camera were used for the collision avoidance behavior where three ground sensors were used for the line following approach. In addition, a GPS was used to obtain the robot's position. The fusion model designed is based on the fuzzy logic inference system, which is composed of nine inputs, two outputs, and 24 fuzzy rules. Multiple membership functions for inputs and outputs are developed. The proposed methodology has been successfully tested in the Webots Pro simulator and with the real time experiment. Different scenarios have been presented with simple, complex, and challenging environments. The robot detected static and dynamic obstacles with different shapes and sizes in a short distance range, which is very efficient in dynamic environment. The distance traveled by the robot was reduced using the fusion model, which reduces energy and computational consumptions and time.

Acknowledgments: The authors acknowledge the reviewers for their valuable comments that significantly improved the paper.

Author Contributions: Marwah Almasri has done this research as part of her Ph.D. dissertation under the supervision of Khaled Elleithy. Abrar Alajalan has contributed to the experiments conducted in this paper. All authors were involved in discussions over the past year that shaped the paper in its final version.

Conflicts of Interest: The authors declare no conflict of interest.

References

1. Tian, J.; Gao, M.; Lu, E. Dynamic Collision Avoidance Path Planning for Mobile Robot Based on Multi-Sensor Data Fusion by Support Vector Machine. *Mechatron. Autom. ICMA* **2007**, 2779–2783. [CrossRef]
2. Hui, N.B.; Mahendar, V.; Pratihar, D.K. Time-optimal, collision-free navigation of a car-like mobile robot using neuro-fuzzy approaches. *Fuzzy Sets Syst.* **2006**, *157*, 2171–2204. [CrossRef]

3. Al-Mayyahi, A.; Wang, W.; Birch, P. Adaptive Neuro-Fuzzy Technique for Autonomous Ground Vehicle Navigation. *Robotics* **2014**, *3*, 349–370. [CrossRef]
4. Anjum, M.L.; Park, J.; Hwang, W.; Kwon, H.; Kim, J.; Lee, C.; Kim, K.; Cho, D. Sensor data fusion using Unscented Kalman Filter for accurate localization of mobile robots. In Proceedings of the 2010 International Conference on Control Automation and Systems (ICCAS), Gyeonggi-do, Korea, 27–30 October 2010; pp. 947–952.
5. Nada, E.; Abd-Allah, M.; Tantawy, M.; Ahmed, A. Teleoperated Autonomous Vehicle. *Int. J. Eng. Res. Technol. IJERT* **2014**, *3*, 1088–1095.
6. Chen, C.; Richardson, P. Mobile robot obstacle avoidance using short memory: A dynamic recurrent neuro-fuzzy approach. *Trans. Inst. Measur. Control* **2012**, *34*, 148–164. [CrossRef]
7. Qu, D.; Hu, Y.; Zhang, Y. The Investigation of the Obstacle Avoidance for Mobile Robot Based on the Multi Sensor Information Fusion technology. *Int. J. Mater. Mech. Manuf.* **2013**, *1*, 366–370. [CrossRef]
8. Kim, J.; Kim, J.; Kim, D. Development of an Efficient Obstacle Avoidance Compensation Algorithm Considering a Network Delay for a Network-Based Autonomous Mobile Robot. *Inf. Sci. Appl. ICISA* **2011**, 1–9. [CrossRef]
9. Rahul Sharma, K.; Honc, D.; Dusek, F. Sensor fusion for prediction of orientation and position from obstacle using multiple IR sensors an approach based on Kalman filter. *Appl. Electron. AE* **2014**, 263–266. [CrossRef]
10. Wang, J.; Liu, H.; Gao, M.; Sun, F.; Xiao, W. Information fusion-based mobile robot path control. *Control Decis. Conf. CCDC* **2012**, 212–217. [CrossRef]
11. Guo, M.; Liu, W.; Wang, Z. Robot Navigation Based on Multi-Sensor Data Fusion. *Digit. Manuf. Autom. ICDMA* **2011**, 1063–1066. [CrossRef]
12. Wang, Y.; Yang, Y.; Yuan, X.; Zuo, Y.; Zhou, Y.; Yin, F.; Tan, L. Autonomous mobile robot navigation system designed in dynamic environment based on transferable belief model. *Measurement* **2011**, *44*, 1389–1405.
13. Wai, R.; Liu, C.; Lin, W. Design of switching path-planning control for obstacle avoidance of mobile robot. *J. Frankl. Inst.* **2011**, *348*, 718–737. [CrossRef]
14. Canedo-Rodríguez, A.; Álvarez-Santos, V.; Regueiro, C.V.; Iglesias, R.; Barro, S.; Presedo, J. Particle filter robot localisation through robust fusion of laser, WiFi, compass, and a network of external cameras. *Inf. Fusion* **2016**, *27*, 170–188. [CrossRef]
15. Chandrasenan, C.; Nafeesa, T.A.; Rajan, R.; Vijayakumar, K. Multisensor data fusion based autonomous mobile robot with manipulator for target detection. *Int. J. Res. Eng. Technol. IJRET* **2014**, *3*, 75–81.
16. Cyberbotics.com, 2015. Available online: http://www.cyberbotics.com/overview (accessed on 23 December 2015).

Permissions

The contributors of this book come from diverse backgrounds, making this book a truly international effort. This book will bring forth new frontiers with its revolutionizing research information and detailed analysis of the nascent developments around the world.

We would like to thank all the contributing authors for lending their expertise to make the book truly unique. They have played a crucial role in the development of this book. Without their invaluable contributions this book wouldn't have been possible. They have made vital efforts to compile up to date information on the varied aspects of this subject to make this book a valuable addition to the collection of many professionals and students.

This book was conceptualized with the vision of imparting up-to-date information and advanced data in this field. To ensure the same, a matchless editorial board was set up. Every individual on the board went through rigorous rounds of assessment to prove their worth. After which they invested a large part of their time researching and compiling the most relevant data for our readers.

The editorial board has been involved in producing this book since its inception. They have spent rigorous hours researching and exploring the diverse topics which have resulted in the successful publishing of this book. They have passed on their knowledge of decades through this book. To expedite this challenging task, the publisher supported the team at every step. A small team of assistant editors was also appointed to further simplify the editing procedure and attain best results for the readers.

Apart from the editorial board, the designing team has also invested a significant amount of their time in understanding the subject and creating the most relevant covers. They scrutinized every image to scout for the most suitable representation of the subject and create an appropriate cover for the book.

The publishing team has been an ardent support to the editorial, designing and production team. Their endless efforts to recruit the best for this project, has resulted in the accomplishment of this book. They are a veteran in the field of academics and their pool of knowledge is as vast as their experience in printing. Their expertise and guidance has proved useful at every step. Their uncompromising quality standards have made this book an exceptional effort. Their encouragement from time to time has been an inspiration for everyone.

The publisher and the editorial board hope that this book will prove to be a valuable piece of knowledge for researchers, students, practitioners and scholars across the globe.

List of Contributors

Yali Zeng, Li Xu and Zhide Chen
Fujian Provincial Key Laboratory of Network Security and Cryptology, School of Mathematics and Computer
Science, Fujian Normal University, Fuzhou 350007, China

Heng Li and and Qun Hao
School of Optoelectronics, Beijing Institute of Technology, Beijing 100081, China

Xuemin Cheng
Graduate School at Shenzhen, Department of Precision Instrument, Tsinghua University, Shenzhen 518055, China

Yi-Xin Guo, Zhi-Biao Shao and Ting Li
The School of Electronic and Information Engineering, Xi'an Jiaotong University, No.28, Xianning West Road, Xi'an 710049, China

Qiang Zhou, Wei He, Dongping Xiao
State Key Laboratory of Power Transmission Equipment & System Security and New Technology, Chongqing University, Chongqing 400044, China

Songnong Li and Kongjun Zhou
State Grid Chongqing Electric Power CO. Electric Power Research Institute, Chongqing 400015, China

Zhaoyuan Yu, Linwang Yuan and Guonian Lv
Key Laboratory of VGE (Ministry of Education), Nanjing Normal University, No.1 Wenyuan Road, Nanjing 210023, China
State Key Laboratory Cultivation Base of Geographical Environment Evolution (Jiangsu Province), No.1 Wenyuan Road, Nanjing 210023, China
Jiangsu Center for Collaborative Innovation in Geographical Information Resource Development and Application, No.1Wenyuan Road, Nanjing 210023, China

Wen Luo
Key Laboratory of VGE (Ministry of Education), Nanjing Normal University, No.1 Wenyuan Road, Nanjing 210023, China
State Key Laboratory Cultivation Base of Geographical Environment Evolution (Jiangsu Province), No.1 Wenyuan Road, Nanjing 210023, China

Linyao Feng
Key Laboratory of VGE (Ministry of Education), Nanjing Normal University, No.1 Wenyuan Road, Nanjing 210023, China

Aitor Álvarez
Vicomtech-IK4. Human Speech and Language Technologies Department, Paseo Mikeletegi 57, Parque Científico y Tecnológico de Gipuzkoa, 20009 Donostia-San Sebastián, Spain

Basilio Sierra, Andoni Arruti, Juan-Miguel López-Gil and Nestor Garay-Vitoria
University of the Basque Country (UPV/EHU), Paseo de Manuel Lardizabal 1, 20018 Donostia-San Sebastián, Spain

Fuming Chen, Chuantao Li, Miao Liu, Zhao Li, Huijun Xue and Xijing Jing
Department of Biomedical Engineering, Fourth Military Medical University, Xi'an 710032, China

Sheng Li
College of Control Engineering, Xijing University, Xi'an 710123, China

Jianqi Wang
Department of Biomedical Engineering, Fourth Military Medical University, Xi'an 710032, China
Shaanxi University of Technology, Hanzhong 723001, China

Francesco Giordano, Gaia Mattei , Claudio Parente, Francesco Peluso and Raffaele Santamaria
Dipartimento di Scienze e Tecnologie, Università degli Studi di Napoli "Parthenope", Centro Direzionale, Isola C4, 80143 Napoli, Italy

Zhenghao Li
Key Laboratory for Optoelectronic Technology and Systems of Ministry of Education, College of Optoelectronic Engineering, Chongqing University, Chongqing 400044, China
Chongqing Academy of Science and Technology, Chongqing 401123, China

Junying Yang, Jiaduo Zhao
Key Laboratory for Optoelectronic Technology and Systems of Ministry of Education,
College of Optoelectronic Engineering, Chongqing University, Chongqing 400044, China

Peng Han
Chongqing Academy of Science and Technology, Chongqing 401123, China

Zhi Chai
Beijing Institute of Environmental Features, Beijing 100854, China

Allan Melvin Andrew, Ammar Zakaria, Shaharil Mad Saad and Ali Yeon Md Shakaff
Centre of Excellence for Advanced Sensor Technology (CEASTech), Universiti Malaysia Perlis, Jejawi, Arau, Perlis 02600, Malaysia

Heng Yuan, Jixing Zhang, Chuangui Cao and Gangyuan Zhang
Science and Technology on Inertial Laboratory, Beihang University, No. 37 Xueyuan Road, Beijing 100191, China

Shaoda Zhang
Pen-Tung Sah Institute of Micro-Nano Science and Technology, Xiamen University, No. 422 South Siming Road, Xiamen 361005, China

Yue Guo and Kyunglok Kim
Department of Electrical Engineering, Stanford University, 350 Serra Mall, Stanford, CA 94305, USA

Yu-Hung Li
Department of Materials Science and Engineering, Stanford University, 476 Lomita Mall, Stanford, CA 94305, USA

Zhiqiang Guo
Department of Mechanical Engineering, Stanford University, 440 Escondido Mall, Stanford, CA 94305, USA

Fu-Kuo Chang
Department of Aeronautics and Astronautics, Stanford University, 496 Lomita Mall, Stanford, CA 94305, USA

Shan X. Wang
Department of Electrical Engineering, Stanford University, 350 Serra Mall, Stanford, CA 94305, USA
Department of Materials Science and Engineering, Stanford University, 476 Lomita Mall, Stanford, CA 94305, USA

Denis Trček
Faculty of Computer and Information Science, University of Ljubljana, Věcna pot 113, Ljubljana 1000, Slovenia

Marwah Almasri, Khaled Elleithy and Abrar Alajlan
Computer Science and Engineering Department, University of Bridgeport, 126 Park Ave, Bridgeport, CT 06604, USA

Index

www.ingramcontent.com/pod-product-compliance
Lightning Source LLC
Chambersburg PA
CBHW061950190326
41458CB00009B/2832